Forest Ecology

Forest Ecology

An Evidence-Based Approach

DAN BINKLEY

School of Forestry, Northern Arizona University

WILEY Blackwell

Registered Offices
John Wiley & Sons, Inc., 111 River Street, Hoboken, NJ 07030, USA
John Wiley & Sons Ltd, The Atrium, Southern Gate, Chichester, West Sussex, PO19 8SQ, UK

Editorial Office
9600 Garsington Road, Oxford, OX4 2DQ, UK

For details of our global editorial offices, customer services, and more information about Wiley products visit us at www.wiley.com.
Wiley also publishes its books in a variety of electronic formats and by print-on-demand. Some content that appears in standard print versions of this book may not be available in other formats.

Library of Congress Cataloging-in-Publication Data

Names: Binkley, Dan, author. | John Wiley & Sons, publisher.
Title: Forest ecology : an evidence-based approach / Dan Binkley, School of
    Forestry, Northern Arizona University.
Description: First edition. | Hoboken, NJ : Wiley-Blackwell, 2021. |
    Includes index.
Identifiers: LCCN 2021001373 (print) | LCCN 2021001374 (ebook) | ISBN
    9781119703204 (paperback) | ISBN 9781119704409 (adobe pdf) | ISBN
    9781119704416 (epub)
Subjects: LCSH: Forest ecology.
Classification: LCC QH541.5.F6 B555 2021 (print) | LCC QH541.5.F6 (ebook)
    | DDC 577.3–dc23
LC record available at https://lccn.loc.gov/2021001373
LC ebook record available at https://lccn.loc.gov/2021001374

Cover Design: Wiley
Cover Image: © Dan Binkley

Set in 9.5/12.5pt Source Sans Pro by Straive, Pondicherry, India

Printed in Singapore
M108057_120721

The development of forests always includes contingent events: if an event happens, such as a fire, windstorm, or insect outbreak, the future of the forest will unfold differently than if the event did not happen (or if it happened in some other way at another time). This book would not be in front of you without the contingent event of Wally showing up as a young professor when I was an undergraduate at the School of Forestry at Northern Arizona University. Wally's engaging curiosity, interest in students, and active research program pulled my interests and future path into the domain of forest ecology. He continued to be a mentor through my grad student days at other universities, and most recently he led us through establishing the Colorado Forest Restoration Institute (modeled on NAU's Ecological Restoration Institute). It's been a good path. Thanks Wally.

Dan Binkley
Fort Collins, Colorado

Contents

Preface

How Do We Come to Understand Forests?

This book supports learning about forest ecology. A good place to start is with a few points about knowledge, followed by a framework on how to approach forest ecology, some key features of using graphs to interpret information, and finally coming around to how to think about questions and answers in forests.

Humans try to understand complex worlds through a range of perspectives. Art tries to capture some essential features of a complex world, emphasizing how parts interact to form wholes. Religions explain how worlds work now, how the worlds came to be, and what will come next. Both art and religion develop from ideas and concepts, originated by individual artists or passed down by religious societies. How do we know if a work of art or an idea in religion represents the real world accurately? This question generally isn't important. Art that satisfies the artist is good art, and religions are accepted on faith.

Art and religion have been evolving for more than 100 000 years, and lands and forests have been part of that development. One of the first written stories is a religious one from the Epic of Gilgamesh, from more than 4000 years ago from the Mesopotamian city of Uruk (now within Iraq). Gilgamesh and a companion traveled to the distant, sacred Cedar Mountain to cut trees. Lines from the epic poem include (based on Al-Rawi and George 2014):

> They stood there marveling at the forest, observing the height of the cedars . . . They were gazing at the Cedar Mountain, dwelling of gods, sweet was its shade, full of delight. All tangled was the thorny undergrowth, the forest a thick canopy, cedars so entangled it had no ways in. For one league on all sides cedars sent forth saplings, cypresses for two-thirds of a league. Through all the forest a bird began to sing . . . answering one another, a constant din was the noise. A solitary tree-cricket set off a noisy chorus. A wood pigeon was moaning, a turtle dove calling in answer. At the call of the stork, the forest exults. At the cry of the francolin bird, the forest exults in plenty. Monkey mothers sing aloud, a youngster monkey shrieks like a band of musicians and drummers, daily they bash out a rhythm . . .

And after slaying the demigod who protected the forest, Gilgamesh's companion laments:

> My friend, we have cut down a lofty cedar, whose top abutted the heavens . . . We have reduced the forest to a wasteland.

What would actually happen if cedar trees were cut on a mountain? Would more cedar trees establish, would the post-cutting landscape provide suitable habitat for the birds and monkeys? Would floods result? Anything could happen next in a story, but understanding which stories about the real world warrant confidence depends on the strength of evidence.

The core of understanding is knowing how one thing connects to another, and if the connections are the same everywhere and all the time, or if local details strongly influence the connections. The seasonal movements of the sun across the sky are consistent across years, but appear to differ from southern to northern locations. Multiple stories might explain the Sun's march with reasonable accuracy. Patterns etched on rocks by ancient artists may line up with key points in the Sun's seasonal patterns, and the movements of the Sun may reliably follow ceremonies convened by a society with the goal of ensuring the Sun's path. With art and religion, people may have understood the movement of the sun through the year was actually *caused* by the etchings on rocks or by ceremonial rites. These ideas may or may not have been true, but stories do not have to be true to be useful. Stories can persist as long as they are not so harmful that a society would be undermined. This idea is the same as genes in a population; natural selection does not aim toward retaining the best genes across generations, it only tends to remove genes that are harmful.

The human drive to understand cause and effect entered a new dimension when the notion developed of trying to figure out if an appealing idea might be *wrong*. Ideas of Newtonian physics and especially relativistic physics not only chart the apparent movement of the sun with more precision than would be possible from rock etchings or ceremonies, they also would be very, very easy to prove to be wrong. A deviation as small as one part in one million could prove the expectations of physicists were wrong. This innovation of science, based on investigating if an idea is wrong, developed very slowly alongside art and religion, and then exploded over the past four centuries to change the world.

Scientific thinking comes with two parts: creative new ideas about how the world works, and tough challenges that find out if the idea warrants confidence. Clearly most of the creative new ideas that scientists developed were wrong, either fundamentally or just around the edges. The ideas that withstood the challenges of testing have transformed the planet, feeding billions more people than our historical planet could have fed, sending machines across the solar system, and giving us an understanding of how our atoms formed in a collapsing star and how those same stellar reactions can be harnessed to obliterate cities. The idea that investigating whether an idea might be wrong has proven to be the most powerful insight humans have ever developed.

Returning to forests, trees and forests continue to be parts of art, religion, and science. When it comes to the scientific understanding of forests, both parts of science are needed: the generation of creative ideas and the challenging of those ideas to see if they warrant confidence. How do creative ideas about forests arise? That complex question has no simple answer, though creative ideas might arouse observation, learning, thinking, and pondering. The second part is more straightforward; once an idea is expressed, the hard work can begin on challenging the idea, to see if it's a better idea for accounting how forests differ across space and time.

A key point in science is being clear on which of these two aspects is being developed. The generation of a creative idea should not be mistaken for a reliable, challenge-based conclusion. Challenging an existing idea is important, though real gains in insights might depend on new ideas and new methods of measuring and interpreting.

How Confident Should You Be?

The confidence warranted in the truth of art or religion does not depend on the strength of evidence. The confidence warranted in scientific ideas always depends on evidence. Some scientific ideas warrant more confidence than others, and a scale of increasing confidence would be:

Weakest: Ideas based on appealing thoughts or concepts;

Weak: Analogies where well-tested insights from another area of knowledge are extended to a new area;

Moderately strong: Ideas supported by good evidence from one or a few case studies or experiments; and

Strong: Evidence-based ideas with robust trends across many locations and periods of time.

This is also a scale representing how surprising it would be to find out an idea was wrong, with the level of surprise increasing down the list. These distinctions may seem a bit dull and uninteresting, but the differences are as important as a person trying to fly on a magic carpet, to fly like a bird, to fly in an experimental airplane, or to fly in airplane certified to be safe with a record of thousands of hours of safe flights. Which approach to flying warrants the highest confidence for arriving safely at a distant destination?

One of the most common sources of creative ideas is making analogies. This tree has fruits that look like acorns, just like oaks have acorns, so this tree belongs with the group of oak species. Another analogy would be that aspen trees regenerate across burned hillsides and so do lodgepole pine trees, so aspen belongs in the group that lodgepole pine belongs in. Analogies may be true or false, but the key is to recognize that analogies represent only an initial, incomplete step of science. An analogy is reliably useful only when challenged by evidence. The acorn example could be challenged in many ways, including comparing other features of the tree with other oak trees, or especially by comparing DNA and genes. The analogy between the aspen and pine is not so obviously useful. If a grouping included trees that do well after severe fires, the trees indeed share useful features. For any other grouping, such as a suitability to feed beavers or mountain pine beetles, they clearly do not.

Creative ideas may begin as concepts or analogies, but gauging confidence depends on taking the next step to list the similarities and differences between the objects or sets of objects. An analogy might have more potential for useful insights when the similarities include major, diverse features. Analogies are less useful (or even harmful) when the list of key differences is substantial (Neustadt and May 1986).

All Forest Ecology Fits Into a Framework and a Method

Whether a creative idea originated in a concept, an analogy or another line of reasoning, science is incomplete if the idea is not challenged by evidence. The challenge needs to include ways that have a chance to show the idea to be wrong. This book raises questions about how forests work, and examines how the ideas have been challenged by evidence.

A good step for thinking about complex systems is developing a framework for understanding pieces of the system, and how the pieces interact. This book uses a core framework that can be used in every forest at all times (Figure A).

The core framework is structured by three simple questions. "What's up with this forest?" leads to familiar methods of measurement. "How did the forest get that way?" can be investigated with a variety of approaches for finding and interpreting historical evidence. "What's comes next?" usually can have only fuzzy answers because the future is not yet written for forests.

FIGURE A The ecology of all forests can be approached with a core framework and a core method, each asking three questions. The questions apply very generally, while the answers always depend on local details.

This set of core framework questions leads to a second trio of method-related questions that develop the necessary details to answer the core questions.

1. What is the central tendency (or mean) for this set of objects? (The set could be trees in this forest, forests across this landscape, or the forests at this location across millennia.)

2. How much variation occurs around that central tendency? (Do all trees increase in growth rate across all ages, or do some decline?)

3. What factors help explain when cases fall above (or below) the central tendency? (Are suppressed trees likely to decline in growth rate while large trees continue to increase?)

The core framework and core method may seem a bit awkward or unclear, but they should become clearer (and more useful) as the book uses them to investigate how forests work.

A Picture May Be Worth 1000 Words, But a Graph Can Be Worth Even More

A graph is full of answers, and the only work a reader needs to do is to bring the right questions, and know how to interrogate the graph. A good place to start exploring a graph is to apply a few questions:

1. What exactly does the horizontal X axis represent?

2. What exactly does the vertical Y axis represent?

3. When X increases, what happens to Y?

Reading information from graphs becomes easier with practice, and a few points about the graphs in this book might be helpful. Some of the points are obvious, but others will take some thought before insights can be pulled out of graphs.

The graphs in Figures B and C come from a massive dataset for tropical forests around the world. Taylor et al. (2017) compiled information on rates of wood growth, with basic information on each location's annual precipitation and average temperature (this example shows up again in Chapter 2). We know that trees need large amounts of water, because water evaporates (transpires) from leaves as the leaves absorb carbon dioxide from the air. We might have an idea that forests with a higher water supply should grow faster. This idea may or may not be true for forests, so checking the evidence from many studies lets us determine if this idea is worthy of our confidence. Indeed, forests with the highest amounts of precipitation grew more stemwood than those with a moderate supply. A few key features are worth pointing out. The X axis for precipitation spans a sevenfold range, from 1000–6000 mm yr^{-1}. The Y axis for stem growth also spans a large range, but the line in the graph goes only from a bit more than 4 Mg ha^{-1} yr^{-1} to something less than 8 Mg ha^{-1} yr^{-1}. Note that a sixfold change in the X axis (precipitation) gives at most a twofold range in stem growth, so the response is not as dramatic as if a twofold difference in precipitation gave a sixfold change in stem growth. Water matters, but not as much as we might have expected.

A second point for the first graph is that the average across all the studies follows a simple trend: a given increase in precipitation gives about the same amount of increase in stem growth, regardless of whether we look at the dry end or the wet end of the spectrum. We might have guessed that a small increase in water for dry sites would have a bigger value for tree growth than the same increase on a site that is already wet, but the available evidence would not support that generalization.

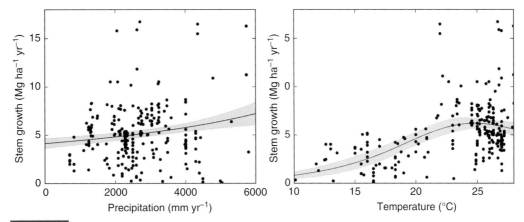

FIGURE B Stem growth in tropical forests is higher for sites with higher precipitation (left). The association with temperature is more complex, increasing at low temperatures and declining at the highest temperatures (**Source:** from data in Taylor et al. 2017).

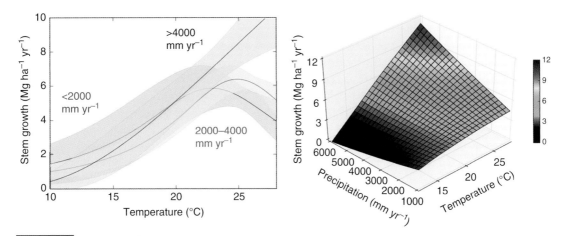

FIGURE C The influence of both factors can be examined together by examining the response of growth to temperature for three precipitation groupings (lower left). Putting sites into precipitation groups leaves out some of the information in the full dataset, and a 3D graph makes use of all the information. 3D graphs can work well for illustrating the overall trend surface, but specific details may be easier to read on 2D graphs (**Source:** from the database compiled by Taylor et al. 2017).

Just because the line in a graph goes up does not mean the trend warrants high confidence. If we chose a set of three numbers at random, the odds are good that the average trend would go up or down, rather than be flat. But if we choose a set with 100 random numbers, the odds are very high indeed that the trend would be flat (as there was no chance of the value of Y being related to X). This third point is illustrated with the shaded band around the line in the first graph, which represents the 95% confidence interval around the line. This means the evidence says the true trend would fall within that band about 95% of the time if the sampling were repeated. If we plotted a line with 240 random points for stem growth for random levels of precipitation, the line would be close to flat. If a flat line was placed into Figure B, it would fall outside the shaded bands of the confidence interval, so high confidence is warranted that sites with high rates of forest growth also tend to have higher precipitation. The odds are less than one in one thousand that growth and precipitation are unrelated (a flat line; this is the "P" value in statistics, which was <0.001 for this trend). This relationship of course does not prove that having more water is the key to producing more growth, but it does show the idea is not counter to the available evidence.

A third point about the first graph is that the dispersion of points around the line is broad indeed. Two tropical forests that have the same amount of precipitation might easily differ by twofold in growth rates. Even if confidence had been warranted in an average effect of precipitation, the average trend would not give a strong prediction for any single observation: half of all observations are always above average, and half below, no matter how much confidence is warranted in the overall trend.

This dispersion of points around the average trend is the fourth important story in the first graph. In statistics, the dispersion is the variance of the sample, and a number is often applied to trends in graphs that characterizes this dispersion around the average. The correlation coefficient (r) describes how tightly the data clump near the average trend, and the square of the correlation coefficient (r^2) tells the proportion of all the variability in Y values that relates to the level of the X values. In the first graph, the r^2 for stem growth in relation to precipitation is 0.04 (4%). We can be strongly confident that stem growth on average increases with precipitation, but that knowledge accounts for only a very small part of the full distribution of growth rates of tropical forests. The idea seems likely to be true, but it gives very little power for precise predictions.

The second graph in Figure B shows the growth rates of the same forests, but in relation to the average annual temperature. The confidence band is a bit tighter in this case, and the dispersion of points around the trend is not as large. The probability that random noise would account for the pattern is quite small (less than one in a thousand), so high confidence is warranted in the association between stem growth and temperature. The average trend with temperature accounts for about 23% of all the variation in growth rates among sites ($r^2 = 0.23$). Does higher temperature directly cause higher growth rates of forests? Possibly, but the association between two things does not mean that one causes the other. It's possible that soil nutrient supply is the real driver of growth, and soils in warmer parts of the Tropics have higher nutrient supplies. Confidence in whether one thing actually drives another depends on further evidence (and often direct experimentation).

The growth of a forest with a given temperature could depend on water supply. The range of sites could be divided into three groups: sites with less than 2000 mm yr^{-1}, 2000–4000 mm yr^{-1}, and more than 4000 mm yr^{-1} (Figure C). The trends between temperature and stem growth are similar across these three groups at temperatures below 23 °C, but at higher temperatures growth seemed to decline more on drier sites than on wetter sites. This breakdown of the temperature relationship into three precipitation groups increases that amount of variation accounted for in growth to 31%, and very high confidence is warranted that predictions of temperature responses of growth differ among the precipitation groups.

Separating the sites into precipitation groups actually throws away some information that might be useful. For example, a site with 1950 mm yr^{-1} precipitation would be tallied in the driest group, and one with 2050 mm yr^{-1} would be separated into the medium group. Yet these two sites would be more similar to each other than the 2050 mm yr^{-1} site would be to another in the medium group with 3950 mm yr^{-1} site. Another version of the analysis could be done with all the data from each site allowed to influence the trend, and then a full three-dimensional pattern can be developed. The second graph in Figure B has two horizontal axes. The temperature axis increases to the right, and "backward" into the 3D space. The precipitation axis goes the other way, increasing to the left and also going backward into the space. This graph shows how any given level of temperature, and any level of precipitation, connect to give an estimate of the expected rate of stem growth. Keeping all the information on precipitation included (rather than lumping into three groups) increases the variation accounted for to 34%. A key difference is that this full-information analysis shows that growth continues to increase at high temperatures if the precipitation is high, but levels off (with no decline) on drier sites. This might seem like a small improvement in the pattern, but the improvement does warrant very high confidence.

It can be challenging to read the values for stem growth on the 3D graph, compared with straightforward 2D graphs. The grid lines give some help for visualizing how the overall trend changes, and the use of colors helps peg a value to any given point on the surface. Overall, 3D graphs can be very useful for illustrating overall trends, but 2D graphs might be more useful when the precise values of variables need to be identified.

Why do temperature and precipitation relate to only about one-third of all the variation in stem growth among tropical forests? Two points are important. This analysis used only annual averages, and two sites with similar annual average might differ in important seasonal ways. A given amount of rain spread evenly across 12 months might have very different effects on growth than if all the rain fell during a 4-month rainy season (with no rain for 8 months). The second point is that stem growth depends on a wide range of ecological factors, including soil nutrient supplies, and the genotypes of trees present. Attempts to explain forest growth often go beyond the ability of graphs to capture the relationship, using simulation models and other tools that have a chance to capture variations in growth patterns that go beyond two or three dimensions (Chapter 7).

The Most Important Points to Understand from Figures B and C Are Not About Precipitation or Temperature

The most important point is one that is not found in the graph, but applies to this graph and most others in this book. Graphs plot the values for a variable (such as forest growth) based on another variable (such as precipitation). Even when the association between the two variables is very strong, it's fundamentally important to recognize that evidence of an association is not evidence of a cause-and-effect relationship. The forests that provided the data for Figure B had very different species composition, different soils, different ages, and different local histories of events. Some of these may happen to vary with precipitation, and might be the actual drivers of the trends that relate to precipitation. Similarly, if forest growth tended to decline in the warmest sites, that might result from increased activities of insects (or monkeys) rather than a direct effect of temperature.

Identification of driving causes behind patterns requires other sorts of evidence, especially evidence from experiments. If the addition (or removal) of water changed growth as much as was expected from the geographic gradient, then increased confidence would be warranted in water influencing growth across many locations. If plantations of a single species also declined in growth at high temperatures, then the trend in Figure B may be less influenced by changes in tree species across sites.

This fundamental idea is summarized in the aphorism, "Correlation does not equal causation." All scientists know this, but placing science into sentences can be challenging for both thinking processes and writing processes. It's easy to find examples where scientists forgot this basic point (perhaps even a few places in this book?).

Confidence Bands Around Trends Come in Two Types

Most of this book's graphs have shaded bands around the trend lines, and these represent the 95% confidence interval around the trend. A narrow band means the value on the Y axis was tightly related to the value on the X axis. Other types of bands can also be used, and Figure D shows a band that describes the distribution pattern for all the observations rather than the confidence warranted in the average trend. Both shaded bands in Figure D deal with 95%, but one describes the region where 95% of the observations are likely to be found, and the other the region where 95% of the trends (from repeated experiments) would be expected to occur. A key point is that the variation in the population of forests does not depend on how many samples are taken; a given proportion of forest would be a bit smaller (or much smaller) than average, and another proportion would be a bit larger (or much larger) than average. That variation does not change as the number of forests are sampled from the same landscape of forests. A sample of 24 forests produces about the same light-shaded band as a sample of 71 forests, but the confidence warranted in the trend is tighter when based on a larger number of samples (the dark bands).

The Stories in This Book Have Two Pieces, Told in Three Ways

The subject of forest ecology combines two different types of pieces: information (or evidence) about important details, and frameworks for how to knit pieces of information into understanding how forests work. The framework described above repeats throughout the book, along with many case studies and experiments that fill in information. This two-piece approach shows up in three complementary ways. The section headings state the key points in each chapter; these headings could be grouped together for

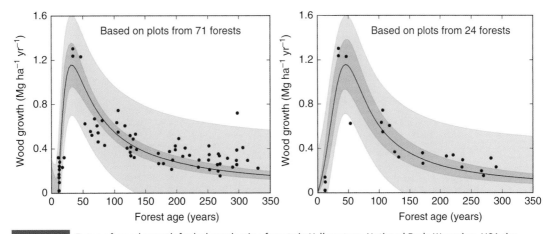

FIGURE D Rates of wood growth for lodgepole pine forests in Yellowstone National Park, Wyoming, USA rise quickly as new forests develop after fires, and then decline more slowly. The left graph shows that confidence in the average trend is warranted within the dark 95% confidence band. The points are dispersed around that average trend, and the lighter band covers the domain where about 95% of the observations would occur. The graph on the right used only a subset of 24 of the plots, and the average trend is similar, but the smaller number of sampled stands leads to a wider 95% confidence band (the darker band) for the trend compared to the full dataset on the left. The light blue band represents where 95% of the observations would be expected to fall, and the breadth of that band is quite similar between the two sampling intensities (Source: based on data from Kashian et al. 2013; see also Figure 9.11). Larger numbers of samples reduce the uncertainty about average trends, but not about the level of variability among forests across a landscape.

a simple overview for each chapter. The text of each chapter lays out the information and framework in detail, while the figures reinforce the headings and text with a third dimension of images and graphs (with detailed captions). Each of these three ways contributes to understanding forest ecology, developing a foundation to be built upon with further conversations, with readings of other books and journals, and especially with curious explorations in forests and across landscapes.

Forests Are Complex Systems That Are Not Tightly Determined

A core idea in this book is that forests are indescribably complex systems, with an uncountable number of interacting pieces under the influence of external driving factors. Simple stories cannot provide high value for specific cases, because the future development of a forest simply is not constrained enough to allow precise predictions. The good news is that evidence from thousands of research investigations around the world does provide a foundation for useful general insights. The next step is to apply general questions – from the core framework and core methods – to any forest of interest to develop strong, locally relevant knowledge. This book tries to clarify some of the basic nature of forests, and how to rely on evidence as a guide for gauging the amount of confidence warranted in ideas about forests.

A different forest ecology book could be written to summarize what we know about the major features of forests: for a very wide variety of questions, what solid answers emerge for each question from the evidence accumulated over the decades and centuries? That approach would provide a strong reference source for describing the general trends for forests, how variable they are, and what factors account for when forests are likely to be above or below the general trends. The focus of this book is somewhat different, though, as it fosters the thinking and understanding that will provide a strong foundation for adding later evidence found in reference books, journals, and other sources.

The future is not yet written for any forest, and that's also true about this book. If revised editions should appear in the future, they would be much improved by feedback provided by readers of this edition. I gratefully invite feedback about typos, mistakes, omissions, and ideas for how the stories could be stronger and warrant more confidence (Dan.Binkley@alumni.ubc.ca).

The final introductory point is that this book could be rewritten with all of the graphs and all of the examples switched out with different examples from other forests in other locations and other times. For example, sal (*Shorea robusta*) is a major, important tree in forests across southern Asia (Figure E), but this is the only sentence in the book that mentions sal. Each reader can make use of the book's questions and perspectives by adding local information for other forests types, other places, and other times.

FIGURE E Sal is a major species across southern Asia, just one species of 700 among 16 genera in the Dipterocarp family. Sal wood is valued for lumber, its leaves are used for various purposes (including plates for food), and oil extracted from its seeds are used in food and mechanical applications (**Source:** photo by Anand Osuri).

Acknowledgements

It takes a virtual village to write a book, and I want to acknowledge and thank the villagers whose insights shaped this book, and my career. Each of my university advisors contributed new dimensions to thinking about forest ecology. Wally Covington at Northern Arizona University set the stage, and brought out interests in connecting chemistry and ecology. Hamish Kimmins at the University of British Columbia broadened my experiences, and patiently endured my skeptical challenges of so many ideas. Kermit Cromack's curiosity and enthusiasm across a broad range of ecology and science was infectious – and persistent. Ed Packee at MacMillan Bloedel provided the questions, insights, funding and free reign that were so important early in my career. Colleagues and students at Duke University's School of Forestry and Environmental Studies could not have done a better job of sustaining the momentum provided by earlier members of the village. The worldwide community of scientists in forest ecology and forest soils provided collaboration, ideas, and education over the following decades. Some of the most generous villagers were Tom Stohlgren, Mike Ryan, Peter Högberg, Bill Romme, Bob Powers, Dale Johnson, Jose Luiz Stape and Bob Stottlemyer. Most of my career developed at Colorado State University, with world-class colleagues and students. I cannot conceive of any village that would have been more fun, more supportive, or more productive; this tree acknowledges coauthors across four decades, with font sizes proportional to how many works we wrote together. Thanks to all of you – and to Mason Carter and Jane Higgins for working through and polishing these chapters. There are many ways a forest ecology textbook could be written; this is the one I could write.

CHAPTER 1

The Nature of Forests

To see a World in a Grain of Sand,
And a Heaven in a Wild Flower,
Hold Infinity in the palm of your hand,
And Eternity in an hour.. . .

William Blake (1757–1827)

William Blake's poetic approach of seeing the general in the specific is a useful approach, two centuries later, for launching into the ecology of forests. The biology of a single tree in a single hour connects outward in time to the course of the tree's development from a seed, back through evolutionary time for the genes comprising the tree's genome. The environmental influences on the tree also connect outward in space, with strong similarities to the forces shaping trees around the world. The value of this literary approach to describing and understanding forests has limits: trees comprise the greatest part of the living matter within a forest, but the vast majority of organisms and species in forests are not trees at all. The biodiversity of forests resides primarily in understory plants, animals, and especially very small organisms such as arthropods, fungi, and bacteria.

The ecology of forests can be explored using Blake's approach of starting very small to begin to understand very large and complex systems. The hourly, daily, and annual story for a single tree can be expanded outward to encompass the other trees in a forest, a landscape, and the forest biome, just as an hour can be expanded to a day, a year, a millennium, and evolutionary timescales.

Forest Ecology Deals with Individual Trees Across Time

A tulip poplar in the Coweeta Basin of eastern North America will be the launching point for developing insights about forests. This particular tree (Figure 1.1) would be over half a meter in diameter (at a height of 1.4 m above the ground, a common point for measuring) and over 30 m tall. Eighty years of biological processes have led to an accumulation of more than 1000 kg of wood, bark, leaves, and roots. The actual living weight of the living tree would be about twice this mass of the biomass, because trees typically contain as much water as dry matter.

The crown carries about 75 000 leaves, with a total mass of about 25 kg (not counting the water). This is enough leaves to provide more than four distinct layers of leaves above the ground area below by the tree crown. The multiple layers of leaves are displayed to capture 90% of the incoming sunlight. A sunny afternoon might have 1000 W of sunlight reaching each m² of ground area.

Many Processes Occur in a Tree Every Hour

Over the course of an hour, the tree leaves would intercept about 140 MJ of sunlight, and about half of the light arrives at wavelengths that can be used in photosynthesis. Perhaps 10–15% of the light reaching leaves reflects back into the environment, with no effect on the leaves. About one-third to one-half is converted to heat, warming the leaves, which then lose heat to the surrounding air (especially if the wind is blowing). Most of the rest of the intercepted energy is consumed as water evaporates from moist leaf interiors into the dry air, also cooling the leaves.

Forest Ecology: An Evidence-Based Approach, First Edition. Dan Binkley.
© 2021 John Wiley & Sons Ltd. Published 2021 by John Wiley & Sons Ltd.

FIGURE 1.1 The Tree. This tulip poplar is a typical tree for temperate forests. The tree may live for a few centuries, integrating daily, seasonal, and yearly fluctuations in environmental conditions to turn carbon dioxide and water into wood (and thousands of types of chemicals).

A few percent of the radiant energy hitting leaves is harnessed to drive photosynthesis, producing about 30 g of sugar in this tulip poplar in an hour. The carbon contained in the newly formed sugar enters the leaves as carbon dioxide (CO_2) during the same hour as the light interception. Small, adjustable openings (stomata) in the underside of the leaves allow CO_2 to diffuse into the interior of the leaves as photosynthesis depletes the concentration of CO_2 inside leaf cells. The rate of diffusion from the air into the leaf depends on the difference in concentration between CO_2 in the atmosphere and inside the leaf. The air has about six times the concentration of CO_2 that would be found inside photosynthesizing leaves, providing a steep gradient for the movement into the leaves. The remarkable biochemical processes in the leaves depend on the presence of more than a dozen elements in the tree, including 500 g of nitrogen (N) and 50 g of phosphorus (P). The bulk of these nutrients were taken up from the soil earlier in the season, but a sizable portion came from reserves that were recycled from last year's leaves and stored over winter in the wood.

The 30 g of sugar produced during an hour would be associated with a release of about 30 g of oxygen (O_2), as oxygen is released when water is split as part of photosynthesis. It may seem that this oxygen could be an important source of oxygen for the atmosphere, but it isn't. As with all accounting in ecology, half a picture might lead to the wrong conclusion. The sugar produced by the tree may be "respired" fairly soon to support the growth of new cells or to maintain old cells, and oxygen is consumed (reforming water) in this reaction. Some of the sugar ends up in longer-lived cells, but even these tend to be oxidized back to CO_2 over years or centuries. Unless the carbon content of a forest increases across generations of trees, the generation of oxygen in photosynthesis is matched by consumption during respiration and decomposition, leaving no extra oxygen in the atmosphere.

Some of the sugar produced by photosynthesis is consumed within the leaf to produce and support the metabolic needs of cells in the leaf. More than three-quarters of the sugar is loaded into the phloem and sent to flowers, twigs, branches, stems, roots and symbiotic root fungi (mycorrhizae).

Exposing the moist interiors of leaves to the dry air allows for uptake of CO_2, but also allows water to be pulled into the dry air. The production of one molecule of sugar entails an unavoidable loss of hundreds of molecules of water. The production of 30 g of sugar in an hour would be accompanied by a far greater loss of water, perhaps 10 liters (10 kg) of water. The water transpired by the leaves during an hour of photosynthesis would have been found lurking in the soil a day earlier, and may have been in the atmosphere a day or a week before.

The tree has tremendous surface area developed within the soil to facilitate uptake of water and nutrients. The surface area of fine roots may be in the order of 100 times the surface area of leaves in the crown, and the surface area of mycorrhizal fungi that colonize roots contribute more than 10 times the surface area of roots. This vast surface area of absorbing roots and fungal mycelia collects water (and nutrients) that move up through the sapwood of the tree. The sapwood is comprised of xylem vessels, each measuring about 0.1 mm in diameter by 1 mm in length. The water passes through more than 1000 vessels for every meter of tree height, taking half a day or a day to move from all the way from roots to leaves.

Lifting water from the soil to the crown requires energy to overcome gravity, about 300 J for 10 liters. This is a tiny amount of energy compared to energy consumed as liquid water in leaves becomes water vapor in the atmosphere (about 2400 kJ for each liter, or 24 MJ for 10 liters). All the energy consumption and dissipation by the tree crown result in a deep shade beneath the tree. The air temperature in the shade may be a degree or two cooler than the air above the crown, but the shade will feel much cooler to a person sitting under the tree because of the greatly reduced energy load from the incoming sunlight.

Tree Physiology Follows Daily Cycles

Over the course of a day, the tulip poplar responds to the changing environment through a daily cycle just as strongly as an animal would. The uptake of CO_2 (and loss of water) begins as the sky brightens across the hillside in the morning, increasing as the intensity of sunlight increases (Figure 1.2). Rates may be highest near noon, decreasing if clouds develop, or if the air becomes so dry that the tree tightens the stomata to avoid losing too much water. Increasing temperatures in the afternoon drive up the capacity of air to hold water, resulting in a climb in the vapor pressure deficit. This deficit is a key force driving the water use by the tree. The tulip poplar would produce about 250 g of sugar on a sunny summer day (more than the average mentioned above for all days of the growing season), when the soil was moist, and transpiration could total 70 liters of water.

Not all processes in the tree shut down when the sun sets. Chemical reactions inside cells continue to renew thousands of biochemicals, generating and expanding new cells, and actively absorbing nutrient ions (such as nitrate and phosphate) from the soil. All of these processes require energy, most of which is supplied directly or indirectly from the sugars formed by photosynthesis. The oxidation of the sugar leads to substantial release of CO_2 from the tree; this "respiration" in all the tissues of a tree may equal half of the total photosynthesis that occurs on a sunny day.

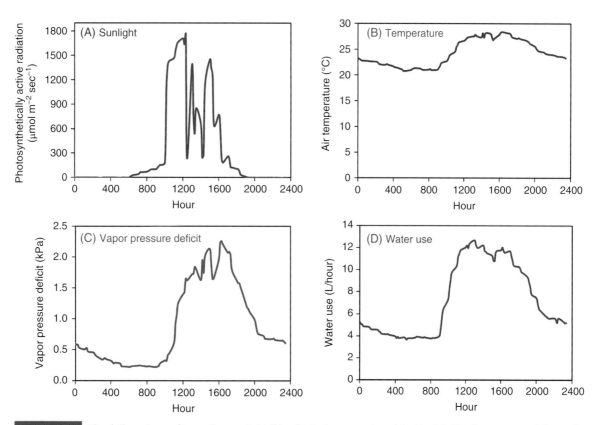

FIGURE 1.2   The daily pattern of incoming sunlight (A) reflects the geometry of the Earth's tilt, the aspect and slope of a hillside, and the passing of clouds through the day. Temperature patterns (B) are driven in part by incoming sunlight, moderated by winds and evaporation of water (which cools the air). The combination of temperature patterns determines the capacity of air to hold water, and the vapor pressure deficit (C) tracks the difference between the current humidity of the air and the saturation point of the air. All these factors influence the rate of water use by the tulip poplar (D), though the connection to vapor pressure deficit is the most direct. **Source:** Data from Chelcy Miniat.

Trees Must Cope with Seasonal Cycles Through Each Year

The environment surrounding the tulip poplar changes through the course of a year. The daily cycles of temperature swing between 5 and 10 °C, while the coldest and warmest day of the year may differ by 25 °C or more (Figure 1.3). Incoming sunlight in winter averages about half the level experienced in summer, as a result of shorter days, lower sun angles, and fewer clouds. These environmental changes lead to regular, predictable patterns in the phenology of the tree.

The tulip poplar begins an annual cycle of flowering and growth with the initiation of root growth late in the winter, followed by flowering in April and May. The tulip-shaped flowers are pollinated by bees and other insects, with 10 000 seeds raining from the crown in autumn. The leaves of the crown also expand in April and May, from expanding buds that were set the previous year. The initial burst of leaf growth depends on stored sugar, but the leaves rapidly provide new sugar for their own growth, and for the growth of all parts of the tree. The growth of a new leaf requires only about one-week's production of sugar; the rest of the span of the leaf's existence contributes to the growth and maintenance of other tissues.

Over the course of the growing season, the tulip poplar may produce 50 kg of sugar. Respiration would consume half, and the growth of short-lived roots and leaves might consume another quarter. Less than 25% of the annual production from the tree's leaves would be found in new stem wood, increasing the diameter and height of the tree. The annual growth of the tree might use more than 8000 liters of water.

Dry periods during the summer lower the rate of photosynthesis in two ways. Low supplies of water in the soil lead to closure of leaf stomata, restricting both the gain of CO_2 and the loss of water. The tulip poplar might also respond directly to the dryness of the atmosphere, and days with very high vapor pressure deficits may have low rates of photosynthesis, even if the soil is moist.

Trees Grow and Reproduce at Times Scales of a Century

Tulip poplar trees originate from seeds that develop following pollination of a flower by a bee or other insect. The flower may have developed on a parent tree as young as one or two decades, or as old as one or two centuries. Seeds develop over a period of five or six months, and then fall to the ground within a radius equal to a few times the height of the parent tree. A single seed

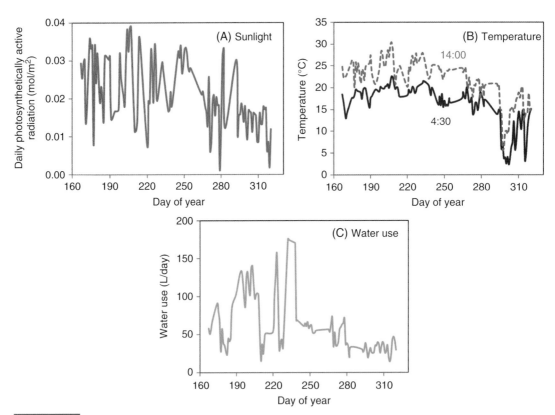

FIGURE 1.3   Seasonal trends in incoming sunlight (A) lead to almost twofold differences between summer and winter. The difference might be larger if not for the frequent cloud cover in summer. Patterns in incoming light lead to both daily and seasonal patterns in air temperatures (B). These environmental driving forces combine with the biology of the tulip poplar to determine the seasonal course of water use by the tree (C). **Source:** Data from Chelcy Miniat.

may germinate the following summer, or several summers later. The vast majority of seeds may never germinate, or if they germinate may not lead to a successfully established seedling. New seedlings require a great deal of luck to establish, including obtaining enough water and nutrients from the soil to support the development of leaves, and enough sunlight (perhaps 10% of full sun) to drive photosynthesis. The full intensity of sunlight may dry out a seedling, or overwhelm the photochemistry of new leaves.

A tulip poplar stem may not be the "first generation" of the "tree." A tree stem may die (from a wind storm breaking the stem, or a saw harvesting the tree), and a new stem may develop from dormant buds in the stump. The early growth and development of sprouted stems is faster and more assured than the tenuous development of a new seedling.

Weather differs a lot from one year to the next, and the growth of the tree during favorable periods may be double the growth rate for droughtier times (Figure 1.4). This response of an individual tree is the outcome of several factors, including the direct effects of the environment on this tree's physiology, and the indirect effects of how fluctuations in the environment change the competition between trees in the forest.

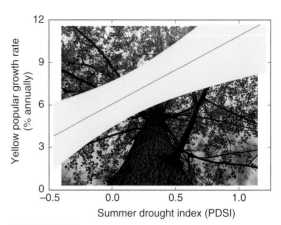

FIGURE 1.4   Growth of yellow poplar trees is low in drier summers (a negative value for the Palmer Drought Severity index), and increases with increasing summer moisture. **Source:** Data from Kardol et al. 2010.

The tree is larger than its local neighbors, and this "dominance" provides a twofold advantage. The tree obtains a higher supply of light, water, and nutrients than its neighbors, driving faster growth. Faster growth then leads to a positive feedback that increases the tree's capture of resources, allowing its growth to increase at the expense of neighbors.

The Story of Forests Is More than the Sum of the Individual Trees, Because Interactions Are So Strong

The tulip poplar tree is enmeshed in a complex ecological system (Figure 1.5). The tree provides habitat for an intricate community of insects and other arthropods. Each kilogram of leaves supports a total arthropod community of about 1 g (Schowalter and Crossley 1988), so the total leaf mass of the crown of 25 kg would support about 25 g of arthropods. Some of these invertebrates feed on the tree, eating leaves (or the insides of leaves), sucking sap, and boring into the wood of branches and the stem. Occasionally the populations of tree-feeding insects might increase to the point where much of the forest canopy is eaten; in most cases trees survive defoliation by native insect herbivores and form new canopies in the same season. Forests have other arthropods that feed on the species that feed on trees, forming complex food webs that include small mammals, a dozen or more species of birds, and even fish in streams and ponds that feed on arthropods from within the forests.

Does the tree benefit from neighbors, or is competition for resources the major effect of neighbors? Competition between trees is very important in all forests, but some possibilities exist for interactions between trees that actually benefit neighbors. One example is having a nitrogen-fixing black locust tree as a neighbor. The tulip poplar would compete with the locust for light, water, and other nutrients, but it might benefit from the enrichment of the soil N supply by the locust. Dozens of species of plants in the understory also compete with overstory trees for soil water and nutrients.

The diversity of plant species may be impressive, but this diversity is overshadowed by the diversity of invertebrates. Each square meter of soil contains about 60 large invertebrates with a total mass of about 1 g (Seastedt and

FIGURE 1.5   The dominant tulip poplar tree in the center of this springtime photo is part of a complex ecological system that includes other tulip poplars, other trees from more than a dozen species, several dozen species of understory plants, hundreds of species of arthropods and other invertebrates, and a soil that is itself a complex system with a level of biodiversity that dwarfs the diversity of the rest of the forest.

FIGURE 1.6 Although this looks like a topographic map of the Coweeta Basin, the colors actually represent the amount of water available for use by trees (hot colors are droughty sites, cool colors are wetter sites), and for draining into streams. Higher elevations receive more rainfall (and snow) than lower elevations, but water also flows downslope through soils, enriching lower parts of landscapes. **Source:** Map provided by D.L. Urban.

Crossley 1988). The number of small invertebrates would be on the order of 10 000 individuals (from hundreds of species) in each square meter; most of these feed on soil fungi.

Each kg of the upper mineral soil contains about 1 or 2 g of fungi, bacteria, and Archaea (Wright and Coleman 2000). The microorganisms are responsible for the majority of the processing of dead plant materials, returning carbon dioxide to the atmosphere, releasing inorganic nutrients into the soil, and altering soil structure and aggregation in ways that protect some organic matter from decomposition for decades, centuries, and even millennia. The small size of the soil microorganisms is matched by an almost unimaginable diversity of "species" or taxonomic units (as the concept of species does not apply well to many microbes). A 10 m by 10 m patch of soil likely contains more than 1000 species (or taxonomic units) of Archaea, another 1000 species of fungi, more than 10 000 species of bacteria, and 10 000 varieties of viruses (Fierer et al. 2007). This biocomplexity remains a largely unexplored frontier in the ecology of forests.

No two locations in the Coweeta Basin have exactly the same forest structure and composition, because local details (such as small variations in soils, or legacy of historical events) always shape local forests. Some broad forest patterns do repeat across the landscapes, as a result of patterns in topography. Precipitation increases by about 5% with each 100 m increase in elevation, rising from about 1500 mm yr^{-1} at 700 m elevation to more than 2200 mm yr^{-1} at 1500 m. Local topography modifies this elevational pattern, as wind flow near ridges can lead to 30% less precipitation falling below the ridgelines than would be expected based on elevation alone (Swift et al. 1988). The water available for use by trees (and flow into streams) depends heavily on local topography. Forests on ridgelines receive water from precipitation, and lose water through evaporation, transpiration by plants, and seepage downhill. Forests lower on the landscape receive water not only as precipitation, but also as water draining from higher slopes. Although more rain falls at higher elevations at Coweeta, some forests at lower elevations have access to more water because of this downhill flow (Figure 1.6).

Temperature also changes with elevation, falling by about 0.5 °C for every 100 m gain in elevation; moist air shows less temperature change with elevation than dry air. The landscape pattern in temperature is also strongly influenced by slope and aspect; the amount of incoming sunlight can vary by more than a factor of two from south-facing slopes to north-facing slopes, generating temperature differences of several degrees. Steep slopes receive more light than flat areas if the aspect points toward the sun, or less light if the aspect faces away from the sun.

These patterns in soil water, sunlight, and temperature lead to predictable patterns in forest structure and composition. Concave slopes (coves) have abundant supplies of water and deep soils, with large forests dominated by tulip poplar, black birch, and eastern hemlock. Dry ridges and convex slopes have smaller forests of oaks and pitch pine. Uniform slopes at lower elevations have mixed-deciduous forests dominated by white and red oaks, hickories, and nitrogen-fixing black locust. Uniform slopes at higher elevations are typically dominated by northern hardwood forests, with sugar maple, red oak, and beech.

Differences in species with elevation and topography also lead to differences in forest diversity and size. Lower elevation forests in the Coweeta Basin average about 18 tree species in a hectare, with diversity declining to about 14 tree species ha^{-1} at upper elevations (Figure 1.7). Diversity shows no trend with topography, as concave locations (coves) have about the same number of species ha^{-1} as convex (ridge) locations. The largest forests occur at middle elevations, and in concave locations.

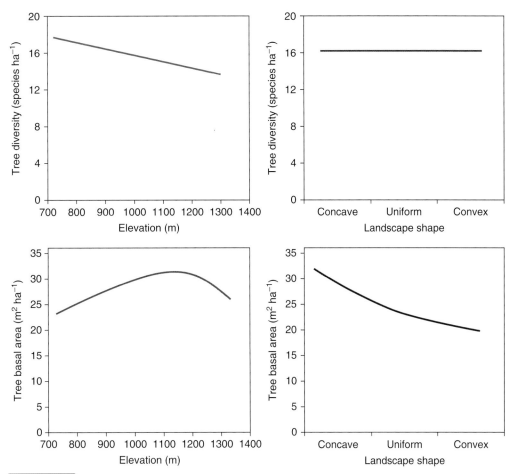

FIGURE 1.7   Forest patterns commonly vary with elevation and with local topography. The number of tree species occurring in a hectare at Coweeta declines slightly with increasing elevation (upper left), whereas tree diversity shows no pattern among concave (cove) locations through to convex (ridge) locations (upper right). The basal area of trees tends to be highest at middle elevations (lower left), and in concave slope locations. **Source:** Data from Elliott 2008.

The Coweeta Forests Aren't the Same as Two Centuries Ago

Forests with large, old trees may give an impression of an unchanging system that seem to be stable for decades and centuries. Some temperate forests may fit this image, but most are quite dynamic. If we could visit a forest before and after 50 years of changes occurred, we would likely find that many of the small trees had died (perhaps replaced by others), along with some of the medium- and large-size trees. The overall size of the forest, in terms of height or mass of wood in living trees, may have increased, but typically this increase in the size of larger trees comes in part at the expense of smaller trees that died.

Forests also change more rapidly, as a result of rapid events that alter the typical year-to-year progression of changes. The forests at Coweeta experienced massive changes in the past two centuries (Figure 1.8) as a result of direct human impacts and unintended, indirect impacts.

The most noticeable change in the forests in the Coweeta Basin is the loss of the formerly dominant tree species, American chestnut. Long-lived, large chestnut trees were the most notable part of the forest in 1900. About half the trees in the forest were chestnuts, and chestnuts comprised about half of the forest biomass. An exotic fungal disease from Asia, chestnut blight, killed almost all the mature chestnuts in forests of eastern North America within a few decades. Not all the mature trees were killed outright, as the fungus creates a canker on the stem that topples the tree. Surviving root systems continue to send up hopeful shoots, but these also form cankers when the stems are few meters tall.

What did the demise of chestnut mean for the forest? Given that competition is so important in the interactions among trees, the loss of chestnut led to a dramatic increase in the biomass of other species, particularly oaks, red maple, and tulip poplar. These species responded not by increasing the number of trees in the forest, but with accelerated growth of the already-present stems.

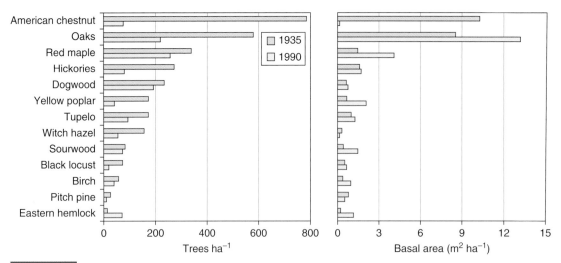

FIGURE 1.8 Forest composition in the Coweeta Basin in 1935 and in 1990. The total density of trees (left) declined from 3000 trees ha^{-1} to 1200 trees ha^{-1}, while basal area (right) increased slightly from 27 to 28 m^2 ha^{-1}. The decline in tree density is a common feature of growing forests; as dominant trees increase in size, many smaller trees die. However, the trend in this forest was largely influenced by the drastic decline of chestnut. This formerly dominant species was decimated by the exotic chestnut blight disease. **Source:** Data from Elliott 2008.

Another major event reshaped parts of the forests of Coweeta after the vegetation survey in 1990 summarized in Figure 1.8. Before this time, eastern hemlock was a major tree species in wetter locations, such as coves and valley bottoms. The hemlock woolly adelgid is an exotic, invasive insect that has killed most of the eastern hemlocks across much of eastern North America. About half the hemlocks died within the first 5 years of the adelgid's arrival (Ford et al. 2012), with 90% or more dead after 15 years (Ford et al. 2012; Abella 2018). The loss of large hemlocks led to drops in the number of trees in the forests by about half, and long-term changes may include expansion of both trees of other species and understory woody plants (such as rhododendrons).

Continuing back in time, the most notable event of the nineteenth century was the extinction of the passenger pigeon. This large (30 cm) bird was a major consumer of large tree nuts, including acorns, beechnuts, and chestnuts (Halliday 1980; Johnson et al. 2010). Huge flocks contained tens of thousands (perhaps even millions) of birds, with large impacts on dispersal of tree seeds and nutrient cycling (with concentrated feces beneath favored roost and nest trees). Passenger pigeons were the most dominant species of bird in eastern North America, and perhaps the most numerous bird species in the world. Over the course of a few decades, this species spiraled to extinction as a result of massive hunting, and perhaps the effects of changing forest cover and even exotic diseases. How did the forests change in the absence of passenger pigeons? This important question is easily asked, but probably cannot be answered without stronger historical records.

Two other human-related events marked the 1800s in the Coweeta Basin. The second half of the century saw Euroamerican settlers occupying the land. Their major impacts included some logging, agricultural cropping on a few hundred hectares, widespread grazing of pigs and cattle, and hunting of wildlife for food. The prior inhabitants were Cherokee Indians, forcibly removed in the 1830s. Cherokee influences on the forests included some agriculture (maize, squash, and beans), extensive hunting (primarily deer, turkeys, and bears); food collection (including tree nuts); and frequent use of fire to clear the forest understory (Van Derwarker and Detwiler 2000; Gragson and Bolstad 2006).

Across Dozens of Generations of Trees, Almost Everything Changed at Coweeta

The past 10 000 years have seen dozens of generations of trees and forests come and go in the Coweeta Basin, in response to fluctuations in climate, events such as hurricanes, and probably sizable fluctuations in populations of humans and other animal species that influence forest dynamics. The frequency of fires may have increased as people ignited forest fires (intentionally or unintentionally). Fires may have burned the tulip poplar/mixed broadleaf stands every 200 years or so over the millennium before European settlement (Fesenmeyer and Christensen 2010). Some notable events include a near-disappearance of eastern hemlock throughout its range, between about 5500 and 6500 years ago (Calcote 2003; Heard and Valente 2009), followed by recovery. The cause of the decline is unknown, and speculations include some sort of novel disease. This also happened to be one of the coolest times in the past 10 000 years, so multiple factors may have been involved.

Continuing back to 12 500 years ago, the continent (and much of the globe) was undergoing rapid warming as the most recent Ice Age ended. Temperatures in the Coweeta Basin would have risen by more than 5 °C from conditions that prevailed for 100 000 years. Under colder conditions, the forests in the Basin would have resembled forests that are currently found farther north, with pines and spruces dominating even the lower elevations. During some periods, the assemblages of trees species across the region included combinations that have no modern analog in local forests, or in forests now found farther north (Jackson and Williams 2004). Assemblages of tree species change in response to interactions among temperature, precipitation, and biotic factors. Unlike organisms, the genotypes of forests change routinely as species come and go.

The most notable difference in the forest at the end of the Ice Age would have been the presence of many large species of mammals in the region. The list of now-extinct species includes tree-browsing American mastodons; grass and tree-browsing Columbian mammoths; woody-plant browsing stag moose; tree-eating giant beavers more than 2 m in length; and large predators such as dire wolves, sabretooth cats, and massive short-faced bears. The now-extinct mammals would have been joined by at least one now-extinct tree species, Critchfield spruce (Jackson and Weng 1999).

The Futures of the Tree and the Forest Will Depend on Both Gradual, Predictable Changes and Contingent Events

The future is largely unpredictable for individual trees, but some predictions may have a high probability of coming true. The dominant situation enjoyed by the tulip poplar featured in this chapter would generally predict steady growth into the future. Growth might even increase as neighbors are suppressed. Dominant trees of this species may live for more than two centuries, and such a long lifespan provides opportunities for dispersing millions of seeds.

A long lifespan also increases the odds that the tree will experience rare weather events. For example, a severe drought with a probability of occurrence once in 100 years might severely challenge a tree's survival. A tree that lives only about five decades would have a 60% chance of never experiencing a 100-year-magnitude drought (if weather is random), whereas a tree that lived two centuries would have an 87% probability of experiencing at least one 100-year drought.

A host of other future factors are more difficult to assign probabilities. The death of a neighboring tree may suddenly increase the supplies of resources available to this tree, or the falling neighbor may collide and uproot this tree as well. Lightning tends to kill large trees more often than smaller trees. Outbreaks of insect populations and fungal diseases influence the long-term development of many forests. The climate experienced by this tree (and its ancestors) may not continue into the future. Novel pests may arrive in the forest, as a result of widespread transport associated with world-wide travel by people and materials. The future of the tree may also depend very heavily on choices made by people; a large tulip poplar tree can be transformed into thousands of dollars-worth of furniture and other products.

Some changes in a forest tend to be cyclic, with repeating patterns of species and growth rates following major events. The major recolonizing species will have predictably high tolerance for full sunlight and rapid early growth, whereas trees that remain after two centuries will likely grow slowly and the community will include trees that thrive under shady conditions. Other changes are clearly not cyclic, and lead us to expect that the future forests in the Coweeta Basin will not be simple analogs of past forests (Jackson and Williams 2004). The development of forests responds to changes in climate, and climatic patterns (and the responses of trees and species to these patterns) have long legacies (Kardol et al. 2010). Changes in future climates may have modest effects on the forests compared to novel insects and diseases. The chestnut blight removed the dominant tree species from the Coweeta forests, and the hemlock wooly adelgid decimated the population of eastern hemlock trees. What will be the legacies of the loss of almost all the chestnuts and hemlocks trees from Coweeta's forests? Might we be able to predict the response of surviving species to the disappearance of hemlock, based on the patterns from 6000 years ago when eastern hemlock experienced another decline, or will other factors (such as changing climate) limit the ability of the past to illuminate the future? We might speculate about how other species will take advantage of reduced competition from these species, but the actual impacts will include the ecological legacies of changes in soils and in animal communities. Forests often respond to more than one event; future forests develop from the combined legacies of historical events (such as losses and gains of species) in combination with current conditions. Warming climate, rising atmospheric concentrations of carbon dioxide, and other factors will influence future forests, shaping the legacies of the losses of chestnuts and hemlocks. Will new species of exotic insects arrive and remove other tree species from Coweeta's forests?

The future development of a tree, and of a forest, derives from the gradual accumulation of routine changes, such as annual increases in height and mass of stems. Over limited periods, these gradual, expected trends are punctuated by contingent events that are largely unpredictable, such as hurricanes and invasions by exotic pathogens. Humans are another force for change in forests, through direct management (typically favoring some species over others, often limiting the opportunity for old trees to develop) and indirect activities (such as nutrient enrichment of rain, air pollution, and climate change).

Given all these forces of change, how can we predict future forests? The short answer is simply that we cannot predict future forests with much confidence. The longer (and more useful) answer is that we can indeed develop insights about the likely forests of the future, if we understand some of the basic features that have shaped forests in the past, and how ecological interactions will combine to shape future forest.

Ecological Afterthoughts: Is a Forest an Organism?

A variety of traits and processes characterize all organisms: they process high-energy sources from the environment (such as sunlight or organic compounds) into low-energy byproducts (such as heat), and they grow, reproduce, and die. Forests do these same things. So are forests like organisms? We have a strong tendency for using analogies to make sense of the world, and sometimes we go beyond analogies to use metaphors, where one is not simply like another, but is essentially the same. Ideas about forests have arisen commonly from analogies, and sometimes even from metaphors. For example, an influential ecologist asserted a forests-are-organisms metaphor a century ago:

> The unit of vegetation, the climax formation, is an organic entity. As an organism, the formation arises, grows, matures and dies... The life-history of a formation is a complex but definite process, comparable in its chief features with the life-history of an individual plant... Succession is the process of reproduction of a formation, and this reproductive process can no more fail to terminate in the adult form than it can in the case of the individual.

(Clements 1916)

Our ideas about forests can shape what we can see in forests, and the belief in the organism-nature of ecosystems led this ecologist to strong confidence in untested ideas, simply because he was seduced by the beauty of the organism metaphor:

> It can still be confidently affirmed that stabilization is the universal tendency of all vegetation under the ruling climate, and that climaxes are characterized by a high degree of stability when reckoned in thousands or even millions of years.

(Clements 1936)

A metaphor that was true might be very useful, but a poor metaphor may be useless or even harmful. An untested metaphor could be a good starting point for science, but could not be a reliable conclusion. If forests were the same as organisms, the future composition, structure and function of forests would be largely predictable. Any deviations in that progression would risk the continued persistence of the forest. If forests are quite unlike organisms, such a belief would befuddle our ability to see the forest and the trees.

The "ecological afterthoughts" in later chapters are open-ended invitations to apply ideas from the chapters to specific situations. The afterthoughts are not intended to convey information or answers, but just to raise questions. This first chapter goes a bit further, highlighting how the afterthoughts might be used for insights.

A listing of similarities and differences would immediately show this metaphor of "Forests are Organisms" would be weak at best, and maybe harmful if taken too seriously. Forests clearly differ from organisms in fundamental ways (Figure 1.9). A tulip poplar seed can only lead to a tulip poplar tree, with growth rates and forms that are shaped by environmental factors and the genes of the tree. A tulip poplar that deviated from normal structure and function would soon be a dead tulip poplar, with no chance to send more of its genes into future generations.

A forest that contains tulip poplar trees is much less constrained in its future development. Unlike organisms, forests routinely gain and lose genes as members of species enter and leave the forest; there is no single way for a forest to be, and no single path that must be followed if a landscape will remain dominated by trees. If we believe forests are organisms, the loss of major components should be expected to endanger the whole. The death of an organism is an event that encompasses all its parts. The "death" of a forest is always a matter of perspective; major events kill some trees, plants and animals, leading to greater opportunities for the surviving trees, plants and animals. Forests persist through the gains and losses of individuals and species; organisms generally don't persist through the gains and losses of organs (aside from seasonal senescence of leaves and fine roots). A tree that lost its leaves and never regrew a new set of leaves would die; a forest that lost all members of a given tree species (with that species never returning), would remain quite viable.

Several of the dominant species formerly found in tulip poplar forests disappeared from the landscape in recent times, including chestnut trees, passenger pigeons, wolves and mountain lions, while the human influences shifted from Cherokee to European cultures. Nevertheless, forests that contain tulip poplar trees continue to exist and change, as individual organisms and species shift in response to changing stresses and opportunities. The forests of the future will not be the same as those in the past, and change over time is a normal aspect of forests.

FIGURE 1.9 As with the tulip poplar and tulip poplar forest examined in this chapter, all *trees* have very limited scope in their development from seed to mature tree, and all *forests* have very broad scope in their composition and structure over time. This ponderosa pine tree (left) developed from a seed that contained genetic material from two parent trees. The potential future states of this plant were determined and constrained by these genes. The actual development of the tree depended on climate, weather events, fires, and interactions with a huge variety of microorganisms, animals, and other plants. No two ponderosa pine trees are identical, yet the range of differences among ponderosa pine trees is miniscule compared to the range differences in the composition of forests that contain ponderosa pine trees. Unlike the tree, the forest where this tree grew did not have a particular beginning; major events such as fires killed many trees, but many survived (including the root system of aspens that sent up a new generation of stems from ancient root stocks), as did many of the understory plants. The dynamics of the forest were not determined or constrained by the genes of a single species, and the composition of the forest shifted over the decades in response to climate, rapid events, and management. The forest continues as the death and birth of individuals subtracts and adds genetic possibilities for the future.

This flexibility in the genetic composition of forests ensures that the organism metaphor confuses rather than enlightens. Complex, changeable forests have far greater capacity for change than organisms, ensuring that the future development of forests is anything but precisely predictable.

CHAPTER 2

Forest Environments

Forests change across landscapes, and some of the changes result from gradients in environmental factors such as temperature and available water. Patterns of forest change across landscapes relate broadly to patterns of environmental factors, such as typical changes in the dominant species as elevation increases in mountains (Figure 2.1). Some broad generalizations may explain general patterns: temperatures are lower at higher elevations, and precipitation is generally (but not always) higher. The underlying details that influence both the broad patterns and specific cases are a bit complex and highly interacting. This chapter develops a foundation for understanding the fundamental physics that influence tree occurrence and growth. Later chapters expand to examine how ecological interactions (including plants, animals and microbes) and environmental factors combine to shape forests.

All the chemical and biological processes within a forest are driven by physical processes: environmental temperatures, water regimes, and solar radiation. In turn, the chemical and biological processing of solar energy shape the microenvironmental conditions found in forests. Interception of sunlight by trees not only drives photosynthesis, it leads to evaporation of water from leaves, cooling the leaf and increasing the humidity of the air. Evaporation of water from leaves (transpiration) physically lifts masses of water and nutrients from the soil up to the canopy, and cools leaves heated by sunshine. Energy "fixed" through photosynthesis cascades down a series of chemical reactions, driving the forest's chemistry and biology until all the energy has been dissipated as heat. All the interesting biology and ecology that occur in forests take place within this physical system.

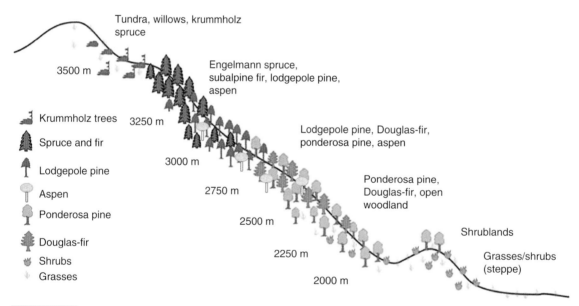

FIGURE 2.1 The vegetation in the Front Range of the Rocky Mountains in northern Colorado, USA grades from primarily grass/shrub domination at low, dry elevations, through open pine-dominated woodlands, to aspens mixed with a variety of conifers, to forests of spruce and fir and then to tundra (alpine) at the highest elevations (**Source:** based on a diagram from Laurie Huckaby). The patterns depend on direct effects of environmental factors on tree physiology, and also environmental influences on fires, insects, and other forest-shaping agents.

Climate Influences Where Forest Occur, and How They Grow

The distributions of forests can be mapped across the world, showing regions where major forest types occupy most of the land-scapes (Figure 2.2). Forests essentially occur everywhere there is enough water for trees to cope with evaporative demands driven by temperature and atmospheric humidity. The two exceptions to this generalization would be areas that are too cold, and areas that have enough water but high frequency of fires leads to greater success of grasses and fire-tolerant shrubs.

The productivity of the world's vegetation has been mapped by satellites, where the wavelengths of light reflected from vege-tation across time periods are combined with simulation models to estimate growth rates. Such global estimates may come close

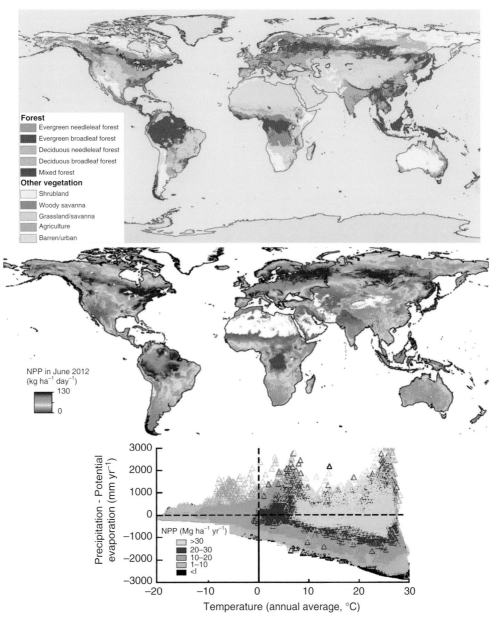

FIGURE 2.2   The distributions of major types of forests can be mapped across the world (upper), along with typical rates of aboveground net primary production of forests and other vegetation types (middle; for more information on production, see Chapter 7; **Source:** maps based on Pan et al. 2013). The spatial patterns can also be examined in a functional way, where aboveground net primary production is plotted relative to annual temperatures, and the balance between precipitation and the energy available to evaporate water (wetter conditions occur above the 0 point on the Y-axis; **Source:** Based on Running et al. 2004.

to pegging the actual global averages and totals, but the accuracy for any single region or landscape would need to be verified for confidence (as with any map).

The same information can be examined in relation to environmental factors rather than spatial locations (Figure 2.2). It's not surprising that the most productive forests (and other vegetation types) tend to occur in warmer locations where precipitation is close to (or higher than) the water that could be evaporated by available energy. Is it possible to be too warm for high growth rates, even where plenty of water is available? This is not clear from Figure 2.2, but the hottest tropical forests do appear to have lower growth rates than somewhat cooler forests (see Figure 2.18).

Warmer Forests Have More Species of Trees

Patterns in forests across environmental gradients can also be examined in terms of structure and species composition. Trees tend to be taller under favorable environmental conditions of moderate-to-warm temperatures and high rainfall (Figure 2.3). Part of the geographic pattern in tree heights relates to the physics of supplying leaves high above the ground with sufficient water to meet the evaporative demand of the atmosphere; dry air requires more water than the tallest trees could transport (see Chapter 4). The number of species found in forests is low at higher elevations, and high at lower elevations and in warmer areas.

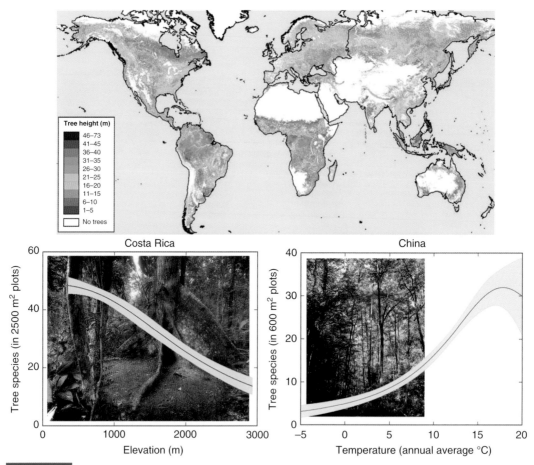

FIGURE 2.3   The tallest trees in the world occur in cool, wet locations, and most tall trees are limited to relatively wet areas. More severe growing conditions limit average tree heights (**Source:** upper map, from Pan et al. 2013/Annual reviews,Inc.). The diversity of tree species in a tropical forest in Costa Rica dropped by about half with a 1000 m increase in elevation (lower left, **Source:** Data from Veintimilla et al. 2019, photo by Cristian Montes). Across eastern China, the number of tree species that co-occur in 600 m² (=0.06 ha) plots increases rapidly with temperature, but levels off (and might even decline) on the hottest sites. **Source:** Based on Wang et al. 2009.

Chemical and Biological Reactions Go Faster with Increasing Temperature

Extreme temperatures tend to stop biological processes. Little biological activity occurs below the freezing point of water, and at very high temperatures organs fail and organisms die. Between these extremes, most biological processes respond strongly to temperature, often changing by a factor of two or more for a 10 °C change in temperature (Figure 2.4). Chemical reactions may follow a simple trend, but the response of biological processes derives from a suite of interacting factors. Some biological reactions such as respiration (the oxidation of carbon compounds to release energy and CO_2) may show an exponential temperature trend, as the effect of temperature on breakdown of carbon compounds is similar to simple chemical reactions.

The temperature effect on more complex biochemical reaction rates includes any effects on the breakdown and regeneration rates of enzymes and other proteins. Biological reactions also change as temperatures affect the supplies of reactants. Higher temperatures lead to greater evaporative stresses, which reduce the supply rate of carbon dioxide flowing into leaves. The products of reactions can suppress rates of further reaction if they accumulate in cells, which can be a problem at high temperatures. Rates of photosynthesis do not increase with a simple exponential trend like rates of respiration (Figure 2.3). The temperature effect on rates of photosynthesis might have a humped shape, reflecting the balance between carbon gains and losses in leaves, as photosynthesis declines in response to rising photorespiration (the diversion of energy to producing water rather than sugar).

The temperature of an object, such as a tree leaf, represents the thermal energy contained within the object. Thermal energy is the kinetic energy of molecules; the molecules of nitrogen in a volume of air move at velocities of about 450 to 500 m per second. A molecule of nitrogen might move about 100 nm before colliding with another air molecule, which translates into billions of collisions for each molecule each second. Molecules move faster as temperatures rise, leading to more collisions and exchanges of energy that we measure as temperature. Molecules are packed more densely in liquids and solids, leading to higher rate of collisions (and transfers of heat).

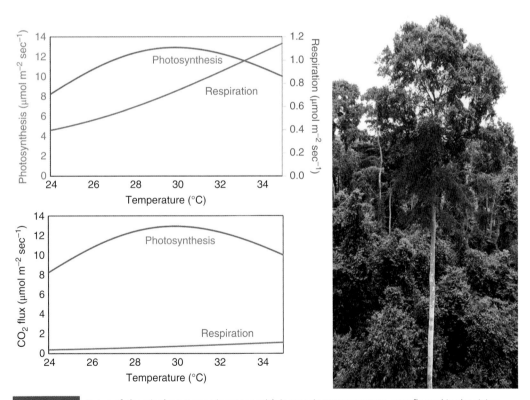

FIGURE 2.4 Rates of chemical processes increase with increasing temperature, as reflected in the rising curve for respiration of trees in a rainforest in Costa Rica (**Source:** Based on Cavaleri et al. 2010). The rate of total (gross) photosynthesis by fully illuminated leaves in a rainforest in Panama also rises with temperature to an optimum near 30 °C, and then declines (**Source:** Based on Slot and Winter 2017). The two graphs present the same data, with the upper graph focusing primarily on how each process responds to temperature, while the lower graph uses a single Y axis to give show how much larger photosynthesis is relative to respiration (across the temperature gradient).

Temperature is the Balance Point Between Energy Gains and Losses

A leaf increases in thermal energy when exposed to radiant energy from the sun, and when exposed to hot air. Absorption of solar radiation raises leaf thermal energy, raising leaf temperature. The temperature of the leaf continues to rise until the gain of energy is offset by energy losses. Leaves can lose energy by exposure to cooler air, by evaporating water, and by emitting (shining) radiant energy.

Objects such as tree leaves and bird feathers, lose energy to the air by conduction and convection if the air is cooler than the object (Figure 2.5). The rate of energy transfer into the air depends on the gradient in temperatures, and how well the object is "coupled" to the air. Large surfaces, such as beech leaves, have stable layers of air molecules comprising a thick boundary layer, slowing energy transfer from the leaf into the air at large. Small surfaces have thinner boundary layers allowing faster transfer of thermal energy; the boundary layer for the spruce needles was only one-fourth that of the beech leaves, explaining a large portion of the difference in temperatures between species. The coupling of surfaces to the atmosphere is improved by winds, as the thickness of the boundary layer declines with increasing wind. A wind of a few meters per second (a few kilometers per hour) can lower leaf temperatures by a few degrees, and increase the delivery of CO_2 to stomata along with loss of water to the air.

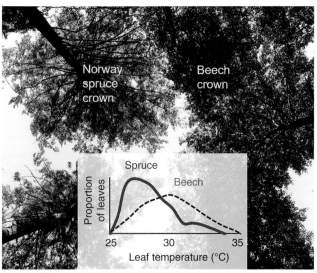

FIGURE 2.5 On an afternoon when air temperature was 25 °C, the temperatures of leaves of beech and spruce were sustained at higher levels as a result of heating by sunlight. The temperatures of leaves covered a range for each species, because of varying angles of leaf exposure to the sun's rays, and the variable levels of shade in the crowns. Beech leaves were hotter (most were about 30 °C) than spruce leaves (most were about 27 °C), indicating beech leaves had higher energy gain (absorbing more light), or lower energy loss (lower transpiration, or poorer coupling with the cooler air). **Source:** Data from Leuzinger and Körner (2007).

The evaporation of water requires large amounts of energy, about 2.4 MJ/l evaporated. This energy flow is referred to as "latent heat," because the temperature of the water molecules remains unchanged as the phase changes from liquid to gas. The amount of energy removed from an object such as a leaf depends of course on the rate of evaporation, and rates of evaporation depend on the water status of the plant, the dryness of the air, and the presence of wind to reduce the boundary layer. Evaporation from a leaf exposed to dry air without any wind may lower leaf temperature by about a degree; the addition of a light wind can increase evaporation enough to cool leaves by 3–5 °C.

All Objects Shine; Hot Objects Shine Brightly

Radiation from the sun can be felt by holding out a hand in bright sun; the hand absorbs the solar radiation and warms up. We can see the hand illuminated by sunlight, because some of the light reflects from the hand and reaches our eyes. Another radiation story is also occurring, invisible to our eyes. The hand is also "shining" like the sun, emitting radiation to the environment around it. The cooler temperature of the hand means the wavelengths of radiation emitted are much longer than sunlight (a long wavelength is the same as a low frequency). The emission of radiation by the hand removes energy, and would lead to cooling of the hand unless another source of energy kept the hand warm.

The radiant energy emitted by the sun is called "shortwave" radiation, because of the short wavelengths (between about 400 and 700 nm for the visible portion of the solar spectrum). Objects at temperatures commonly encountered in forests "shine" or emit radiation at longer wavelengths, on the order of 10 μm. Longwave radiation may be sensed by skin as warmth, even though the weak radiation cannot be detected by our eyes.

The differences in the wavelength of light that shines (or is "emitted") from an object are associated with very large differences in the amount of energy transfer. A very hot object emits shorter wavelengths of light, along with much higher amounts of energy (Figure 2.6). An object at room temperature (about 20 °C) shines out about 420 W of energy for each m² of surface area. If the same object warmed to 37 °C (typical human body temperature), the energy transfer would rise to 520 W m⁻². The loss of this emitted energy cools the objects, which in turn lowers the rate of further losses of energy. The energy loss may be counteracted by any energy being added to the surface from the environment (in the form of radiation, or hot air), and temperatures stabilize when energy gains from the environment match the energy losses.

Incoming Sunlight Decreases in Winter and at Higher Latitudes

The emission of light from the sun is essentially constant through a year, and the distance of the Earth from the sun differs by only a few percent (a bit closer during winter in the Northern Hemisphere). The strong patterns of variation in incoming sunlight with latitude and season of the year result from a tilted planet doing an annual revolution around the sun. Summers are warmer because incoming light is several-fold greater than in winters; the tilt of the Earth is toward the sun in summer, giving high-angle incoming light that lasts for more hours in the day (Figure 2.7). The incoming light for a flat site at 23° latitude in mid-summer is 2.5 times the amount received in January. At higher latitudes the difference between mid-winter and mid-summer much larger (eightfold at 43 °).

The intensity of incoming light also depends on the angle of a surface relative to the incoming angle of the sunlight. Incoming sunlight is at a maximum on a perpendicular surface, and the intensity lessens as a surface is tilted away to more oblique angles. High sun angles in mid-summer bring almost as much sunlight to a wide range of slopes, as high sun angles combine with the long path of the sun through the sky to mostly even-out the differences among aspects (which direction a slope faces; Figure 2.8). North-facing aspects receive little sunlight (in the Northern Hemisphere) because of fewer hours of direct sunlight, and lower angles of the sun relative to the slope.

These graphs of incoming radiation assume that all locations are fully exposed to the sun, but forested slopes are often shaded partially by surrounding topography. Shading by adjacent hillsides can be important, varying of course through a day and across a season in relation to the angle of the sun in the sky. A set of watersheds in northern Idaho, USA shows that the "blockage" of incoming light ranges from 0 to more than one-third when averaged across a year (Figure 2.9). This landscape shading has large effects on soil temperatures, the accumulation and duration of snow cover, and of course on vegetation.

How important are these differences in incoming sunlight? The differences are important enough that the typical elevation for a given species may be a few hundred meters lower on N-facing aspects (in the Northern Hemisphere) than on S-facing aspects. The latitudinal range of species may reach hundreds of kilometers farther south when N-facing slopes are available as habitat. Why does the incoming light make so much difference? It might seem that the simple answer would deal with the supply of light to drive photosynthesis, but two other factors are likely more important. The first is the seasonality of temperatures that favor growth. South-facing aspects are warmer throughout the year, which might benefit some species in the spring and autumn. Incoming solar energy is a major driver of evaporation, and S-facing aspects experience higher evaporative demands that may dry soils while soils on N-facing aspects remain moist. The apparent dryness of S-facing aspects is not a difference in precipitation inputs; the difference is in the drying effect of the extra radiation.

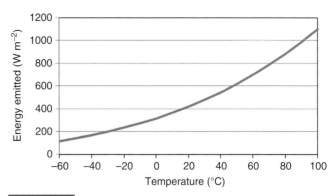

FIGURE 2.6 All objects emit radiation to the environment, and hotter objects emit more energy (for a given surface area) than cooler objects. If the temperature of a soil surface increased from 20 to 40 °C, the emission of energy from the soil into the air would increase from 420 to 550 W m^{-2} (based on Stefan-Boltzmann Law, and assuming that the objects are very good emitters ("black body" emission rates).

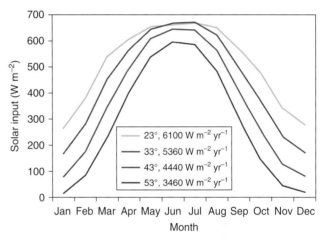

FIGURE 2.7 The total potential sunlight (not accounting for clouds) at 23° latitude (6100 W m^{-2} yr^{-1}) is almost double that at 53° latitude (3460 W m^{-2} yr^{-1}). The latitudinal differences are much larger outside the summer, because in summer the low sun angles that extend through very long day lengths give similar totals to sites with higher-angle sun for shorter days at low latitudes. **Source:** Based on spreadsheet by Nicholas Coops.

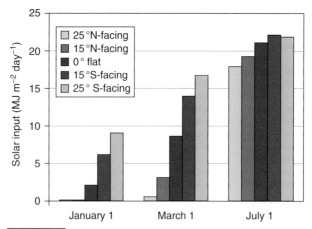

FIGURE 2.8   The daily amount of incoming sunlight depends on the aspect of a site. In the Northern Hemisphere at a latitude of 43°, a hillside that faces toward the north at an angle of 25° receives almost no direct-beam sunlight in winter, when a 25° hillside facing to the south receives 9 MJ m⁻². The effect of aspect is small in mid-summer. **Source:** Based on a spreadsheet by Nicholas Coops.

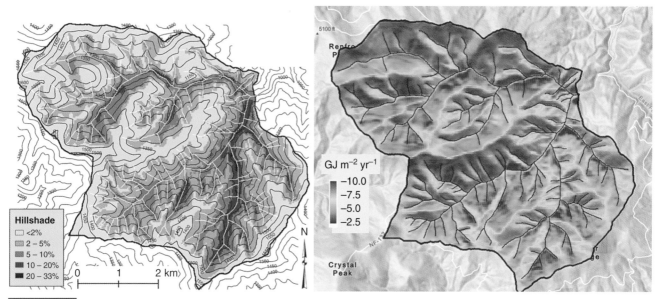

FIGURE 2.9   The amount of incoming radiation received by a site depends not only on latitude, slope angle and aspect, but also on whether nearby hillsides block sunlight (left). The shading effects in a mountainous landscape in northern Idaho reduces incoming sunlight by only a few percent on south-facing slopes, but by an average of 30% on north-facing slopes. The effect of slope combined with shading from surrounding hills resulted in a fourfold range of incoming solar radiation annually (right; **Source:** Wei et al. 2018 / Elsevier).

Forests Receive Shortwave Sunlight, and Shine off Longwave Radiation

The sun is so hot that the wavelengths of light emitted are strong enough to activate the light sensors in our eyes. Cooler objects, like surfaces in forests, are so cool that the radiation they shine is too low in energy to activate our eyes. These radiation patterns are illustrated in Figure 2.10 for a clearcut forest in Oregon, where the incoming solar radiation hits the soil surface. Most of the shortwave solar radiation is absorbed in the system, though 13% reflects away (which is handy for us, or our eyes could not see to

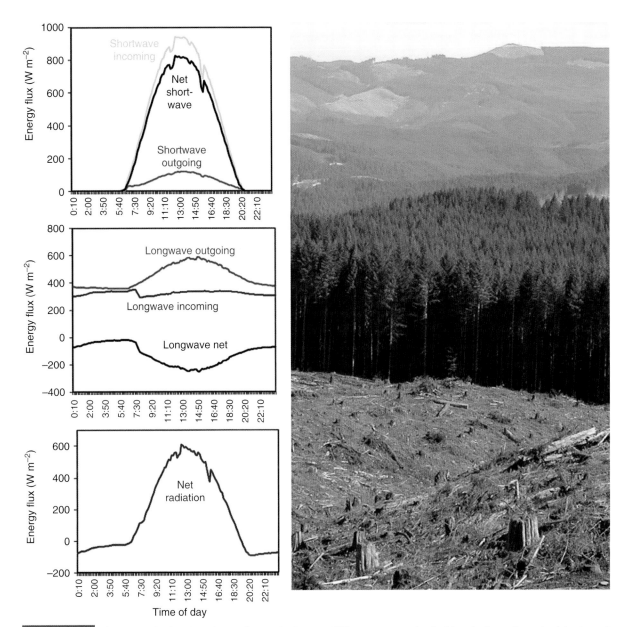

FIGURE 2.10   The energy budget for a forest clearcut in Oregon, USA on a summer day is driven by incoming solar (shortwave) radiation. About 13% of the light is reflected, with no effect on the forest. The rest of the solar energy is either absorbed by the forest (warming it), or evaporating water. The forest itself "shines" at long wavelengths, and the intensity of the emission depends on the daily trends in temperature. The warm air also emits longwave radiation to the forest, about $300\,\mathrm{W\,m^{-2}}$ through the day and night. The emission of longwave radiation from the forest increased by about 50% through the day, as absorption of shortwave solar radiation increased the thermal energy stored in the forest. The balance between incoming and outgoing fluxes of shortwave and longwave radiation determines how much energy is available to evaporate water or change the thermal energy storage (and temperature; **Source:** data from Mike Newton and Liz Cole).

walk across the site). The site also receives some longwave radiation, mostly from the warm air. The soil surface shines longwave energy back the sky, and in fact the soil is hotter than the air at midday and the longwave emissions remove more energy than the sky returns as longwave emissions. When the incoming shortwave radiation from the sun stops in the evening, and the emission of radiation from the soil leads to cooling.

Combining both shortwave and longwave budgets results in a large net gain of energy ($14\,\mathrm{MJ\,m^{-2}}$) to the soil across 24 hours. What does a net gain of $14\,\mathrm{MJ\,m^{-2}}$ mean for the site? It's possible that the energy moves deeper into the soil, contributing to the gradual warming of the soil over the summer. Some of the energy also leaves the site in the form of heated air that moves away. If the soil is moist, a large amount of energy could go into evaporating water (a latent heat loss): $14\,\mathrm{MJ\,m^{-2}}$ could evaporate about 5 l of water, or 0.5 cm of water across $1\,\mathrm{m^2}$.

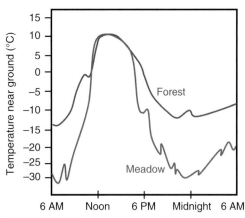

FIGURE 2.11 The temperature of the air in the forest in northern Arizona, USA, remained above −15 °C on a winter's night, compared with −30 °C in the meadow. Some of the difference could result from cold-air drainage from upslope, and better mixing of the air column by tree crowns, but the largest portion resulted from the high emissivity of the ground exceeding that of the air, with a greater energy loss from the soil chilling the air at the ground surface. **Source:** Based on Kittredge 1948, with data from the US Forest Service.

At this point, a technical detail about temperature and emission of radiation (and loss of energy) needs to be mentioned. From the description above, it sounds like there would be no net loss of energy by longwave radiation from a soil that had the same temperature as the air. The losses of energy from the soil to the air would match the energy emitted from the air to the soil. This would be true if the soil and air were perfect "black body" emitters of radiation. Most objects are more like "gray" bodies, emitting less energy than this ideal maximum. Soils and plants are almost-black box emitters of radiation, emitting about 90% or more of the black-body maximum. Air is a poorer emitter, especially if humidity is high, emitting only about 70% of the maximum. This means that a soil at 8 °C will emit more energy to the air than the air at 8 °C will emit back to soil, leading to greater cooling of the soil. This difference in emissivity can lead to quite large differences in temperatures at the soil surface compared to the air a few meters above the ground. The presence of trees can increase turbulence, preventing a layer of very cool air from establishing at the ground surface. In the absence of trees, the difference in emissivity between air and ground can chill a layer of air on windless nights by 10 °C relative to a few m above the ground (Figure 2.11; Holtslag and de Bruin 1988), with dire consequences for seedlings. The presence of clouds or high humidity can prevent this severe cooling. The emissivity of water is very high, so the emission of energy from moist air toward the ground would be about the same as the ground's emission, when air and ground start at the same temperature.

Temperatures Decline with Increasing Latitude

The patterns of temperatures around the Earth include familiar trends of hot tropical conditions near the equator, and chilly boreal conditions to the far north and south. Empirical measurements of temperature of course support this general trend, and traveling northward for 550 km from a starting point of 35° latitude reduces the average annual temperature by 2.5 °C when 40° latitude is reached (Figure 2.12). But why do temperatures follow this pattern? The air at any given latitude does not sense its location, so latitude is a covariate with temperature, not a direct driver. A more causal explanation can be plotted with temperature in response to the annual amount of incoming solar energy. The statistical fit for the data is the same, whether we use latitude or solar radiation on the X axis. This is because the amount of solar radiation relates strongly to latitude because of the geometry of a round planet with a tilted axis moving around the sun. The key difference between these two "explanations" of average temperatures is that one provides a good predictive answer, and the other provides both a predictive answer and a good explanation for why.

We can be very confident in the general trend of temperature in relation to latitude or incoming solar energy, as the 95% confidence intervals around the average trends are relatively tight. The actual temperatures for some locations fall substantially outside this confidence band, because the confidence band relates to the average trend, not to the dispersion of sites around the trend. What might explain why one site falls above the average trend, and another falls below? Temperatures also tend to be colder at higher elevations, and the third graph in Figure 2.12 shows that adding information on elevation can improve the prediction of temperature compared to latitude alone. Other factors are important too, including distances from oceans (which tend to moderate temperatures) and mountains (which limit the ability of oceans to affect temperatures).

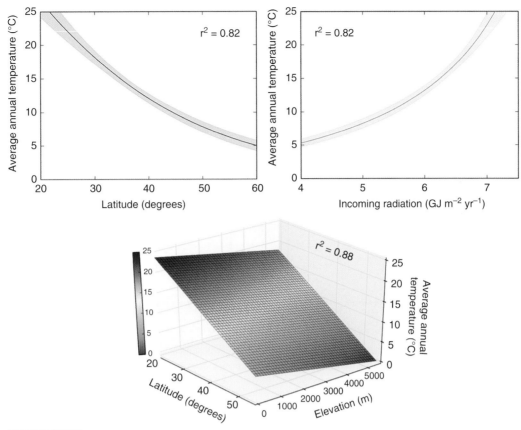

FIGURE 2.12 Average annual temperatures for sites around the world decline with increasing latitude (distance N or S from the Equator (top left). Latitude is a good predictor of temperature, but this is only a correlation not a process-based explanation. The annual amount of incoming solar radiation (top right) provides a strong explanation, and has the benefit of relating directly to a process influencing temperature. The variation of individual sites around the general trend can explained in part by considering other factors, such as elevation (bottom).

Temperatures Increase at Lower Elevations

What we measure as temperature depends on the collisions of molecules, with temperatures rising as the number of collisions increase. The mass of air molecules in $1\,m^3$ of air at 3000 m is about 0.8 kg, rising to 1.2 kg at sea level (about a 50% increase for a 3000 m loss of elevation; or in the other direction, a 33% decrease for a 3000 m gain in elevation). The increasing density of air means more collisions between molecules, which means a higher temperature. The temperature increase depends on the amount of moisture in the air, and a typical rate would be an increase in temperature of about 1 °C for each 100 m drop in elevation (Figure 2.13). This pattern is referred to as adiabatic heating (or cooling), because the overall energy among the molecules doesn't change even though the temperature changes with air density (or pressure).

The general relationship graphed in Figure 2.13 can be examined for specific cases, and for details that cause some deviations from the central pattern. Figure 2.14 compares daily high and low temperatures for a valley in the Rocky Mountains of Colorado, USA with corresponding temperatures in the foothills about 50 km away. Rising 1500 m in elevation lowered air pressure by 17%, reducing the summertime average daily highs by about 8 °C, and the average daily lows by 10 °C. The differences in daily lows were

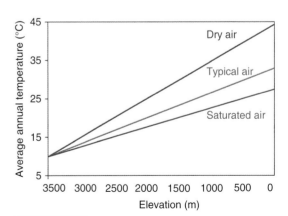

FIGURE 2.13 Air temperature increases with decreasing elevation because increasing air density leads to more frequent collisions among air molecules. This adiabatic (no change in energy content) heating depends in part on the moisture content of air, as evaporation or condensation of water moderates the trend that would occur in dry air.

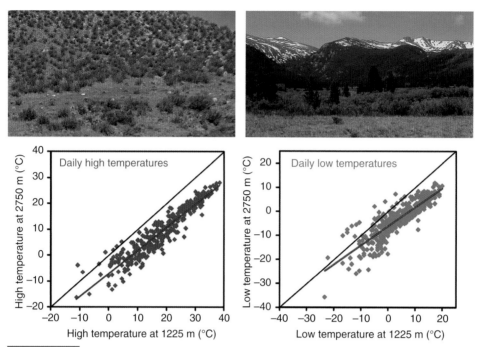

FIGURE 2.14 Daily comparisons of high and low temperatures at locations that differ by 1500 m in elevation showed a general pattern that also included specific days that were higher or lower than the trend. Some of the variation related to the moisture content of the air, especially on colder days, and other variation relates to cloudiness and shifting weather systems. **Source:** Data from Steven Fassnacht.

smaller in winter, only about 2–3 °C. During the warm season, the lower humidity of the air allowed for a stronger elevation effect than occurred during cool periods when condensation of water moderated temperature changes. Some days in winter were actually warmer at the higher elevation, illustrating the occasional importance of shifting weather systems and cold-air inversions.

Temperature Variation Over Time, and Across Space, Strongly Influences Forest Ecology

The complex physics of radiant energy and heat transfers come together to determine the temperatures that influence the development and success (or failure) of organisms. For example, the differences in incoming sunlight between N- and S-facing aspects of a glacial valley in Colorado, USA resulted in an average difference in air temperature (2 m above the ground) of 1 °C, the equivalent difference associated with a 200-km difference in latitude (on flat terrain). This difference was also large enough to affect the accumulation and retention of snow, which in turn influenced soil temperatures. Snow is a good insulator, and soils beneath deep snow may stay near 0 °C in winter, compared to lower temperatures in soils unprotected by snow. In December, a small amount of snow on the cooler N-facing side of the valley kept soil temperatures a few degrees warmer than across the valley on the S-facing slope (Figure 2.15). In May, the greater accumulation and slower melting of snow on the cooler N-facing side kept the soil near freezing while the soil on the S-facing warmed with spring sunshine.

Patterns that result from seasons, latitudes, and aspects are further modified by the structure and dynamics of forests. Fires can reduce the tree canopy, shifting the absorption of solar energy from leaves to soils. Forest harvesting similarly alters how radiation influences temperatures, and even how the convection of air influences temperatures at the soil surface. Harvesting a forest at high elevation in Montana, USA led to much hotter temperatures at the soil surface (Figure 2.16). Perhaps more surprising is the effect on nighttime low temperatures. Without a forest canopy, a thin layer of very cold air may settle near the soil surface, with higher emissivity of the soil (compared to the air) accentuating cooling. The presence of tree crowns promotes more mixing of the air, limiting the ability of cold air to settle on the soil (Löfvenius 1993), and provides for a closer match in emissivity between soil and tree canopies (compared to soil and air). This is one reason that harvesting prescriptions may retain some live trees (green tree retention, or shelterwood designs).

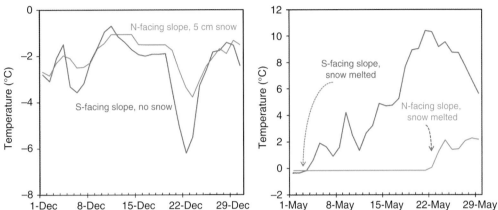

FIGURE 2.15 The temperatures of the soil (10 cm below the O horizon) differed between north- and south-facing aspects in this glacial valley in Colorado, USA (2750 m elevation), in part because of the effect of aspect on snow accumulation, retention, and time of melting.

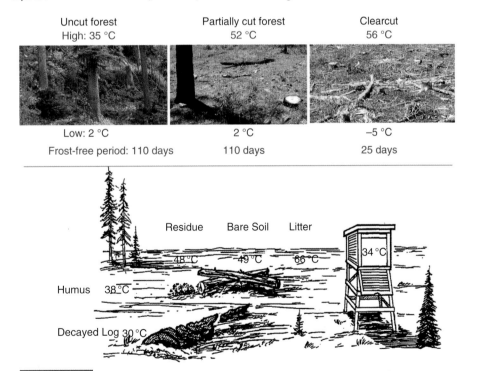

FIGURE 2.16 Removing some or all of the tree canopy from a high-elevation forest in Montana, USA (upper) resulted in much higher mid-day temperatures at the soil surface on a day in August, and the clearcut site actually experienced nighttime lows below freezing. Just as temperatures change through a day (and year), they also differ over space. Patterns of temperatures within a clearcut site vary substantially with the soil surface materials, with very large implications for microsites where new plants might establish. **Source:** Based on Hungerford 1980.

Large differences in temperature also occur among locations within a forest, especially after the tree canopy has been reduced or removed. An afternoon in the clearcut site in Montana had an air temperature of 34 °C, but temperatures of dry organic matter at the soil surface exceeded 50 °C (Figure 2.15), hot enough to kill tree seedlings. Bare mineral soil was somewhat cooler, because mineral soil conducts energy deeper into the soil, reducing the peak temperatures at the surface. Decaying logs were the coolest, as water accumulated in the wood evaporated and consumed energy. These spatial patterns of temperatures of course vary through a day and across seasons, and they illustrate some of the detailed interactions that are always important in determining local temperatures.

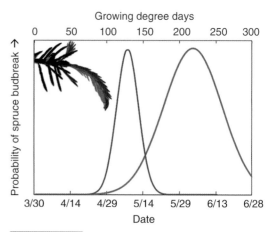

FIGURE 2.17   Buds on spruce trees in northern Sweden burst open and expand new needles near June 3 each year, but warm springs may see budburst come in mid-May, and cold springs may have budburst delayed till late June. The timing of budbreak is less variable when examined based on the accumulation of warm days; the average of 130 growing degree days (with a baseline temperature of 5 °C) shows relatively narrow variation across years. **Source:** Based on Langvall et al. 2019.

Temperature Strongly Influences Phenology and Growth

Most forest organisms have distinct cycles through seasons and years. Plants germinate, grow, blossom, and produce seeds. Temperatures are fundamentally important in the timing of these aspects of physiology (called "phenology"). Buds on spruce trees in Sweden may withstand winter temperatures of −40 °C, and then buds break and new needles form in the spring. At a site in northern Sweden, the average date of budbreak is June 3 (Figure 2.17), but budbreak may occur a couple weeks sooner in warm springs, and a couple weeks later in cold springs. Budbreak depends more on temperatures than on the days on a calendar, and a common way to gauge the overall warmth of a season is with growing degree days. The idea of growing degree days is that the biological response to several days of moderate temperatures might equal the response from only one or two days of high temperatures. The cumulative tally of warmth first requires a baseline temperature, such as 5 °C for an average daily temperature, which is considered too cold to lead to a biological response such as budbreak. The average daily temperatures that are higher than 5 °C contribute to the tally. Three days of 10 °C would add 15 (=3×5) growing degree days, and three days at 15 °C would add 30 growing degree days. The timing of budbreak of spruce trees in Sweden shows much less variation from year to year when examined in terms of the sum of growing degree days than when based on calendar days.

Cold temperatures slow the growth of plants, and many animals too (especially those whose body temperatures reflects surrounding environmental temperatures). Chemical reactions are slow in the cold, and cold locations have shorter growing seasons that also limit growth. It's easy to imagine that trees in a northern boreal forest would likely grow more slowly than trees in a temperate or tropical climate, but the effects of temperature are also large across smaller gradients. Tropical climates are generally warm throughout the year, but of course some locations are warmer than others. Across the Tropics, an increase of average annual temperature from 20 to 22 °C is associated with a 22% increase in wood production (Figure 2.18; revisiting one of the studies used in the Preface to discuss statistics). This is a very sensitive response compared to differences in rainfall. For example, increasing rainfall from 500 mm yr^{-1} (a dry tropical forest) to 2000 mm yr^{-1} (a tropical rain forest) is associated with only a 14% increase in growth if temperatures are the same. The value of extra water for forest growth increases substantially, though, with increasing temperatures. This description of the pattern in Figure 2.18 implies that because the trend in forest growth is associated with the amount of precipitation that the difference is driven by precipitation. In fact, correlation patterns should not be assumed to show cause and effect; and implication of a causal connection needs to be done carefully (as noted in the Preface). Is it possible that the soils differ substantially along with precipitation, and that soil differences are the actual causes of the pattern in growth? Or is it more likely that rainfall drives the pattern in growth?

The influence of temperature on forest growth may be even stronger on plantations where intensive silviculture reduces the influence of other factors that might limit growth, such as competing understory vegetation, low nutrient supplies, and genetics that do not optimize growth. Eucalyptus plantations in Brazil are managed on short rotations, and high rates of growth (trees may exceed 30 m height in six years) achieve forest volumes that would take decades in other regions (Chapter 7).

Forests Use Very Large Amounts of Water

Earth's atmosphere contains vast amounts of water, so much in fact that water is the most important greenhouse gas that keeps the planet from freezing. The rates of precipitation and evaporation are equally huge, and the atmosphere contains only enough water to supply Earth with 10 days of rain. About 90% of the evaporation that refills the air comes from open bodies of water, and most of the rest comes from water released from the insides of plants (mostly trees).

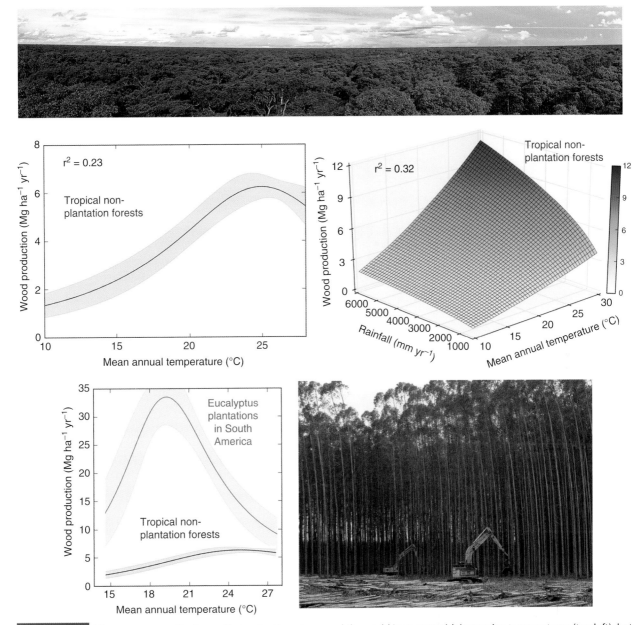

FIGURE 2.18   The average growth of wood in tropical forests around the world increases with increasing temperatures (top left), but declines as average annual temperature exceeds 25 °C. Temperature accounts for 23% of the observed variations in wood growth, so other factors (such as rainfall, species composition, forest age, etc.) lead to situations that are well above and below the average trend. The combined effect of temperature and rainfall accounts for 32% of the of the variation among studies, so the importance of high rainfall appears to be higher on warmer sites (top right, **Source:** Based on of Taylor et al. 2017, Cleveland 2017). Intensively managed plantations of eucalyptus grow much faster than unmanaged tropical forests (bottom figures, plantation in picture is six years old), with an even stronger apparent influence of temperature. **Source:** Based on Binkley et al. 2020.

The units used for precipitation and evapotranspiration are typically one-dimensional (mm). The one-dimensional number can be converted to volume with multiplication by an area (such as 1 m² or 1 ha). This approach also works for streamflow; dividing the total amount of water transported in a stream by the land area of the contributing watershed gives a one-dimensional measure (mm) of streamflow.

The rain and snow that fall on forests do not accumulate from one year to the next; the water either reenters the atmosphere or runs off into groundwater and streams. Evaporation in forests may be divided into three categories. Some water sitting at the surface of the soil may evaporate, but this evaporation is a small part of the water budget. Much larger fluxes are evaporation of water from the outer surfaces of leaves (canopy interception), and water from within leaves (transpiration). The sum of these three vectors is total evapotranspiration. Forest soils can retain several cm of water, but high intensity storms (and melting of snow) can exceed the soil's capacity to store water, leading to recharge of ground water and streams. Forest evapotranspiration accounts for virtually all the water falling onto forests in dry regions (Figure 2.19). In some cases, riparian (streamside) forests use even more water than falls from the sky. Evapotranspiration in dry areas comes close to matching precipitation, leaving very little water for streamflow.

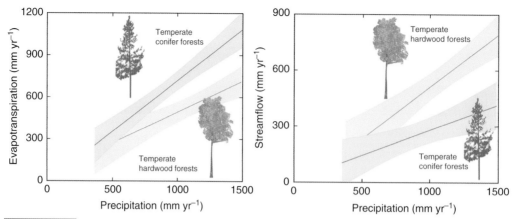

FIGURE 2.19   Forests occur in temperate regions where precipitation is typically greater than about 500 mm yr^{-1}. Almost all the incoming precipitation is lost in evapotranspiration in drier regions (left), whereas streamflow accounts for almost as much water loss in wet regions. Conifers and broadleaves have similar evapotranspiration rates in dry areas, but conifers use more than broadleaves in wetter areas. **Source:** Based on Ford et al. 2010.

A broadleaved forest in the Coweeta basin of North Carolina (Chapter 1) received 2160 mm of rain in a year (Ford et al. 2010). Evaporation from the soil surface was minimal, as the top part of the soil O horizon was often dry. The evaporation of water from the outsides of canopy leaf surfaces (interception) was 145 mm (7% of precipitation), and of course most of this happened during the growing season because of the low surface area of trees without leaves in winter. Transpiration loss of water from within leaves summed to 200 mm yr^{-1} (9% of precipitation), again mostly during the growing season. The sum of these losses (346 mm yr^{-1}) accounted for 16% of precipitation, leaving 84% (1800 mm yr^{-1}) to leave as streamflow.

Scientists at the Coweeta Hydrological Lab planted two entire watersheds with white pine trees to provide insights on water use by conifer forests versus native hardwood forests. The higher surface area of the pine needle canopy led to an interception loss of 280 mm yr^{-1} (13% of precipitation), and transpiration of 420 mm yr^{-1} (19% of precipitation), for a total evapotranspiration of 700 mm yr^{-1} (30% of precipitation) and streamflow of 1460 mm yr^{-1}.

Water Flows Down Gradients of Potential, Which Sometimes Means Going Up

The physics of water transport through soils and trees to evaporate into the air depends on water potential. Water potential can be thought of as a gradient, similar to a gradient in elevation. A drop of water sitting at rest in a puddle would have a low potential; it would not be possible to obtain work from movement of the water, and its potential could be defined as zero. But if the water could follow the gravitational gradient into the soil, that movement might have an opportunity to do work (though not much!). In this case, a potential for the water at the soil surface would be zero, and the potential of water deeper in the soil would be less than zero (a negative value). Movement along gradients goes from higher potential to lower potential, and zero is higher than negative numbers.

Water at the soil surface might move into the soil along a potential gradient that does not relate to gravity. A key feature of water molecules is an imbalance in electrical charge from one side of the molecule (slightly positive) to the other side (slightly negative). This polar aspect of water makes molecules line up with each other, providing surface tension to water drops. It also causes water to adsorb (stick) onto surfaces such as soil particles. Indeed, the potential for water being adsorbed onto surfaces of soil particles is very low (a large negative value), which means water in a puddle can be "sucked" into dry soil, faster than movement from gravity alone.

The sizes of mineral soil particles are important for influencing water infiltration into soil, movement through the soil, and storage between wetting events (see Chapter 6). The smallest particles are clay-sized, meaning <2 μm. One gram of clay has more than 1 m^2 of surface area to interact with water, so most water molecules are close enough to clay surfaces to slow their mobility. Water molecules that interact with surfaces have a lower (more negative) water potential than free water. The story of potentials also explains how water moves up trees, ascending tens of meters (or even one hundred meters) upward against the pull of gravity. Dry air has a tremendously negative potential compared to the insides of leaves, so water is sucked from leaves into the air, driving a potential gradient that goes against the gradient provided by gravity (Chapter 4).

Wind Shapes Trees and Forests

Forests have a complex relationship with wind. For example, wind moving past a leaf cools the leaf in two interacting ways. Leaves have a "boundary layer" of air that restricts the exchange of energy, water and CO_2 with the atmosphere at large. Winds strip away the outer portion of the boundary layer, making it thinner and facilitating more transfer of energy and matter. How much does this matter? A leaf exposed to sunshine with no wind might have a temperature about 3–5 °C cooler in a breeze than in still air (as noted earlier in this chapter). If the tree is well supplied with water and the air's vapor pressure deficit is not a problem, the leaf would be another 3 or 4 °C cooler, for a total temperature difference of about 7 °C (Knoerr and Gay 1965). This difference in temperature in relation to wind (with interacting effects on transpiration) would be large enough to change photosynthesis and respiration in a leaf by 10–50% (Figure 2.4).

Moving from the scale of leaves and minutes up to trees and years, wind affects the way trees form stemwood. Trees may experience high wind if they grow in a windy location, or if they're at the edge of a forest with few sheltering neighbors. Tree crowns act like sails on ships, catching the wind and enduring very large forces that bend stems. Trees that are chronically exposed to strong winds develop a strong taper, broad at the base and thin at the top. Trees that are less exposed to wind have less taper, with tall slender stems that decline less in diameter going up the tree. Some of the windiest environments occur at high elevation, where crowns appear to be pushed to the downwind side of trees (flagging), or even pressed down close to the ground (krummholz, from German, "twisted wood"). The upper-elevation limit of tree occurrence may relate as much to severe winds as to short growing seasons. High winds contribute great stresses in these extreme locations for trees, including wind-blown ice crystals that damage leaves.

Trees experience "low-to-average" wind conditions most of the time, but severe wind events can bring forces on tree crowns that are higher than any time in the previous decade or century. When the severe event is not too far beyond the typical range, individual trees topple over, leaving gaps in the forest canopy and to some extent in the soil (Figure 2.20). Trees that blow over and create gaps typically are uprooted, with a large amount of soil raised up and a pit left behind. In other cases, the

FIGURE 2.20   Severe winds that are not too extreme may topple individual trees within a forest, leaving most of the canopy intact. High winds toppled many beech trees, and a few spruce trees, in this mixed forest in central Germany. Treefalls create gaps in the canopy that allows more light, water, and nutrients to be available for neighboring trees and understory plants (including small trees). Abrupt edges between harvested and unharvested forests can expose trees on the edge to higher winds, breaking off stems or uprooting trees (bottom left, Vancouver Island, Canada). Severe storms can level entire forests, breaking off stems of trees that are too solidly anchored to uproot, as in this former Scots pine forest in Poland (picture from one year after the storm. **Source:** Skłodowski 2020 used by permission.

stem may resist the force of the wind less well than the roots, and the stem snaps off above ground. Severe wind events that greatly exceed typical storms, such as tornadoes, hurricanes and typhoons, can uproot or snap-off most trees across large areas (Chapter 10).

Events and Interactions Are More Important Than Averages and Single Factors

For learning purposes, it may be reasonable to talk about patterns and processes of energy, temperature, water and wind separately. Forests are influenced and respond to all of these all at once, with complex interactions that have legacies that can last for centuries. We can enter a forest and look around, employing knowledge and measurements to understand what's going on currently in the forest, including temperatures, water use by trees, and connections with growth. The current composition and structure of the forest resulted from a legacy of environmental factors in the past, especially the big events of storms, fires, and insect outbreaks. These historical events often leave traces that can be unearthed with careful study. The future of any forest will depend on the current operations of environmental factors, on the historical legacies of past major events, and on big events that may or may not happen soon. Any forest can be examined with these three questions (what's up with this forest, how did it get that way, and what's next?), but the answers always require an inconvenient amount of local detail. And when it comes to "what's next?", only broad, hazy insights are possible because big events just can't be predicted with much clarity.

Fires Depend on Temperature, Water, Winds

Most forests on Earth have been shaped by fires in the past, and the behavior and impacts of the flames depend on the fuel structure of the forest, on the water content of the fuels, the temperature and humidity of the air, and the speed of the wind. No single one of these factors would be very useful in understanding fires. Many forests experience fires multiple times within the lifespan of dominant trees, including ponderosa pine forests of western North America. Repeated fires lead to ecosystem structures with low densities of trees among small meadows (Figure 2.21). During hot, dry periods, fires may burn readily through the meadows and around the bases of the trees, and a few trees may even have fires reach into their crowns. Strong winds enable fires to race through canopies, but when canopies are very patchy, the necessary wind speeds would be so high that trees would topple over before burning. If fires are absent from such systems for the span of a human lifetime, the forest structure and fuels change so much that surface fires may be less likely, and only moderate wind speeds would be needed to fan flames from tree crown to tree crown. The details of interactions of forest structure, winds, temperature, and humidity would be different for other forests, but the importance of understanding these interactions would be important for all types of forest fires.

FIGURE 2.21   What will happen in this forest on a windy day in June? Back in 1928, the forest was a mosaic of trees within a matrix of small grassy meadows (**Source:** photo by H. Krauch, US Forest Service photo 16974A). A partially logged forest would require a windstorm of over 200 km hr^{-1} for a fire to spread from crown to crown across the forest (which might be strong enough to topple the trees). With heavy cattle grazing and fire suppression for eight decades, the grass meadow matrix was replaced by a high density of closely packed trees (white arrow points to the same rock; **Source:** photo by Andrew Sanchez-Meador), and a wind of only 50 km hr^{-1} could spread a crown fire. For a related, spatially explicit example, see Figure 11.14. (**Source:** D.W. Huffman, J.D. Bakker, D.M. Bell, and M.M. Moore, unpublished).

Droughts Affect Trees, Beetles, Forest Structure and Fire Intensity

Insects have coevolved with tree species, and favorable environmental conditions can change insect populations from low background levels to incredibly high numbers that overwhelm the ability of trees (and other vegetation) to cope. Most conifer species are hosts to one or more specialized species of bark beetles, and fungal spores spread by the beetles can kill trees (Chapter 10). The success of bark beetles may be higher in forests with trees that have experienced a strong drought, or winds that toppled large numbers of trees (where dying trees succumb to the beetles). These changes in forest structure result from interactions of drought, wind, and insects also change the potential of fire to affect the forest, with legacies that last for decades. In the Sierra Nevada mountains of California, droughts increase stress in pine trees, increasing the success of beetles. The beetles' fungal symbiont plugs the trees' water-conducting sapwood, and needles dry and die. Dry needles ignite more easily and release more energy when burned than green needles, and forests with large numbers of recently killed pines have a high potential for running crown fires (Figure 2.22). The dry needles fall off within a few years, and the reduction in crown fuels reduces the potential for crown fires. After the woody material of dead trees falls to the ground, the high fuel loads can support high intensity surface fires.

The actual "story" of the environmental drought effects on the trees and forests would be only partially about water supply and tree physiology, and more about beetles, fires, and all the interacting legacies that shape forest changes over time. Most forests do not burn soon after beetle outbreaks, so the variety of post-beetle forests that develop across landscapes can be quite broad.

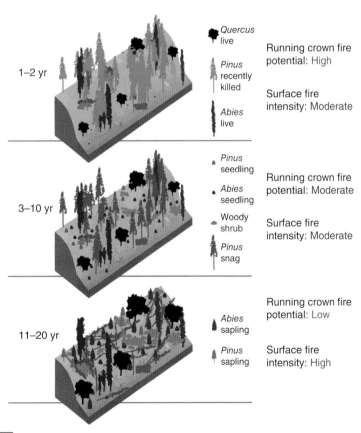

FIGURE 2.22 Severe droughts can foster outbreaks of bark beetle populations, with legacies that last for decades or centuries. If a fire occurs shortly after the trees are killed, the dry needles can support a fire that leaps from crown to crown (upper), killing many (or all) of the surviving trees. A few years later, the dead needles have mostly fallen, and the risks of a spreading crown fire declines. Dead tree stems fall after a decade or so, just as small trees and shrubs are increasing in biomass. The accumulation of fuels near the ground can support surface fires that burn at high intensities. **Source:** Stephens et al. 2018/ Oxford University press.

Weather Events Can Matter More than Averages

The two examples above underscore the importance of weather extremes in shaping forests. The year-to-year ecophysiology of trees in response to daily and seasonal weather is important (see Chapter 4). However, the longer-term changes in forests often result from big events that may, or may not, happen in any given year. And once a major event has happened, the responses of vegetation and animals can set up patterns in forest composition and structure that continue to shape forests in ways that diverge from trends that would have occurred if the major event had not.

Ecological Afterthoughts

The major tree species of the Rocky Mountains in the USA can be plotted in two-dimensional space of average annual temperature, and average annual precipitation (Figure 2.23). The points represent the environment where each species currently shows its highest frequency of occurrence, and the oval clouds represent the full range of species occurrences (based on 9500 plots across the region). The overlap of ranges makes it hard to see separate species, but it paints a picture of how similar (but not identical) the distributions of species may be. The environment where each species has its peak of occurrence is not in the middle of the species' oval (lower graph). Plots like these represent how two features covary, but not whether one or both factors actually drive the pattern (correlation should not be confused with strong evidence of causation). Do the species ranges likely result from direct effects of temperature and precipitation? If not, what other factors might be important? The ranges for the species could be coupled with climate simulation models, with the output showing the geographic locations where each species might be expected to be found. What limitations would be important in how well such stories would actually map onto real landscapes? How would the factors on such a list of limitations interact, and what would the implications be for the interactions?

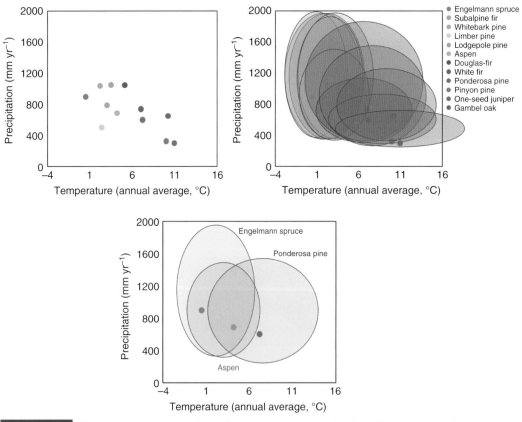

FIGURE 2.23   The major tree species in the Rocky Mountains, USA can be plotted in relation to where they are most frequent on axes of temperature and precipitation (upper left, based on 9500 locations). The full range of occurrence can be plotted as overlapping clouds (which are difficult to decipher for any single species in the upper-right graph). The points of maximum frequency for each species does not fall in the middle of each species' range (lower graph; **Source:** Data from Martin and Canham 2020).

CHAPTER 3

Evolution and Adaptation in Forests

What's in a Name?

This question holds implications for most of the topics covered in this chapter. Prior to the late 1800s, Carl von Linné (1707–1778) was the most influential person in the history of biological thinking. His contribution was the creation of a rational, carefully designed system for giving names to species of plants and animals, and grouping similar kinds into hierarchies. His nomenclature system was based on giving each species two names in Latin, one for its general type (genus) and one for its specific type (species). The choice of Latin for names reflected the value and importance of communicating across cultures. Indeed, Carl decided to give himself a Latin name as well, and history generally refers to him as Carolus Linnaeus.

The development of names and relationships launched opportunities for cataloging the diversity of life on the planet, and for thinking in more concrete ways about factors that led to the sorts of patterns displayed by the geographic ranges of species. One of the first fruitful questions was: why are some species found in certain types of locations but not others? This question, along with sub-questions that follow from it, might account for half of all the ecological investigations since the time of Linnaeus.

Linnaeus' system "worked" because species really do seem to exist as separate categories, with varying levels of relationships across categories. Spruce trees and pine trees clearly belong to different categories, yet they have more similarities to each other than spruces or pines have to oaks or maples. Linnaeus assumed (with no glimmer of doubt) that the categories were real because species were created by a god. Not much contrary evidence was apparent. Some individual specimens in the same species might show some small differences, but that could be the result of the local details of where the individual developed. The boundaries between some species seemed indistinct, with gradations of differences rather than sharp distinctions. But that might reflect the imperfections of the state of botanical and zoological knowledge rather than an imperfection in the solid, immutable reality of species.

All this rigid thinking changed when Charles Darwin and Alfred Russel Wallace offered the revolutionary paradigm of evolution by natural selection. Well, that's not what really happened. The rigid thinking of concrete species, each created separately by a god, did not in fact evaporate in the light of this novel idea and overwhelming evidence. But the idea of evolution by natural selection did rapidly change the views of most scientists (and some naturalists, including Henry David Thoreau). The seductive appeal of familiar, entrenched stories held back the evolution of thinking by many other scientists, religious scholars, and others with strong beliefs.

The consideration of species and evolution in this book can only provide a brief overview, illustrated with a few useful examples. This introduction to the longer-term and larger-domain issues of forests around the world can be followed up with many fabulous books to stoke curiosity and convey insights, such as Richard Dawkins' *The Greatest Show on Earth* (2010), and Colin Tudge's *The Tree* (2005).

Forest Ecology: An Evidence-Based Approach, First Edition. Dan Binkley.
© 2021 John Wiley & Sons Ltd. Published 2021 by John Wiley & Sons Ltd.

The Core Idea of Evolution Is the Combination of Variation, Failure, and Innovation

Darwin's long path toward developing ideas about evolution by natural selection began with the observation that offspring are not identical copies of either parent, or even to their siblings. In a random world, any particular offspring would be as likely successful in becoming a parent in turn as any of its siblings. But the world is not quite random, and some offspring have traits that are more likely to be eliminated from the clan of future parents. An acorn that lacks chemicals that deter nut-eating rodents may not last long enough to germinate, and an oak that lacks the ability to endure severe frosts in November may not live to reproduce. The production of variation in each generation is coupled with environmental "filters" that have a non-quite-random likelihood of removing individuals from a population, leading to changes in a population over time. This might be thought of as "survival of the fittest," but in functional terms it might be more productive to think of "higher death rates for the less fit." Favorable innovations have no guarantee of persisting into future generations, as individual organisms die from a broad range of causes, and any one of those causes could cut short the legacy of even the best innovation. This is of course a scheme of very low efficiency indeed, but efficiency is not a goal of evolution (which of course has no goals in any case).

This short-hand description of natural selection could shift the frequency of characteristics within a species, but it would not account for any innovation in characteristics. How can a novel trait appear in a population, a trait that no ancestor had before? Innovation needs to arrive either from the outside (new genes entering the population), or be newly created within the species. The appearance of innovation within the gene pool of a species can develop from mutation (a mistake in the replication of DNA), or from shifts in genetic details of how genes are transcribed (or not) into proteins. Evolution by natural selection depends on variation among related individuals, environmental filters that remove some variants, and various sources of novel characteristics.

Darwin Could Not Explain Why Variations Occurred, or Why They Were Passed on to Offspring

In some ways, Darwin's boldest act was offering a theory he found to be very compelling despite a total lack of any mechanism that could make it happen in the real world. It's easy to imagine that one oak tree is by far the best oak in a forest, but after that oak's flowers are pollinated by inferior neighbors, its acorns would be only half as superior as it was. Another generation would lead to another dilution of the original oak's superiority, and this clearly could not lead to the improvement of a species or the creation of a new species.

The mechanisms that allowed evolution by natural selection to work became clearer over subsequent decades, with identification of inheritance happening in "packages" rather than gradations (the idea of genes, which cannot be infinitely diluted in a population), and the full suite of biochemical insights related to genes, DNA and protein synthesis. All the links are now well known, with many fascinating details being developed every year.

Does Selection Work on Species or on Genes, or Is This Only a Chicken-and-Egg Question?

The division of the perceived universe into parts and wholes is convenient and may be necessary, but no necessity determines how it shall be done.

Gregory Bateson, 1979

It may seem obvious that species are the domain where evolution occurs. Evolution cannot take place at the level of a whole "genus" because the multiple species that constitute a genus evolve separately. Evolution cannot be captured at the level of a single organism either. A single organism can reproduce, or fail to reproduce, but the organism itself cannot evolve. Therefore, it seems that evolution happens at the domain of species (not individuals, or genera). This idea seemed obvious until Richard Dawkins (1976) introduced the upsetting idea that selection really operates at the level of genes, not species. This perspective recognizes that organisms are coalitions of genes, which influence the success or failure of the organism. Closely related organisms share more of the same genes, but genes are dispersed across populations, species, and broader domains of life. The biochemistries and genetics of dinosaurs and humans have quite a lot in common, with many of the same genes shaping both lineages. And the microworld of bacteria and Archaea has even less correspondence with the concept of a species, as genes can move laterally from one strain of a microbe to another.

The focus on species in biology can also be limiting when the reality of breeding fails to follow clear distinctions. In the Western US, narrowleaf cottonwood trees can interbreed with plains cottonwood trees, and the offspring can be reproductively viable (unlike the mules that come from mating donkeys and horses). It might seem that these two species do not meet the definition of being separate species (which are usually defined as reproductively isolated, either genetically or geographically), yet each "species" does tend to segregate to different parts of a landscape, with somewhat distinctive morphology, function and ecological interactions (Floate et al. 2016). These two groups (narrowleaf and plains cottonwoods) do represent different suites of characteristics (in anatomy and ecology), yet the two closely related groups can mix genes too.

The somewhat nebulous idea of "species" is quite useful in forest ecology, even though a species is "only" a temporary vehicle passing genes along into the future. When ancestral species evolve into new species, the pool of DNA and genes continues mostly intact. The sorts of interactions that occur in the field of forest ecology deal strongly with the convenient idea of species, though in the long run the real stories take root deeper in the pools of genes.

Biology Operates from a Simple Story of DNA to Incredible Complexity of Proteins and Biochemistry

The story of life on Earth has a remarkable thread that reaches from relative simplicity at the base level of DNA, to infinite complexity at the level of organisms and ecosystems. In the second half of the twentieth century, the structure and role of DNA was revealed to be based on the patterns of just four nucleotides: adenine, cytosine, guanine, and thymine. A sequence of three of these provides the code for one of 20 or so amino acids, and the assembly of amino acids provides the information base for enzymes and complex proteins which lead to the full range of thousands of biochemicals found in each cell. A sequence of DNA that encodes the amino-acid sequence for a particular protein is a gene, and massive chains of genes comprise chromosomes that are visible (with a microscope) in the nucleus of plant and animal cells.

The number of chromosomes varies among species, and often within species for plants. The division and recombination of chromosomes in plants typically leads to diploid offspring with one set of chromosomes from each parent. Sometimes recombination leads to polyploidy, where offspring have more than the usual set of two chromosomes (with an extra set from one or both of the parents). Aspen in the Rocky Mountains of the USA are commonly either diploid (a set of 19 pairs of chromosomes, one from each parent) or triploid (a third extra set of chromosomes). Triploid aspen are more common in the southern ranges where clonal reproduction dominates, and diploid aspen dominate to the north with reproduction by seeds (Greer et al. 2018). Redwood trees, the tallest trees on Earth, are hexaploid, with six copies of each chromosome! The number of chromosomes varies among species, and the size of the genome (the number of nucleotides in DNA) varies even more. The poplar genus has 19 chromosomes, with 38 chromosomes in a diploid individual, and a genome of about 400 million bases (nucleotides), forming about 45 000 genes. The eucalypt genus has only 11 chromosomes (22 in a diploid individual), and a genome of about 640 million bases forming 36 000 genes. Redwood trees have 11 chromosomes, but the 6 copies of each leads to an astounding 27 000 million bases. For comparison, humans have 23 pairs of chromosomes, with a modest genome of about 3000 million bases forming 20 000 genes.

Cells in animals also have chromosomes in mitochondria, the legacy of a distant symbiosis when a bacterium took up permanent residence within another cell to create the lineage of eukaryotic organisms. The formation of animal embryos brings together genes on chromosomes of both parents, but the mitochondrial genes are passed on only by mothers (with some rare, notable exceptions).

Cells in plants have chromosomes in the nucleus, in mitochondria, and also in chloroplasts. The formation of seeds also entails genes on nuclear chromosomes from both parents, with mitochondrial genes coming only from mothers. The genes in the chloroplasts of most angiosperm seeds come only from the mothers, whereas chloroplasts genes in seeds of gymnosperms come only from fathers (which can be quite useful for tracing lineages of mothers and fathers).

Much of the DNA present on chromosomes is faithfully passed along through the generations, but never "read" to create any proteins. Some genes have DNA sequences that do code for proteins, but the genes may be turned off (or on). The field of epigenetics focuses on the activation and deactivation of genes, which can provide great variations in biochemistry, physiology, behavior, and ecological interactions.

The biochemical factories in cells produce thousands of different compounds, and cells differentiate over generations to produce the array of cells that comprise the diverse types of tissues of animals and plants. These tissues in turn need to interact in clearly defined, mutually supportive ways for the continued survival of an organism. The coordination involves simple chemistry (such as the carbon dioxide levels in animal blood influencing pH which regulates binding and release of oxygen by hemoglobin), biochemistry (such as terminal shoots producing auxin that inhibits lateral buds), and neurochemistry (such as the adrenaline response of a hare in the presence of a lynx).

A key insight from this six-paragraph sprint through biology is that the viability of an organism is the result of almost inconceivable numbers of interacting parts and functions. A list could be prepared of "things that could go wrong" and lead to the death of an organism, or the failure of the organism's genes to make it to the next generation. That list would have almost exactly no overlap with what can happen at the level of ecosystems. The vast assemblages of genes in forests come in packages of organisms and species, and the total gene pool of a forest ecosystem changes routinely as organisms (and species) come and go. The failure of any species is typically a positive opportunity for the competing organisms (in contrast the failure of an organ in an animal: no mouse ever found that its heart was happier when its kidneys failed). Most of the interactions we examine in forest ecology are the interactions among competitors, though symbioses are also fundamentally important among different types of organisms (such as plants and mycorrhizal fungi). This sort of competition characterizes a chaotic system rather than a tightly constrained, organized systems. Chaos does not imply randomness; a system shaped by competition can lead to high predictability (such as big trees are more likely on average to outcompete smaller trees). At the scale of a hectare, a forest may be pictured as an "ecoanarchy" more than a tightly regulated ecosystem. The exception to that statement comes when considering the fundamentally important interactions between co-dependent species.

Mycorrhizae and trees have a symbiosis that entails a sort of competition (how much C does the tree send to the mycorrhizae versus retain for its own growth?), but neither survives without the symbiosis. Words, analogies and metaphors are at the heart of how we think (including scientific thinking), so a clear awareness of hidden implications and expectations behind these words is vital to understanding forest ecology.

Why Are There Only Two Species of Tulip Poplar, and Why Are They 12 000 km Apart?

The nearest relatives to the tulip poplars at Coweeta (Chapter 1) are found about 12 000 km away (20 000 km is as far apart as two points can be on Earth's surface). Why are they so far apart? We could speculate that both species shared a common ancestor in Europe somewhere, and then seeds migrated east (across the Atlantic Ocean) to North America and west (across the Himalaya Mountains) to China. The distances would still be vast, with a mystery remaining of why there were no tulip poplars to be found in Europe anyway.

Fortunately, we aren't limited to speculation. A combination of geology, paleontology, and genetics provides the evidence needed to explain the observed distribution of tulip poplars. The first key from geology is that 100 million years ago the continents were not located where they are now, and North America connected to northern Asia. The connection wasn't across the Bering Sea between Alaska and Siberia, but actually along what is now the northern edges of each continent. The spread of tulip poplars did not have to contend with crossing either the Atlantic or Pacific Oceans (these oceans didn't exist). The paleontology (fossil) record shows tulip poplar species were common across what would become North America, Europe and Asia (Figure 3.1). Comparisons

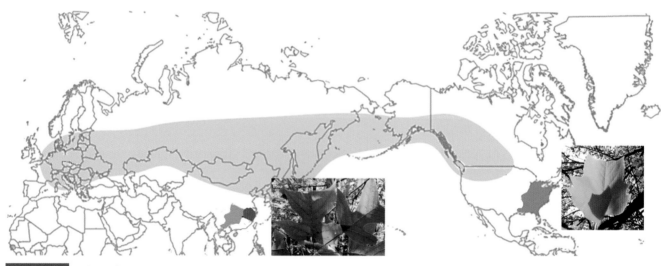

FIGURE 3.1 Millions of years ago, tulip poplars were found across much of what are now the continents of Eurasia and North America (tan zone). These areas were not spread so far apart in the time of the tulip poplar ancestors. The current genus of tulip poplars has just two species, found 12 000 km apart in eastern China and eastern North America. Without the evidence from geology, paleontology, and genetics, the origin of the two species in such distant places would be a curious mystery. (**Source:** map based on Chen et al. 2019, photo of Chinese tulip poplar leaves from Haishan Dang).

of genes and DNA shows two distinct varieties of tulip poplar in China (though usually classed as a single species), and those two varieties are only slightly more similar to each other than the North American species is to the eastern China tulip poplars. If tulip poplar species existed in Europe, why did they go extinct? One possibility is a lack of refugia to the south during glacial periods. That idea is consistent with a wider gene pool of Chinese tulip poplars, with had diverse refugia available in southern Asia, compared to the more limited peak-glacial warm sites for tulip poplars in North America.

All tree species have similarly fascinating histories, across time spans when continents were far across the Earth's surface from their current locations. The roots of gymnosperm species date back farther than tulip poplar's ancestry; the diversification of gymnosperms began when all our present continents were massed together in Pangaea.

Tall Growth Requires Strong Stems

The evolution of ancestral tulip poplars followed a split from the broader group of magnolias about 100 million years ago. That of course is a hugely long time; dinosaurs were rampant and mammals were small. But 100 million years is a short portion of the time since Earth formed; the tulip poplar story comprised only 2% of Earth's history. Indeed, almost 90% of the story of Life on Earth could be told without mentioning plants that grow on land, because land-based plants didn't evolve until 4 billion years had passed out of the Earth's total age of 4.5 billion years (Morris et al. 2018). The innovations needed for life on land included ways to conserve water when exposed to air, and ways to cope with gravity without the buoyancy of water. About 100 million years of evolution of plants on land gave rise to plants with internal conducting systems that could move water from the soil to leaves, and this ability supported the next step of vascular plants in growing tall: the formation of lignin "glues" to cement fibers (such as cellulose) together and provide rigidity to stems.

The first plants to form woody material and raise their photosynthetic tissues high above the ground were related to ferns, reproducing by spores rather than seeds. (Spores are single cells that can generate a new plant, and seeds are complex, multicellular structures that include stores of compounds to support the early growth of the new plant.) Fossils of *Wattieza* and *Archeopteris* from 350 to 385 million years ago indicate the trees reached 5–10 m in height, with frond-like leaves.

The First Trees from Seeds Were Gymnosperms

The clade of gymnosperms traces back about 350 million years, followed by the appearance of the first conifers about 300 million years ago. The conifers diversified into about 20 families dominating much of the planet for 100 million years or so, but only eight of those families remain today (Farjon 2018). Extinction of conifer lineages has left only about 600 species, or about 1% all current species of trees (Botanic Gardens Conservation International, 2020; Tudge 2005). The low number of species does not prevent conifer forests from covering vast areas of Earth, with conifers dominating about one-third of the Earth's forest area. Conifers also provide the majority of the supply of wood and wood products to people.

The seeds of conifers are borne in cones, and cones are either male or female. Some species have both sexes of cones on individual trees (monoecious), and others have separate male and female trees (dioecious). Seeds develop without being enveloped inside an ovule or fleshy fruit (such as in apples), though this trait of naked seeds (gymnosperms) may seem a bit confused for some species that do develop fleshy coats for the seeds (such as junipers). The fertilization and development of conifer seeds can span more than a year, and the number of cones and seeds carried to maturity varies substantially across years (often in response to periods of favorable and unfavorable weather). The spread of conifer pollen relies on wind, and most conifer seeds are dispersed by wind. Some conifer seeds are eaten by birds and mammals, providing longer-distance dispersion away from parent trees.

The most varied family of conifers is the Podocarpaceae, with at least 19 genera occurring in the Southern Hemisphere, with a few reaching into Central America and Southeast Asia. Podocarp species include small understory plants and towering trees (Figure 3.1) that can exceed 60 m in height. The cones of many species are fleshy, leading to dispersion of seeds by animals.

The Araucariaceae (Figure 3.2) were widespread across the planet during dinosaur times, but now the family has only three genera, all in the Southern Hemisphere. Species of the genus *Arauracaria* are found in South America, and on islands in the Western Pacific that are remnants of the former supercontinent, Gondwana. The island of New Caledonia has 13 endemic species of *Arauracaria*, along with great diversity of other conifers and angiosperms. Another impressive member of this family is the kauri genus of New Zealand, with trees that exceed 50 m in height and 4 m in diameter. The third genus is *Wollemia*, which was thought to be extinct until a few dozen individuals were discovered in the 1990s in the Blue Mountains of Australia.

FIGURE 3.2   Some examples of trees from the major conifer families (beginning top left). Sources: Podocarpacae: *Podocarpus totara*, totara, 31 m tall, 2.6 m diameter, about 1000 years old, photo by Tomas Sobek from the Peel Forest, Canterbury, New Zealand. Araucaraceae: *Araucaria araucana*, monkey puzzle tree, photo by Vicente Fernández Rioja, Conguillio National Park, Chile Pinaceae: *Pinus sylvestris*, Scots pine, coast of Öland, Sweden Cupressaceae: *Thuja plicata*, western red cedar, photo by Brew Books, Lake Quinault, Washington, USA.).

The Pinaceace family has the most species of all conifer families, with over 200 grouped into 11 genera. All the species are native to the Northern Hemisphere, and include pines, spruces, firs, larches, and hemlocks. The western coast of North America has the tallest Pinaceae, with Douglas-fir trees approaching 100 m, Sitka spruce exceeding 95 m, and sugar pines exceeding 80 m in height. Many species of Pinaceae are important commercially, and pine plantations cover more of the earth than those of any other genus. The oldest documented trees come from the pine family, with bristlecone pines reaching 5000 years. Some other species may have older individuals, such as a Norway spruce in Sweden that survived more than 9000 years. In this case the currently living spruce stem is a younger shoot, connected to a root system with decaying portions with a very ancient date based on ^{14}C dating.

The Cupressaceae includes cypresses, junipers, and towering redwoods. The family holds several world-wide records: the most widespread conifer is the common juniper shrub, the tallest trees are coastal redwoods (the tallest known tree is 115 m), and the most massive trees are giant sequoias (over 500 Mg for a single tree).

Collaboration with Insects Helped Angiosperms Take over the Planet

Over the past 100 million years, broadleaved tree species (angiosperms) rose to outnumber the species of conifers by 100 : 1. They occur across the planet, in every environment that is suitable for trees. The remarkable diversification of broadleaved species derives from the remarkable alliances developed with insects for pollination of flowers. Conifers are primarily pollinated by wind (as are some broadleaved species), and most broadleaved species rely on insects to move pollen from tree to tree. Some trees have conspicuous flowers, including familiar fruit trees, and trees in the magnolia family (including the tulip poplars introduced in Chapter 1). Tree flowers evolved to be conspicuous to insects, by sight and by scent, with nectar or other food encouraging visits to multiple flowers by each insect. Sometimes the mutual benefits have evolved into highly specialized symbioses. There are more than 800 species of fig trees in the mulberry family, and most have co-evolved with a single species of wasp for fruit pollination. A high degree of specialization between plant and pollinator can be very efficient.

Angiosperms (including the broadleaved trees) have a type of double fertilization for the production of seeds. Pollen arrives to flowers carrying two gamete (reproductive) cells: one fuses with the egg to form the plant embryo, and the other fuses with two other cells in the flower's ovule. This triploid ovule system (one set of chromosomes from the father, two from the mother) develops into the rich store of carbohydrates, fats, and proteins that provision the initial establishment of the newly germinating plant.

Seeds of some broadleaved trees are dispersed by wind, including the tiny seeds of willows and poplars, and the larger winged-seeds of maples, ashes, and tropical dipterocarps. Many species have evolved with animal dispersal of seeds, encasing seeds within large, nutritious fruits. Learning the details of broadleaved trees and forests would take more than a lifetime. Five cases are described below, just to give a taste of what evolution has developed.

The Highest Diversity Is in Tropical Rain Forests

A diverse temperate forest might have as many as two dozen species of trees in a hectare. An average hectare of tropical forest in Africa would have 75 tree species (Figure 3.3), while twice that number would be common in tropical forests of South America and Southeast Asia (Sullivan et al. 2017). Some tropical rainforest patches have more than 200 species of trees/ha. The tropical diversity

FIGURE 3.3   The view from 30 m up: Tropical rain forests are famous for their diversity of tree species, made possible by favorable environments for growth and intricate biological interactions (this old-growth forest in Ghana likely has about 100 tree species/ha). Some tropical environments are not so favorable, and few species may be able to cope with features such as flooding in the Pantanal of Brazil (where almost all trees in this picture are a single species of *Vochysia*).

numbers are especially astounding given that all of Europe has only about 250 native species of trees, and North America only about 600. The diversity of tree species is only the tip of a massive pyramid of diversity, with the tree species vastly outnumbered by smaller plant species and animals, which are vastly outnumbered by the biotic community within the soil.

Do all Trees Need to Have Trunks?

Tropical forests commonly have vines, or lianas, that root in the soil and climb up trees to reach sunlight. Some species of fig take this approach a step further. The figs wrap stems around unrelated trees, raising their own canopies up into the light (Figure 3.4). These "strangler" figs compete for nutrients and light with the supporting tree, and sometimes this might hasten the death of the

FIGURE 3.4 The young fig in Kerala, India in the upper two panels is wrapped around a supporting tree, with fig branches spreading out in the light above. The fig in the lower two panels outlived its supporting tree, and all that remains is the hollow interior among the fig stems that trace the outlines of the former tree. The lattice-like structure of the old fig trees supports crowns very well, an interesting variation compared to the typical solid stems of other tree species.

supporting tree. In other cases, the fig simply outlives the supporting tree. The eventual decay of the supporting tree can leave the fig standing with its own macramé-like stem, with a hollow center where the supporting tree once stood.

Some Broadleaved Trees Make Fertilizer Out of Thin Air

The growth of almost all forests in the world is limited by low supplies of one or more essential nutrient, and nitrogen limitation is very common. The crowns of trees are awash in nitrogen gas (N_2), but no tree has evolved an ability to directly tap this unlimited pool. Some species have evolved symbioses with bacteria that synthesize enzymes that break the bonds between the two N atoms in N_2 and then build amino acids. The symbiosis includes growth of nodules on roots, where the bacteria are protected from high concentrations of oxygen and supplied with carbohydrates. Nitrogen fixed within the nodules is transported into the tree (Figure 3.5), and then into the broader ecosystem when tissues on the trees die (including annual litterfall). Most of the nitrogen-fixing trees are found in the legume family, but the symbiosis has also evolved in some species of the birch family (such as red alder), and casuarina family. The rates of nitrogen fixation are often very high, with annual inputs similar to what farmers might apply in agricultural fields.

What's the Largest Tree in the World?

Simple questions often have complex answers. The largest living stem of a tree is a giant sequoia in California, USA, with almost 600 Mg (megagrams, metric tons, 1,000,000 grams) of mass. Some coastal redwood trees measured in the twentieth century might have been even a larger. It's not unusual for trees to send up more than one shoot from its root system, and sometimes the number of multiple stems can be phenomenal. The largest identified clone is an aspen in Utah, USA. This single genetic individual spreads across about 40 ha, with more than 40 000 stems summing up to about 6000 Mg (or about 10 times the size of the largest giant sequoia; Figure 3.6). The ages of the stems are younger than 200 years, but how old would the clone itself be? One way to estimate the age would be to use ^{14}C dating on the roots, as was done to determine the age of the Norway spruce described above. The aspen clone probably has no surviving root material from the original tree; the roots grow, die and decay in these clones, just like the stems. Another way would be estimate how long it would take for a clone to spread across so much area. A 40-ha circle would have a radius of about 600 m, and the edge of aspen clones might expand outward at rates of 0.5 m yr^{-1}. A quick calculation would suggest the clone might be 1200 years old, which sounds reasonable but of course is

FIGURE 3.5 Red alder trees form a symbiosis with *Frankia* bacteria, housed in nodules on the alder roots. The bacteria convert N_2 from the air into amino acids, which are used by the trees. The "fixed" nitrogen then cycles in the forest as tissues of the alder die and decay. **Source:** forest photo by Walter Siegmund.

FIGURE 3.6 The 50 or so aspen stems visible in this picture are only half of 1% of all the tree stems in the single individual of the Pando Clone of aspen. **Source:** photo by John Zapell.

only a rough guess. Stems within a clone that are far apart might accumulate genetic mutations as cells divide. These somatic mutations could be used to estimate total clone age, if the rate of mutation could be established. So far the genetic approaches have not yielded precise estimates of ages for aspen clones, though ages of centuries to a few thousand years would be consistent with the genetic evidence (Ally et al. 2008).

Does it make sense to consider the Pando Clone an "individual"? One stem may actually share some roots with adjacent stems, but it would not be likely that a continuous network of intact roots connects all stems (especially after a road was placed through the middle of the forest). If a clonal forest of aspen might be considered to be one individual, what about clonal trees that have grown from cuttings of a single ancestor tree? The tall, thin Lombardy poplars that characterize landscapes of northern Italy are genetically identical. Would it make sense to sum up the mass of all the Lombardy poplars across the world to get the "largest" tree? This might not seem like a good idea, and some interesting ecological evidence could be used to argue they are not all a single individual. Epigenetics deals with changes which genes are turned on or off, determining which proteins are made. A study of Lombardy poplar used cuttings from trees across Europe and examined their environmental responses in a greenhouse (Vanden Broeck et al. 2018). Stems grown from trees collected in cold climates set buds about five days sooner in the autumn than stems from warm sites, despite the trees being genetically identical and the shoots grown all together in a greenhouse. Epigenetic "memories" like this can arise from methylation of DNA molecules that keep genes from being read. This "clone as the largest individual" idea could be taken to an extreme in Brazil, where single genotypes of eucalyptus are propagated and planted across tens of thousands of hectares, qualifying some clone of eucalyptus as the largest "single" individual.

History Has No Need to Repeat Itself

If environments were unchanging, species might evolve precise adaptations to cope with predictable conditions. Forest environments are always changing across all scales of space and time (Chapter 2), so the "environmental filters" that remove some individuals from populations (and gene pools) are not constant. Highly specialized adaptations for a particular suite of environmental conditions may be unsuited to rapid events (droughts, early frosts, fires). Environmental factors such as temperature and precipitation may have average ranges that we could compute for defined locations and time periods, but environments show long-term changes that shift these average ranges. And of course the success of an individual can end suddenly when eaten or infected by a disease, and the co-occurrence of species that consume and infest organisms change over time.

The Earth's climate cycled into ice ages and back again more than 10 times in the past two-and-a-half million years. This Pleistocene Epoch ended about 12 000 years ago, when the most recent warming launched our present Holocene Epoch. The ranges of species contracted and expanded with these major shifts in climate, and the present distributions of species continue to change. Rates of change will likely increase in response to human-induced changes in climate.

The geographic range of a species is limited by various legacies of history. The dispersion may be limited to short distances from current ranges, life histories (flying animals and light-weight wind-blown seeds may travel farthest), and contingent events that influence availability of areas to colonize (such as volcanoes, floods, fires, and storms). The success of a new tree seedling can depend on animals that might eat it, on mycorrhizal fungi that are necessary for tapping soil resources, and a host of pests and diseases. The ranges where we find species to occur at present is just a single frame in a long-running, continually evolving movie of changing environments and shifting species.

Paleoecologists investigate changes in the ranges of species with a wide variety of approaches. One of the most useful involves sampling pieces of plants that are found in pack rat middens. Pack rats are small rodents that build nests in protected rocky areas, bringing in bits of nearby plants for food or nest construction. Urine from pack rats cements the materials together into middens, and the reuse of the same site by many generations of pack rats leads to a long-term sample of a location's vegetation. The plant macrofossils found in the layers of a midden can be dated with ^{14}C methods, giving a

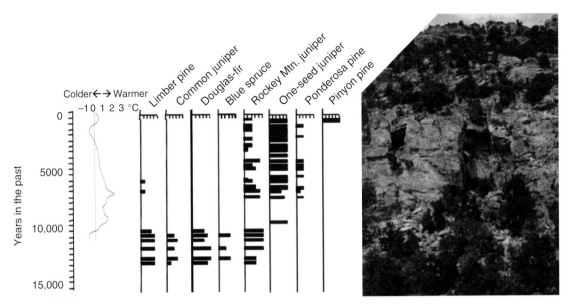

FIGURE 3.7 The history of vegetation at a site may be recorded in the form of macrofossils, sometime encased in solidified deposits in caves. For a few thousand years after the last ice age ended, this site near the boundary of Utah, Wyoming and Colorado, USA was dominated by juniper shrubs and trees of limber pine, Douglas-fir and blue spruce. For the past 7000 years, the vegetation was dominated by two species of juniper trees, with some ponderosa pine. Pinyon pine is currently a dominant member of the vegetation, but it arrived only a few centuries ago. The length of the bars indicates prevalence of macrofossils. Sources: Based on Jackson et al., 2005; temperature trend (present climate = 0, based on Kelly et al. 2013.

historical catalog (Figure 3.7). Hundreds of middens have been examined across the western USA, providing a record of very dynamic shifts in species. By looking across many middens, it's possible to chart the general patterns of changes in the ranges of species.

Regional histories of vegetation have also been developed by identifying types of pollen that accumulate in sediment layers in ponds. The annual deposition of pollen sinks to the bottom of ponds and accumulates in layers with sediments. Cores from the ponds are sorted by depth intervals, and ^{14}C methods identify the pattern of sediment depth over time.

The overriding story from paleoecological studies is that vegetation changes over time, as the ranges where species occur shift in different ways. We might think that a set of species that commonly are found together would have been found together in the past, just in different locations that had suitable climates. This has been true in some cases and some periods in time, but species often shifted ranges very independently, leading to some combinations of species being common in the past, but no longer occurring together today. Three case studies of tree species illustrate how differently tree species responded as the last ice age ended.

Critchfield Spruce Melted Away at the End of the Last Ice Age

The extinction of mammal species after the last ice age was severe by any measure (see later in this chapter), but almost all tree species responded by persisting in areas despite changing climate, or successfully shifting their geographic ranges to follow suitable climate conditions. The fossil record does record one extinction of a tree species in North America that was a dominant part of forests of the southeastern USA: Critchfield spruce.

Why did Critchfield spruce fail to make a transition from the most recent Ice Age into warmer periods? Possible factors can be listed, but evidence is not available to distinguish among them. Perhaps Ice-Age climates provided a combination of temperature and water dynamics that gave a competitive edge for Critchfield spruce relative to red, white, and black spruces that were also in the region. Perhaps that distinctive climate was not repeated geographically as climate warmed, and the other spruce species were more successful at moving into ice-free areas to the north. The success of trees also depends on surviving the impacts of pests (such as bark beetles) and diseases (such as fungi that rot wood), and the interactions between trees, pests, and diseases may shift with changing details of the climate.

The gene pool of a species can also change over time. Serbian spruce now occurs only in isolated places in the Balkan region of southeastern Europe, but it was more widespread during the Ice Age. The current gene pool within locations is somewhat broad, but there has been very little sharing of genes among the currently isolated locations (Aleksić and Geburek 2014). The long-term dynamics of the Serbian spruce gene pool depends on periods of "escape" from the isolated refugia; otherwise, narrowing of the gene pool could lead to higher risks of extinction. Any or all of these may have played a role in the extinction of Critchfield spruce.

Ponderosa Pine Went from Obscurity to Prominence in Just a Few Thousand Years

The home range for ponderosa pine in the Southwestern USA at the end of the last Ice Age was very limited, with macrofossils indicating populations restricted to a few areas near the border with Mexico (Figure 3.8; Norris et al. 2016). The dominant pine across the Rocky Mountains during the Ice Age was limber pine, a five-needle pine that is tolerant of cold and drought. As climate warmed rapidly, the range of ponderosa pine raced northward, reaching northern Colorado about 5800 years ago. If a generation of ponderosa pine trees lasted for 100–200 years, ponderosa pine has been present in these landscapes (Figure 3.7) for only about 25–50 generations. At the northern limits of the present range of ponderosa pine, some of the trees are the first-generation pioneers of the advancing front.

The synthesis of a story like this depends on knitting together large amounts of information, but the story could also be limited by the absence of information. For example, the oldest pack rat midden in the Black Hills of western South Dakota is less than 4000 years old, and ponderosa pine macrofossils were present at that point. This information adds to the story, but without an older layer that documented the *absence* of ponderosa pine it is not possible to say how much earlier the species was present in this corner of its current range. Information on the historical distribution of ponderosa pine farther east is also missing, owing to a lack of pack rat middens to be examined. It's possible that some of the ancestors of current ponderosa pines endured the ice age somewhere to the east (not just in the far south), and these refugia may have provided the seeds for some of the northward (and westward) expansion. The historical information from middens can be supplemented with other sources of information, such as the genetic strains of populations. The genetic evidence would be consistent with some of the northeastern populations of ponderosa pine descending from populations that endured the ice age somewhere to the east. More evidence may come available in the future, providing a stronger version of the Holocene story of ponderosa pine, correcting some inaccuracies and adding new insights.

Each species has a different story of shifting distributions, and three features of the ponderosa pine story are particularly interesting. The first is that the range was very dynamic, taking a species from relative obscurity to dominance across a vast range. If forest ecologists were around 12 000 years ago, they probably would not have predicted this major change in fortune of ponderosa pine. The second feature is that ponderosa pine landscapes are characterized by frequent,

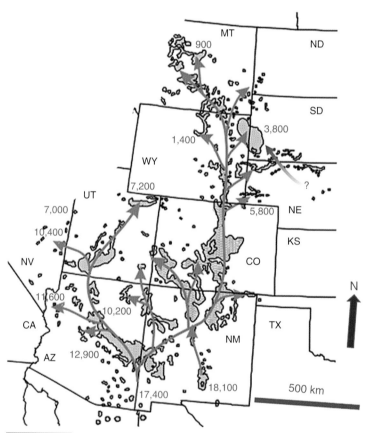

FIGURE 3.8 Ponderosa pine trees were absent from the Rocky Mountains during the last Ice Age, occurring only on mountains the near the US/Mexico border (numbers indicate age of the oldest macrofossils). As climates warmed, the species' range migrated northward, reaching northern Colorado about 5800 years ago. The limit continues northward, with some currently living trees as members of the first cohorts to establish in new territories. Importantly, the southern limit of the species distribution did not change. The red arrow raises the question of whether some populations of ponderosa pine may have endured the ice age to the east (not just the far south), and moved northwestward with warming climates. The lack of middens farther east currently limits the ability to test this possibility, but some genetic evidence in current populations supports the idea. **Source:** Based on Norris et al. 2019.

low-severity fires with fire-free intervals of just a few years or a few decades. Most other forest types in the Rocky Mountains have much longer fire-free intervals. How much is this pattern of fire associated with the climate conditions where ponderosa pine occurs (and so favoring ponderosa pine over other species), and how much relates to the presence of this particular species? The third feature may be the most valuable insight. A simple expectation might be that warming climate shifted the suitable environments from southern locations to northern ones, with northward expansion matched by contraction from southern areas that became too warm. Ponderosa pines still occur in the ranges where they "overwintered" during the last Ice Age (though now at higher elevations), and factors that allow a species to expand into new areas should not be expected to closely match factors that might contract a species' range.

Eastern Hemlock Has Had a Dynamic History of Up and Down

The fortune of species over long time periods may not be all good news or bad news. The past 10 000 years have had major ups and downs for eastern hemlock. This conifer can be a major component of forests that are dominated by broadleaved species, currently ranging from the Mississippi River across southeastern Canada and down to the southern United States. Hemlock moved northward after the last Ice Age, reaching its current limits about 6000–7000 years ago. A massive decline in hemlock occurred throughout its range, beginning about 5000–5500 years ago and lasting about 2000 years (Figure 3.9). No other species showed a sudden and large decline, which might seem to rule out any particular deviations in climate as the causal factor. Perhaps an insect pest or disease ran through the population, and hemlock took 2000 years to resort its gene pool to yield large populations of resistant trees. Would it be possible to find evidence of increased populations of an insect pest, such as the hemlock looper caterpillar? Yes, insect fossils could

be extracted from sediment cores, just like pollen grains. But no, so far the investigation of sediments has not produced evidence to support a hypothesis of an insect explanation for the hemlock decline (Oswald et al. 2017). In some cases, very precise sampling indicates there may have been some large fluctuations in hemlock populations for a few centuries before the major decline (Booth et al. 2012). Perhaps fluctuations in climate (along with insect populations?) combined with varying intensity of competition with trees of other species (less sensitive to droughts?) to drive down the hemlock populations. Whatever the true answer, the pattern of a major decline that lasted for about 2000 years is clear, and the decline of hemlock was followed rapidly by increased populations of other species (Figure 3.9). And now hemlock is undergoing another massive decline throughout its range, this time as a result of tree-killing exotic insect, the woolly adelgid (Chapter 14).

Almost all the Animal Species Are Missing from Temperate and Boreal Forests

The American naturalist Henry Thoreau spent much of his life observing nature and celebrating its beauty. His records of the timing of spring flowering have provided documentation for the ecological effects of changing

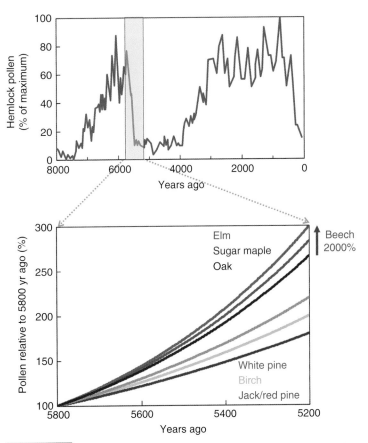

FIGURE 3.9   Hemlock was a major species of the forests of southern Ontario, Canada beginning about 8000 years ago (several thousand years after the glaciers melted). A dramatic hemlock decline occurred about 5500 years ago across the species range (upper), leading to a dramatic increase (2- to 20-fold) in the pollen record in a few centuries for all other species (lower; **Source:** Based on Fuller 1998).

climate (see Chapter 14). Thoreau realized that the nature he was observing was missing many of the major animal species that would have been present two centuries earlier: cougars, lynx, wolverines, black bears, wolves, moose, deer, and beavers. He lamented:

I cannot but feel as if I lived in a tamed, and, as it were, emasculated territory. Would not the motions of those larger and wilder animals have been more significant still? Is it not a maimed and imperfect nature that I am conversant with?. . . I take infinite pains to know all the phenomena of the spring. . . thinking that I have here the entire poem, and then, to my chagrin, I hear that it is but an imperfect copy. . . that my ancestors have torn out many of the first leaves and grandest passages and mutilated it in many places.

Henry Thoreau, March 23, 1853

Some of these animals returned when human hunting pressures decreased after Thoreau's time, but overall the poem remains far more tattered than Thoreau imagined. What he thought of as the all the lines needed for "entire poem" was really only the tattered remnants of the full poem from 12 000 years earlier. The assemblage of large animals that dominated the landscapes of eastern North America (and across the continent, across Eurasia, and across the Southern Hemisphere outside Africa) essentially vanished, leaving only smaller species roaming forests since then.

Hiking across a landscape in North America or northern Eurasia 12 000 years ago would have been fantastic and dangerous. The forests and other landscapes were homes to mammoths, mastodons, huge deer relatives, rhinoceroses, giant ground sloths, long-horned bison, camels, horses and even a beaver species as large as a bear. The list of dangerous predators included short-face bears (larger than brown and polar bears), several species of long-tooth cats, dire wolves (larger than modern gray wolves), and even a large species of lion in North America. These species all survived the Ice Age, as well as several that came before (at intervals of 100 000 years or so), but they disappeared soon after the cold period was over. The Americas and northern Eurasia lost all their very large animals, as well as most of their medium-to-large species (Figure 3.10). What happened? It's difficult to reach iron-clad conclusions about historical events, but much of the evidence supports the idea that human hunting was a key factor.

Every single modern tree species evolved with the presence of this incredible menagerie of animals. How did the extinct animals influence forests? The array of impacts would include major eating of trees (mastodons specialized in eating trees, and they were one of the most widespread of the North American megafauna), eating fruits and transporting seeds (Osage orange trees produce large fruits that no current animals eat). The trees would have evolved in response to the animals, including features such as massive thorns on some stems (Figure 3.10). When the megafauna existed, their influence on forests would have been profound, but their extinction did not lead to the end of forests: forests persisted, though they were not the same.

Climate, Animals and Fire Interact Across Forest Generations

The drivers and responses of forest change over very long time periods typically begin with climate, and the effects are not simple. Warmer climates may benefit two species, but if one benefits more than the other, a warmer climate could limit the success of the less-responsive species. Warmer climates can lead to changes in tree/pest relationships, with follow-on implications for competition among tree species. Fire patterns are very responsive to changes in climate, for both temperature and moisture patterns. It would be easy to come up with a longer list, but the point is that so many interacting factors and responses would ensure that simple stories cannot apply very broadly or for very long. A case study from a landscape in Indiana, USA illustrates how multiple aspects of a forest changed over an 8000-year period, spanning from cold, ice-age conditions to periods that were as warm as today (Figure 3.11). (This location was

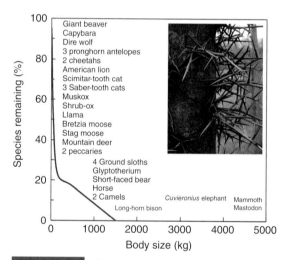

FIGURE 3.10 The current animal community in North America bears slight resemblance to what would have been found over the last 100 000 years. This figure lists the extinct species in relation to their body sizes. After the last Ice Age, extinctions removed all the largest species, and 80% or more of the species that had body weights over 100 kg. Almost all species of small animals (<10 kg body mass) survived (**Source:** Based on Koch and Barnosky 2006). The long-term development of the forests of North America was shaped by large animals that no longer exist, but a few clues remain. Honey locust trees still carry 8–20 cm thorns on their stems, but only up to the height that a mastodon's trunk could reach (**Source:** Photo by Greg Hume).

south of the continental ice sheet, not buried under ice.) Fires became common only after the Ice Age. The spores that come from fungi feasting on the dung of herbivores dropped dramatically after the warming, but in this case the driving factor in the loss of herbivores would be the megafaunal extinction, not a simple climate effect. The distributions of tree species across the landscape shifted. For a few thousand years, the combination of species occurring at this site did not match that of currently forests that occur currently to the north or to the south. No-analog forests are almost as common in historical records as are forests comprised of species that currently co-occur.

Modern Forests Are Changing Faster Than Ever, on a Global Scale

History will certainly not repeat itself as we move into the future of forests on Earth (Chapter 14). Humans have moved species around the planet: eucalyptus trees cover swaths of South America, and radiata pines from North America cover much of New Zealand and South Africa. The gene pools of species shift as they change environments, and they shift even more if humans are involved in nurturing some genotypes more than others. The global expansion of species also includes diseases and browsing animals that can decimate populations of tree species (and animal species) that did not have a chance (yet) to coevolve with the new player in the ecosystem. The shifting biology is occurring at the same time that climate changes, with both subtle and dramatic effects. Changing climates can alter the mortality risk for big, old trees, and somewhat minor changes in climates change fire occurrence and severity in ways that generate massive, long-term legacies. Higher concentrations of carbon dioxide in the air raise the efficiency of water use by trees. The broad historical dynamism of forests in relation to ice ages and long-term evolutionary situations should leave us with confidence in two conclusions: the forests of the world will clearly persist across much of the planet, and many future forests will be "no-analog" forests compared with those of our grandparents (Chapter 14).

Ecological Afterthoughts

The ranges of species shift over time, sometimes by hundreds of kilometers in latitude, and by hundreds of meters in elevation. The burning of fossil fuels is warming Earth by increasing the CO_2 content of the atmosphere, and simulation models try to predict how fast the climate will warm, and how large the changes in precipitation will be. The future will clearly be warmer, but the details of how that warmth will develop across time and space cannot be known perfectly. The responses of the global hydrologic system to warmer conditions are even more challenging to pin down. On average, precipitation will rise in response to increased energy available to drive evaporation. How much will increased precipitation come during extended rainy seasons, in more intense storms, or in gentle rains? Which areas will be wetter, and will there be larger areas of arid conditions?

The ranges of tree species will continue to shift into the future, as this chapter has described the changes through the past. One approach to guessing the future range of species is to map a species' current range in relation to climate, then use simulations of future climates to ask where the current "climate envelope" of the species will exist in the future. A great deal of careful, reasonable thought and calculation lay behind these scenarios, but the intensity of invested effort does not

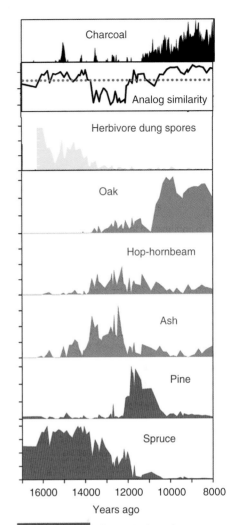

FIGURE 3.11 Change is the only consistent story of forest ecology for the landscape around Appleman Lake, Indiana, USA. Some key stories are the decline and disappearance of spruce as the Ice Age ended about 12 000 years ago, the rise of pines for a 2000-year window, and the varying comings and goings of the broadleaved species. The changes related in part to changing climate, but the decline in "herbivore dung spores" (*Sporormiella* ascospores) documented the extinction of the megafauna, along with their influences on forests. By 11 000 years ago, the accumulation of charcoal in the lake indicated a climate that was suitable for forest fires. The "analog similarity" index charts how similar the assemblages of tree species would be to current assemblages that would be found currently in other places. The 2000-year period where the similarity fell below the dotted line describes combinations of species that do not currently occur together anywhere (a no-analog forest community). The scales of the Y axes vary among the graphs, providing an overall story for the changes in each panel over time. **Source:** based on Gill 2014.

guarantee accuracy or value of the scenarios. The scenarios in Figure 3.12 try to map the future for aspen in western North America. How valuable are these scenarios? That overall question would probably be easier to consider by breaking it down into smaller questions. What would those smaller questions be? Do the smaller questions warrant high confidence in their answers, as a result of clear evidence? If the smaller questions could be answered only in fuzzy ways, what would this mean for putting together the small-question answers to address the original question of whether such scenarios of the future have value?

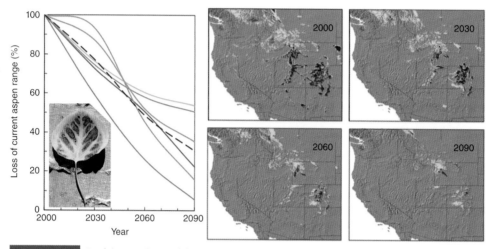

FIGURE 3.12 Earth is warming, and this warming will be accompanied by changes in patterns (seasonal and geographic) in precipitation. The current distribution of aspen in western North America is well known, and it can be related to recent patterns in temperatures and precipitation. Rehfeldt et al. (2009) took the current "climate envelope" for aspen and plotted where that envelope would be found into the future, using six different climate model predictions. All models predicted that areas currently occupied by aspen would become less suitable, with half or more of the current range no longer suiting aspen in 50 to 100 years (left; solid lines are predictions of different models; the dashed line is the average of six models). The maps show areas where the climate envelope would be very suitable for aspen (green), or somewhat suitable (yellow) for four points in time. **Source:** Based on Rehfeldt et al. 2009.

CHAPTER 4

Physiology and Life History of Trees

The development of a single tree progresses from formation of a seed, through germination, successful transition into a seedling, growth into a mature tree, with eventual production of flowers and seeds. The stories of forests assemble from the stories of the life cycles of individual trees, along with the stories of understory vegetation, animals, and microbes. Trees shape the three-dimensional structure of forests, determining how the broad environmental factors of incoming light, winds, and precipitation are modified to create the local microenvironmental conditions within the forest. The structure of forests changes over time as trees grow (and die). The process of photosynthesis, and the growth that it powers, is at the core of forest ecology stories.

Biological Energy Is About Moving Electrons

The discussion of energy in Chapter 2 focused on physical processes of radiation and temperatures, processes that apply to the surface of the Moon as well as the Earth. The biological processes of forest ecology are driven by chemical energy, as stores of energy are created by photosynthesis, stored in organic compounds, and used to fuel the physiological processes of plants, animals, and microbes. The key to biological energy stories is the idea of potentials, similar to the idea of potentials that determine the flows of water in soils and plants. Water moves from zones of higher potential to lower potential, releasing energy. Chemical reactions release energy stored in high-energy compounds as reactions generate low-energy compounds. The flow of energy in chemical reactions entails the flow of electrons: electrons move into higher energy compounds with energy added (as in photosynthesis), and they moving to lower energy compounds with releases of that energy (as in fire). The electrons are not created or destroyed, so the flow of electrons can only occur from a source coupled to a sink. The electron flow is called reduction–oxidation (or redox), and redox reactions are diagramed with two half reactions: one half supplying the electrons, and one half accepting the electrons.

Most redox reactions in forests use oxygen (O_2) to accept the electrons, forming water (with addition of H^+ ions) at a very low energy state. Some other electron acceptors are important in soils (and the guts of some animals), where oxygen is in low supply. Where do redox reactions in forests obtain high-energy electrons for transfer to oxygen? Organic molecules store energy in chemical bonds, and the energy is released as electrons flow from organic molecules to oxygen (forming low-energy water and carbon dioxide). This sequence is called respiration, and it's the same process that occurs in our bodies (and in fires). This is why breathing is called respiration (with oxygen entering, and water and CO_2 leaving).

How much energy is released when electrons flow from organic matter (such as sugar or wood) to oxygen? One square meter in a forest might typically contain about 10–15 kg of organic matter. The energy stored in the molecules would be the equivalent of about 6–10 liters of gasoline. Decomposition of organic matter slowly releases this energy as heat, or fire releases it in a matter of minutes.

Forest Energy Comes from Sunlight; Wood Comes from Thin Air

We all know that plants perform photosynthesis, obtaining energy from the sun to make biology possible. Photosynthesis is commonly represented by an equation:

$$\text{Water} + \text{Carbon dioxide} \rightarrow \text{Sugar} + \text{Oxygen}$$

with the reaction driven forward by sunlight. Photosynthesis might seem to be a process that takes solar energy and charges a battery with high energy, which is a reasonable analogy. But that analogy misses half the story, because photosynthesis also builds massive batteries. Absorption of photons by leaf pigments such as chlorophyll is only the beginning of photosynthesis; the production of sugar completes the system by providing the carbon-based building blocks that assemble into plants.

Even though all students are taught the basics of photosynthesis, this most important of processes is actually grasped by only about one-third of them. For a number of years, incoming students were polled in graduate and undergraduate classes in ecology and forestry, asking "Where does most of the material in this piece of wood come from?" Students were given multiple choice options that included molecules dissolved in water, large soil molecules (humus), inorganic molecules from the soil, and small molecules in the air. Students already knew that forests use lots of water, so molecules dissolved in water was typically the most popular answer. Even though the equation for photosynthesis has "carbon dioxide" as one of the two key reactants, only 30–40% of students chose the correct answer of "small molecules in the air." It might seem unlikely that something so thin as air could provide the materials that comprise very solid pieces of wood, but indeed the carbon that comprises about half the mass of organic molecules (such as wood) does come from carbon dioxide molecules floating around in the air. (For an excellent documentary about how students learn but fail to understand, see: www.learner.org/series/minds-of-our-own/2-lessons-from-thin-air/.) Photosynthesis needs to be understood as having two steps: capturing sunlight into a chemical form, and then using that energy to split water and add H^+ to CO_2 to produce sugar.

Why Are Leaves Green?

The common answer is that sunlight has a spectrum of colors that combine to make white light, and leaves absorb red, yellow, and blue light with green light reflecting from the leaves. This answer is mostly on the mark, but a number of important details strengthen the answer.

The first detail is that light doesn't really come in colors. Light (and other electromagnetic radiation) has wavelengths (or the inverse, frequencies) of oscillations. Some chemicals effectively absorb only certain wavelengths, and that's the case for receptors in our eyes. We have receptors that are activated strongly by light with wavelengths near 440, 540, and 580 nm. The electrical signals sent from our eyes are interpreted as colors in our brains, with millions of variations created by combining the intensities sensed in these bands of wavelengths. The wavelengths themselves possess no property of color. Without eyes, the world has no color.

The second detail is that green leaves actually absorb more than 80% of the green light from sunlight, reflecting only 10–15% of the green light (Figure 4.1). Molecules of chlorophyll do not absorb much of the green light, but other molecules in leaves absorb

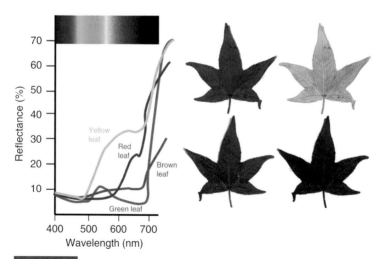

FIGURE 4.1 The apparent color of leaves changes in the autumn as changing chemistry within the leaves reduces the absorption of some wavelengths, leading to higher reflectance. Yellow leaves still absorb green and blue wavelengths, and brown leaves absorb most wavelengths. **Source:** Based on Jensen 2000.

quite a lot. Our eyes interpret leaves as being green because green leaves absorb about 95% of the other visible wavelengths, but a bit less of the green light, so our eyes perceive more photons with a green wavelength than any others.

Leaves Are Not Always Green

When buds burst in spring and leaves begin to form, the colors of leaves tend to be more reddish than green. Red colors in leaves derive from anthocyanins, non-photosynthetic pigments that absorb blue, yellow, and especially green wavelengths of light while reflecting red. A class of compounds such as anthocyanins may have multiple roles within plants. A primary role for anthocyanins seems to be absorbing light (dissipating the energy as heat) and reducing possible damage from intense radiation when the intensity would overwhelm the photosynthetic capacity of chlorophyll (Croft and Chen 2017).

Leaves change color in the autumn because some of the red and blue light that was absorbed strongly by chlorophyll is reflected after chlorophyll breaks down (Figure 2.1). The anthocyanins absorb blue and green wavelengths, but not much red, and anthocyanin concentrations may increase during leaf senescence. Carotenoids are another class of pigments (involved in photosynthesis) that absorb blue and green wavelengths, and especially reflect yellow wavelengths.

Carbon Uptake Is the Second Half of Photosynthesis

The growth of forests relates in part to the supply of water, because the uptake of carbon dioxide (CO_2) from the atmosphere can be accomplished only by exposing the moist interiors of leaves to the dry air. For each CO_2 molecule that diffuses from dry air into the moist interior of a leaf, something like 200–500 molecules of water escape into the atmosphere. When water supplies are low, trees close the pores (stoma = mouth, stomata or stomates for plural) on leaves, minimizing water loss and unfortunately preventing uptake of CO_2. Plants transpire water at high rates when the supply of water is abundant, and when the atmosphere is not too dry (humidity is high; Figure 4.2).

The story of potentials also explains how water moves against gravity, ascending tens of meters (or even one hundred of meters) upward in trees against the pull of gravity (mentioned in Chapter 2). Dry air has a tremendously negative potential compared to the insides of leaves, so water is sucked from leaves into the air. The potential of water inside leaves is lower than in twigs, so the evaporated water is replaced in the leaf with water sucked from the twigs. The twig's water potential is lower than the stem, and the stem is lower than the root, and the root is lower than the water potential of the soil. This chain of potentials moves water from the soil to the

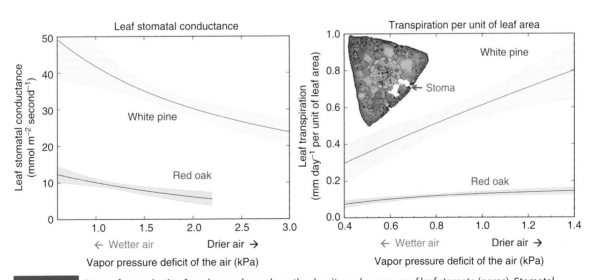

FIGURE 4.2   Rates of transpiration from leaves depends on the density and openness of leaf stomata (pores). Stomatal opening is indexed as stomatal conductance (how many molecules could slip through the stomata present on $1\,m^2$ of total leaf area), and stomatal openings shrink as the air becomes drier. White pine needles and red oak leaves both have stomata that are sensitive to vapor pressure deficits, but the pine needles sustain more open stomata at all humidities (left). The amount of water lost from a leaf depends on stomatal conductance, the water status of the leaf (water potential, Figure 4.3), and the dryness of the air (right). With very dry air, the loss of water can go up even as stomata close down. The pine leaves transpire much more water at any given level of vapor pressure deficit than the oak leaves. **Source:** Data from Ford et al. 2010; cross-section photo of a white pine needle from Paul Schulte, http://schulte.faculty.unlv.edu//Anatomy/Anatomy.html.

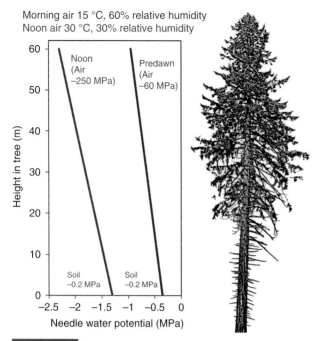

Morning air 15 °C, 60% relative humidity
Noon air 30 °C, 30% relative humidity

FIGURE 4.3   Water flows "down" gradients of potential, from high potentials in soils (which are high, even though the sign on the value is negative) to lower potentials in trees (more negative values). Water flow up this old-growth Douglas-fir tree was along a milder potential gradient just before dawn, when the low water demand of the atmosphere coupled with the leaf water potential for a moderate rate of transpiration as the sun rose. By midday, the vapor pressure deficit of the hot, dry air was extreme enough that the water potential declined in the needles and stomata closed, restricting transpiration even though soil water supply did not change. **Source:** Data from Bauerle et al. 1999.

atmosphere, by virtue of the very strong ability of water molecules to stick to surfaces and to each other. The potentials in soils and plants are typically negative, which means the potential is lower than in a free-standing puddle of water on the soil surface (at a potential of 0 MPa). Going from a higher potential to a lower potential means that water will flow down the potential gradient from −0.2 MPa in soils, to −1 or −2 MPa in trees, to −60 MPa (or lower) in dry air (Figure 4.3).

Growth Happens After Photosynthesis – Sometimes Long After

Plant growth depends on photosynthesis, but not necessarily on recent photosynthesis. Sugar is the immediate product of photosynthesis, and sugars may be used quickly (for respiration needs to maintain leaf cells). Most sugars are converted to starches for short-term storage during the day, with breakdown to soluble sugar at night for transport through the plant. It might seem that photosynthesis should be coupled tightly with the time course of forming new cells in plants (called growth), and this is true generally: trees photosynthesize and grow when the local climate conditions are favorable. The two processes aren't tightly coupled in an immediate sense, because plants have vast storage pools of carbohydrates that are used over longer periods of time to support the growth of new cells and tissues. The term non-structural carbohydrates refers to these storage pools, and they're so important they add up to about 10% of the mass of plants. The importance of stored carbohydrates is apparent in the age of the carbohydrates used to grow new tissues in plants. Most of the C used in leaves and twigs is young, coming from photosynthesis in the current year. Some pools of carbohydrates in roots are several years old, or even decades old (Hartmann and Trumbore 2016).

Trees Do Not Live by Carbon Alone

The products of photosynthesis contain only carbon, oxygen and hydrogen, and the mass of these three elements accounts for more than 90% of the dry mass of trees and other plants. More than a dozen other elements comprise the rest of the mass, each providing a unique biochemical opportunity. For example, carbohydrates are characterized by generally symmetrical chains and rings of atoms. Nitrogen is similar to C in many ways, but different enough in that including N inside carbon chains and rings dramatically alters the three-dimensional shapes of the molecules, changing the chemical and electrical characteristics. Other elements are important for building cell structure, such as calcium (Ca) for plant cell walls. Potassium is a very mobile ion, and is used to control the osmotic potential inside guard cells to open and close leaf stomata. Some enzymes are activated by other elements. One of the most interesting elements is sodium (Na), which is not generally needed by plants, but is a vitally important nutrient for the nervous systems of animals.

Photosynthesis and Growth Depend on Acquisition of Resources

The establishment, growth and survival of plants requires large amounts of light, water and nutrients. Indeed, the quest for obtaining these resources is a root concern for plants on the way to producing seeds and reproducing. Tree growth can raise crowns above competitors, improving light capture. Roots are needed to anchor trees against the massive forces of wind, and to tap stores of water and nutrients in the soil, in part through symbioses with mycorrhizal fungi. Positive and negative feedbacks between resource acquisition and tree growth influence the shapes of trees, the partitioning of growth above and belowground, and the development of soils.

More Leaves Means More Light Capture, up to a Point

Plants use stored carbohydrates (and nutrients) to grow new leaves. It takes leaves about a week or two of photosynthesis to recover the cost of growing them, and then leaves support the plant. This simple story has some complications, including the cost of sustaining leaves (the complex biochemistry of leaf cells consumes a sizable portion of the leaf photosynthesis). Indeed, the "cost" of sustaining leaves can be higher than the gains through photosynthesis if a leaf is too shaded. The number of leaves and their display within tree crowns is determined in part by the presence of light available to be absorbed. The capture of light by leaves depends strongly on the amount of leaf area for a tree or forest, along with the orientation of leaves (mostly flat or mostly vertical) and clumping within crowns.

One Square-Meter of Leaves Has a Mass of 50–150 g

No single design of a leaf would be optimal for photosynthesis in all situations, and leaf shapes range from cylindrical needles of some conifers to broad, thin leaves on many broadleaved tree species. Substantial ranges in leaf morphologies are common within the crowns of individual trees and canopies of forests. Leaves that are brightly lit in the tops of trees are thick, with a high mass per m^2 of two-dimensional leaf area. Leaves in the shady lower crowns may require only half the mass to form $1\,m^2$ of leaf area. Leaves in full light in a rain forest in Costa Rica were thicker than more shaded leaves (Figure 4.4), maybe indicating that fatter leaves are useful in dealing with the full strength of sunlight. Shaded leaves were thinner, perhaps allowing the maximum amount of light to be captured per gram of leaf when light intensities are low. This type of pattern is common in ecology: the trend in one variable (the mass of leaves per m^2 of area) is related to the size of another variable (percent of full sun). We might be pleased when a clear pattern emerges rather than a random scatter. However, the presence of a pattern is not the best evidence for a cause-and-effect relationship. Is it possible that the thickness of leaves is related to another factor, which also happens to relate to the amount of sunlight reaching a position in the canopy? In this case the tightest pattern came from relating leaf thickness simply to height above the ground, regardless of whether a leaf was in a sunny or shady location. Of course this does not prove that leaf thickness results from trees "knowing" how high a given branch is above the ground, but it does indicate that a next step for investigation might focus on something about height above the ground.

How might we gain more confidence that the relationship between leaf mass and area depends on height rather than sunlight? One step would be to examine another case to see if the same pattern emerges. A test of leaf mass and area for eucalyptus plantations in Hawaii matched the rain forest's pattern with height aboveground (Figure 4.5). As the trees grew, the upper crowns of course

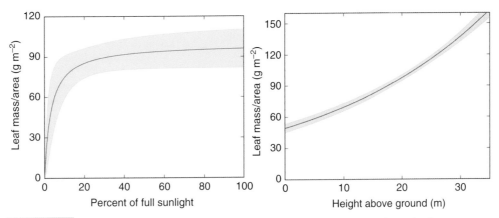

FIGURE 4.4 The structure of leaves in the canopy of a rain forest in Costa Rica showed a clear pattern between the light environment of leaves and the how much mass was necessary to form $1\,m^2$ of leaves (left; the band is the 95% confidence interval based on 12 levels of light measured in 45 profiles through the forest). Leaves receiving full sun had a mass of about $120\,g\,m^{-2}$, whereas leaves receiving 10% of full sunlight had a mass of about $90\,g\,m^{-2}$. The correlation between leaf thickness and light does not mean that the light environment is the key factor determining leaf thickness. Indeed, the variation in leaf thickness related much better to simply how high leaves were off the ground (right, with observations from 45 profiles grouped into 7 height classes; **Source:** Based on data from Cavaleri et al. 2010).

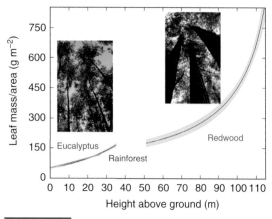

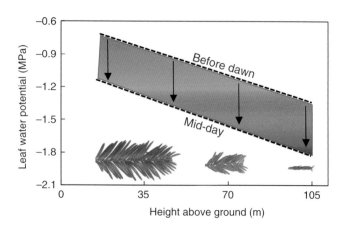

FIGURE 4.5 Eucalyptus plantations in Hawaii also show a clear pattern between leaf mass and area as trees get taller with age, even though the light environment at the tops of the crowns remains the same (left; **Source:** Data from Coble et al. 2014). Needles on tall redwood trees in California also fit the pattern of increasing leaf mass area with height above the ground. Height above the ground influences leaf water potentials (right), both before dawn (after the tree's water content has recovered overnight) and at mid-day, leading to very large changes in needle morphology with height (**Source:** Based on Koch et al. 2004).

remained fully illuminated, but the leaves became thicker as the trees grew taller. Another test of a relationship could be made by going to very extreme situations. The pattern of leaf mass and area for redwood trees in California also demonstrated a very strong pattern with height above the ground. The water potential within trees decreased with height above the ground for basic physical reasons (see Figure 4.3), and the leaf thickness may well be an adaptation to the water potential that relates to height above the ground. The leaves of redwood trees change morphology (as well as physiology), from very large, broad needles lower in the crowns to small, thick needles high above the ground. Needles below 30 m experience a range of water potentials that are high enough for good rates of photosynthesis, but needles higher from the ground have such negative potentials that photosynthesis would be very limited.

The ability of leaves to carry out photosynthesis is strongly affected by the influence of height on leaf structures and ecophysiological function. The photosynthetic capacity of leaves at the tops of trees (under full sunlight) is usually much lower than capacities of leaves closer to the ground (Figure 4.6). Of course the amount of sunlight intercepted by a leaf may be lower in the lower canopy, so actual rates of photosynthesis would depend on both light supply and the photosynthetic capacity of leaves.

Would the pattern in Figure 4.6 represent other situations? This

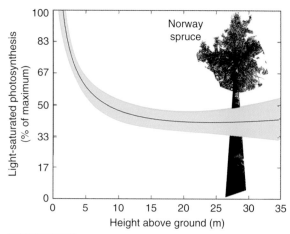

FIGURE 4.6 The photosynthetic capacity of needles (with full sunlight) in a Norway spruce trees in Estonia declined with increasing height above the ground, because of interacting effects of structural differences (resulting in part from differences in water potential) and ecophysiological functioning (**Source:** Data from Räim et al. 2012).

question applies to all patterns in forest ecology, and extrapolations always need to be done carefully. The pattern might well apply to other Norway spruce forests, and perhaps to other conifer forests. The physiology of leaves of broadleaved tree species are more dissimilar, though the challenges of the physics of water with tree height would be similar. If the pattern were examined in a very different forest type and the overall trend was consistent, that would be evidence of stronger, generalizable effect of tree height on photosynthetic capacity. Alas, a tropical rain forest with multiple species of dipterocarp trees showed a trend that was opposite to the Norway spruce pattern, with maximum light-saturated photosynthetic capacity occurring in the tallest trees (Kenzo et al. 2006). The question of how height affects physiology applies everywhere, but local details are very important for answers that would vary across cases.

Each Square Meter of a Forest has Multiple Layers of Leaves above

If all the leaves in a forest were laid flat on the ground, the forest floor would have between two and eight layers of leaves. This means that a m² of forest land area typically has 2 to 8 m² of leaves above it. It might seem that more than one layer of leaves would be redundant, as the incoming sunlight might be captured by just a single layer. The spatial arrangements of tree stems and crowns prevent any single layer of leaves from catching all the light. The actual capture of light by forest canopies depends on the clumping of leaves within crowns, the clumping of crowns across forest canopies, and the geometry of how leaves are displayed (flat, tilted, or vertical) in relation to the sun's apparent movement across the sky.

The capture of light by canopies of eucalyptus plantations at a site in Brazil showed a pattern of increasing light capture with increasing leaf area index (Figure 4.7) across a dozen genotypes. Leaf area index is the number of two-dimensional layers of leaves above each area of ground. The average trend showed that a clone with a leaf area index of three would capture about 70% of incoming light, and those with double the leaf area captured about 90%. The extra light capture by the clones with more leaf area was not very large, illustrating the importance of the angle of leaf display toward the sun, and the degree of clumping (self-shading) within tree crowns and between neighboring trees.

The lower graph in Figure 4.7 looks similar to the upper graph, but each curve represents the capture of light moving down through a canopy. The first layers of leaves in a canopy may catch 20–50% of all incoming light, with the additional capture of light declining with each increasing unit of leaf area index. The amount of light captured for a given leaf area differed by more than a factor of two, again illustrating the importance of crown shapes and leaf arrangements for capturing light.

Large Trees Depend on Large Roots

When strong winds blow, the crowns of trees act as sails that intercept the force of the wind and transfer it to the stems. Sailing ships harness the force of the wind to move through the water, but trees can only transfer that force down the stem and into the ground (as noted in Chapter 2). Well-rooted trees can withstand very high winds, though severe storms can break or topple

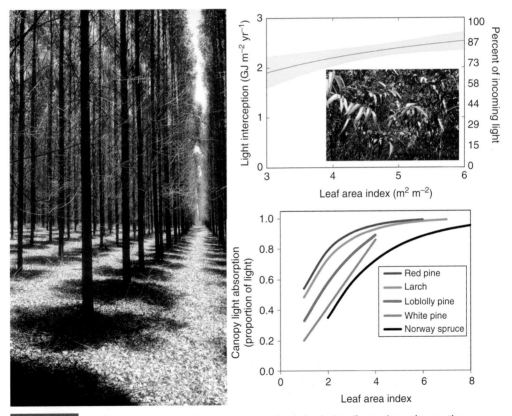

FIGURE 4.7 The tropical location of this eucalyptus plantation in Brazil experienced noon-time sunshine directly overhead, with shadows cast on the ground. The bright areas between the shadows demonstrate how important canopy structure is for capturing (or not capturing) light (**Source:** photo provided by Clayton Alvares). The leaf area index differed by almost twofold among clones planted at the same site in Brazil, but the extra light capture by clones with high leaf areas was small (upper graph, **Source:** Data from Mattos et al. 2020). The light interception patterns differ strongly among species; each line in the lower graph shows the light capture by each layer of leaves in the canopy. The first layer of leaves in the white pine forest captured 20% of incoming light, compared with 50% of incoming light captured by the first layer of leaves in the red pine forest. The upper endpoint of each line indicates the total leaf area, and the total light intercepted (**Source:** Based on Binkley et al. 2013).

even strong trees. The large roots of trees (called coarse roots) typically account for about 20% of the total mass of the trees. The morphology of root systems varies among tree species and in response to soil types. Rocky soils provide options for tree roots to anchor solidly within the matrix of rocks, and wet soils may restrict roots only to the uppermost part of the soil (requiring very broadly branching systems of coarse roots to dissipate wind energy). Some tropical tree species develop massive "buttress" roots, which would help resist wind and disperse the force of wind across are larger area of soil (Figure 4.8).

Coarse roots also provide an important part of the conducting network. Material collected by the finest root hairs come together in fine roots, which come together in larger roots, eventually entering the largest roots on the path into the tree stems and up to the leaves.

Networks of Fine Roots Permeate Soils

On an annual basis, the percentage increase in the size of a tree is usually not very large, ranging from less than 1% in large old trees to perhaps 20% (or even more) in the fast-growing smaller trees. If the mass of coarse roots needed to support a tree would be a constant 20%, then only small-to-moderate increases would be needed each year, with the growth of coarse roots each year accounting for a small part of the total production of a tree (see Chapter 7). However, about one-third to one-half of the total photosynthates produced by trees flows to the belowground system, not for the formation of new coarse roots, but for the production and maintenance of small (fine) roots and their mycorrhizal symbionts. The mass of fine roots may amount to 5–10% of the mass of a tree, but the fine roots typically live for one or a few years and so the cost of replacement accounts for a substantial proportion of a tree's production.

Many sorts of numbers can be used to characterize the fine roots of trees and forests. A 40-year-old forest of Norway spruce in Estonia had about 180 g of fine roots in each m² of ground area, with an astounding 2800 m of fine root length per m² of ground area (Figure 4.9). Each cm of fine roots sprouted about four root tips to explore the soil, giving a total of about one million root tips in each m² of the forest.

Fine roots are typically concentrated in the uppermost soil, where the supplies of water and nutrients are concentrated. The topmost layer of most forest soils is an organic horizon (O horizon), and roots proliferate in O horizons except in dry forests. The depth of the entire soil may be quite shallow, particularly in rocky terrain, steep slopes and areas where soil formation has not progressed far. In deep forest soils, tree roots descend deeply when soil supplies of water and nutrients warrant the investment. Half of the fine roots in a eucalyptus plantation in Brazil developed below 0.5 m, with some extending many meters into the soil (Figure 4.9).

FIGURE 4.8 This onyina (or kapok) tree in Ghana has large buttresses which dissipate the wind force that might topple an unbuttressed tree.

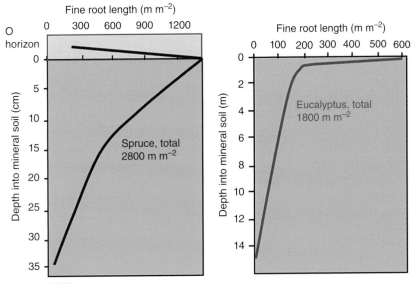

FIGURE 4.9   Fine roots concentrate in the upper soil, including the O horizon in a 40-year-old forest of Norway spruce in Estonia (left, **Source:** Based on Kucbel et al. 2011). Fine roots can extend to great depths wherever water and nutrients can be obtained, such as tropical forests growing on ancient, deep soils (right, **Source:** Data from Christina et al. 2011).

Do Roots Take Up Water and Nutrients?

The obvious answer would seem to be "of course!", but the true answer is "somewhat." The roots of all trees form symbiotic associations with mycorrhizal fungi. Much of the exploration of soil and acquisition of water and nutrients is done by mycorrhizal fungi, followed by transfer to tree roots. Why do trees rely on fungi for uptake of water and nutrients rather than simply use their own fine roots and root hairs? The filaments of the fungi (hyphae) are an order of magnitude smaller in diameter than fine roots, and each gram of the upper soil in a forest may have 200–500 m of mycorrhizal hyphae (Smith and Read 2008). The mass of a 5-cm thick layer of mineral soil would be about 50 kg, so each 1 m^2 of a well-colonized soil would contain an astronomical 10 000 km of fungal hyphae. Trees support these communities of mycorrhizal fungi with carbohydrates derived from photosynthesis in the canopy, and the fungi exploit the full suite of microsites in the soil.

Two major types of mycorrhizae colonize forest trees: ectomycorrhizae and arbuscular mycorrhizae (Smith and Read 2008). Ectomycorrhizae include thousands of species that associate with many conifers (especially in the pine family). They form dense nets of hyphae around root tips, with carbon, water, and nutrients transferring across the outer epidermis of the root tips. Arbuscular mycorrhizae include fewer than 200 species worldwide, but they associate with most broadleaved tree species (and many non-pine conifers). Arbuscular fungal hyphae actually penetrate inside roots to exchange carbon, water, and nutrients.

Trees allocate about 5–20% of their annual photosynthesis to mycorrhizal fungi (Smith and Read 2008), supporting one of the most important symbiotic systems on Earth. A root without mycorrhizae might draw water and nutrients from the soil up to 1 cm away from the root surface. A root colonized by mycorrhizae would extend that distance to 25 cm from the root surface. Importantly, the supply of nutrients is highly localized at tiny scales within soils and the movement of nutrient molecules depends heavily on proximity of the absorbing surface of fungal hyphae. The massive surface area of mycorrhizal hyphae ensures that the primary uptake of immobile nutrients such as phosphorus occurs by the fungi, with later transfer into tree roots.

The symbioses between trees and mycorrhizae benefit both members, but symbioses in biology and ecology should not be expected to benefit each member equally (Yong 2016). Natural selection operates at the level of genes and species, and some coevolution also shapes the genes and species that are most successful in symbiotic situations. However, selection is typically too weak (and flexible) to lead to equal benefits to members of symbiotic systems. Under nutrient rich conditions, trees may have less "reason" to support the mycorrhizal community. Soils with very low nutrient supplies can lead to mycorrhizae accumulating nutrients (for their own use) with only small amounts passed on to trees (Högberg et al. 2003; Näsholm et al. 2013). The fascinating dynamics of tree/mycorrhizal symbioses can be characterized as a complex system similar to economic markets, where the overall market works, but with varying advantages to each participant (Franklin et al. 2014).

Trees (and Mycorrhizal Fungi) Obtain Nutrients by the Interaction of Mass Flow and Diffusion

Water within forest soils has dilute concentrations of nutrients required by plants, and uptake of water provides a "mass flow" of nutrient molecules into mycorrhizal fungi and plant roots. The energy for taking up these nutrients is simply provided by the gradient in water potential that drives water uptake. The mass flow delivery of nutrients is not sufficient to support high rates of plant growth, and additional nutrient uptake occurs through active uptake where energy is spent to accumulate nutrients at higher concentrations within the hyphae and roots than in the external soil. The zone near an uptake surface is depleted of nutrients as a result of active uptake, so more nutrient molecules passively diffuse into the low-concentration zone around the absorbing surface. This diffusion is a slow process, but the restocking of nutrients near the root surface gets resupplied by mass flow of water moving into the roots and hyphae. Diffusion is a key part of the movement of nutrients taken up actively by fungi and roots, but diffusion rapidly depletes the zones near the uptake surface unless mass flow replenishes the nutrient supply (McMurtrie and Näsholm 2017).

Life History Is the Story of Going from Seed to Mature Seed-Producing Tree

It may not be possible to list all the stresses and events that trees cope with through their lifetimes, but even a short list requires very impressive physiological responses. A germinating seed uses carbohydrates and lipids (oils and fats) stored in the seed to produce roots and shoots to obtain water, nutrients, and light. The transition from new germinant to established seedling includes vital steps of gaining carbohydrates through photosynthesis before the stores in the seed run out, and establishing vital symbioses with mycorrhizal fungi. The microenvironmental conditions around a seedling shift through day, with changing sunlight, changing humidity, and changing temperatures. Outside the tropics, seedlings need to cope with large seasonal changes in environmental conditions, including day length and minimum and maximum temperatures. Summertime physiology fosters carbon gain and growth, but winter temperatures would destroy any cells that were still in their summertime condition. Transitioning from seedlings to mature, seed-producing trees is so challenging that the vast majority of seedlings are not successful; the number of seedlings in a forest is often orders of magnitude larger than the numbers of seed-bearing adult trees. All these topics are part of the life history of a species. The idea of a species having a life history is a bit vague, given that variation among individuals, of microclimatic factors, and contingent events that may or may not happen. Life history is sometimes referred to as "silvics" by foresters. Excellent summaries of life histories of North American trees are available to download (Bonner and Karrfalt 2008; Burns and Honkala 1990). Some broad characterizations may be useful for thinking about general patterns, as long as general tendencies are not mistaken for case-specific details.

Tree Seeds Range in Mass from Smaller than a Flea to Larger than a Mouse

Seedlings that develop from large seeds (such as walnuts and acorns) have a great advantage over seedlings endowed with very, very tiny seeds (such as cottonwoods and willows). Successful establishment of tiny-seed species tends to happen only under ideal conditions of high light, high water, and high nutrient supply. For example, cottonwood seeds are so small that it takes about 5000 to add up to a single gram of mass! Successful germinants occur on recent flood deposits (Figure 4.10) that have the requisite light, water and nutrients. Recent flood deposits dry out as the water table recedes, and first-year growth and survival depend on whether seedling roots can descend into the soil quickly enough to tap receding water and nutrient supplies.

Large supplies of carbohydrates in seeds might be a tremendous benefit for new germinants, so it might seem that all tree species should produce large seeds. New germinants of pines benefit from seeds that are much larger than cottonwoods seeds (25 seeds g^{-1}), and oak and walnut germinants have far larger seeds than pines (5–10 g for each seed). To visualize this range of

FIGURE 4.10 The tiny seeds of cottonwoods provide almost no resources to support establishing germinants, so success depends on landing in locations with high supplies of light, water, and nutrients such as this floodplain on the Yampa River in Colorado, USA. The broadleaved seedlings are Fremont cottonwoods, and the narrow-leaved seedlings are invasive tamarisks, which also have tiny wind-dispersed seeds. Successful establishment depends on sinking roots quickly into the soil, and increased supplies of water or nitrogen speed seedling growth and increase the number of survivors. **Source:** Data from Adair and Binkley 2002.

seed sizes, if a cottonwood seed was a sphere the size of the period at the end of this sentence, a large oak acorn would have a diameter of about the width of the writing on this page.

Ecological situations always involve more than one dimension, and trade-offs between different characteristics are important. Large seeds fall only in the vicinity near the mother tree, and given the current occupancy of the site by the mother tree this is a difficult place for successful regeneration. Large seeds may appeal to seed-eating birds and mammals, with the two-pronged probabilities of seeds being destroyed (digested) versus dispersed to favorable sites. Tiny seeds allow for dispersal by wind (and low consumption by animals), with the potential that some seeds will reach favorable sites for establishment (along with billions of unsuccessful seeds).

Why Is the Understory of a Forest a Tough Place for Small Trees to Thrive?

The answer to this question may seem obvious, because we know trees need light for photosynthesis, and the light is usually dim under the canopies of large trees. Indeed, the light passing through the canopy of a forest may amount to only 5% or less of the light hitting the top of the canopy. The obvious patterns of low light being associated with poor conditions for small trees may lead us to conclude a cause-and-effect story. Repeating one of this book's themes: an obviously appealing story may or may not be a true and useful story. Experiments are needed to challenge whether an appealing story warrants our confidence.

Over a century ago, this appealing story of the critical influence of shade on small trees was considered experimentally by a Finnish forest scientist, Viktor Aaltonen (1919). He noted that openings in some forests had good regeneration of trees, but only at distances of more than 5–10 m from surviving trees (Figure 4.11). The light supply was high at closer distances, but small trees did not thrive. Therefore the supply of light could not be a sufficient explanation for the success (or failure) of small trees. He inferred that regenerating trees were probably limited by low supplies of soil resources (probably nutrients). He tested this next "story" by examining growth of understory trees on poor soils and rich soils. Understory trees were more abundant and faster-growing on the higher-productivity sites, despite greater shade. Abundant light on poor sites did not foster success of small trees, but abundant soil resources were sufficient to support small trees on rich sites.

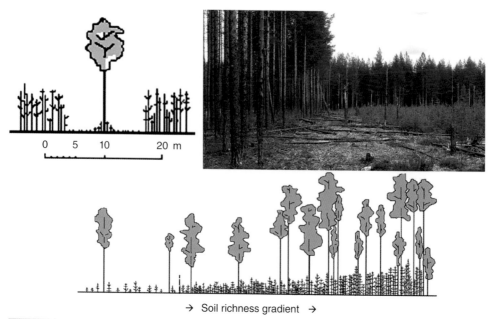

FIGURE 4.11   Aaltonen (1919) concluded that light alone did not explain the growth of understory trees, because areas close to mature trees had abundant light, but poor-growing understory trees (upper, showing empty zone with well-illuminated ground but few regenerating trees; **Source:** photo by Mona Högberg/Taylor & Francis). If competition for soil resources were the key, then understory trees would be expected to grow well on richer soils, even if denser overstory canopies created deeper shade (lower; **Source:** based on Kuuluvainen and Ylläsjärvi 2011 / Taylor & Francis; see also Figure 8.17).

Aaltonen's views were challenged in the 1930s with trenching experiments under intact forest canopies. Small plots were trenched to remove competition for water and nutrients with large trees, while leaving the light regime unaltered. In most cases, trenching led to strong increases in understory plant cover and growth, supporting Aaltonen's story. A more recent experiment evaluated the influence of light as well as soil resources. Petriţan et al. (2011) measured height growth of Douglas-fir seedlings planted under the canopy of a Norway spruce forest in Germany. Seedlings (about 0.5 m tall at the beginning of the experiment) showed much greater height growth when planted within trenched plots than in untrenched areas (Figure 4.12). The strongest response to trenching occurred with the highest levels of incoming light, when evaluated as cm of height growth. When evaluated as relative response (height growth in trenched plots/growth in untrenched plots), the effects were highest under low light conditions. This sort of experiment across a resource gradient provides clear insights into trends. For example, seedlings in untrenched plots receiving 40% of full sunlight grew about 50% better than those receiving only 20% of full sunlight. This effect was smaller than the 80% increase in growth that came from trenching. Light clearly mattered, but

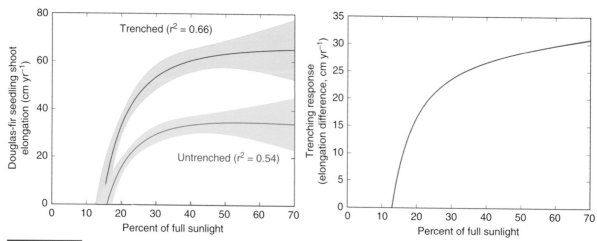

FIGURE 4.12   Douglas-fir seedlings planted in the understory of a Norway spruce forest in Germany grew about 30 cm in height when receiving about 1/3 or more of full sunlight (left). For any given level of light, seedlings in trenched plots that did not have to compete with overstory trees for water and nutrients grew about 80% faster (right; more than 15 cm with 20% or more of full sunlight). **Source:** Data from Petriţan et al. 2011.

not as much as soil resources. This experiment illustrated a common pattern in ecology: the growth of the seedlings was limited by more than one resource (by both soil resources and light), with non-linear interactions between the resource limitations.

For any question in forest ecology, we may ask about the general direction and size of an effect, how much variation to expect around the general trend, and what factors may explain some of the variability. These questions are often addressed in review articles, where the authors evaluate all the relevant investigations and summarize the findings. A review of woody plants led to the conclusion that soil resources commonly limit growth of understory plants as much as light supply does, and that general expectations of simple cause and effect may not be very useful (Coomes and Grubb 2000). The forestry community refers to these complex realities with the term "shade tolerance" even though soil resources are as commonly important as shade (and may or may not covary with shade). Species are described as shade-tolerant if they cope well with large neighbors, and shade-intolerant if they thrive away from large competitors. Once a term becomes entrenched in our vocabularies (and minds), it's hard to replace it with a more accurate and powerful term, and we need to be careful not to be lazy and think "shade tolerance" is a story about light. A useful phrase to keep in mind is that shade tolerance is not entirely *un*related to shade.

All Good Summers Come to an End

Temperate and boreal forests have tremendous swings in seasonal climates that challenge the ability of plants to move between fast-paced physiology in the growing season to suspended animation during severe winters. The influence of temperature on the revitalization of trees and understory plants in spring was described in Chapter 2 in terms of growing degree days. The reverse suite of processes that shut down plants in the autumn begin with "signals" of declining day lengths and chillier temperatures. The biggest challenge is for tissues to transform from warm cells with thousands of reacting chemicals dissolved in liquid water into non-functioning, torpid cells that can endure severe cold without dying.

Many broadleaved trees (and a few conifers) drop leaves for winter. A series of steps begins with the withdrawal of compounds (some derived from the breakdown of chlorophyll) from leaves into twigs, followed by the formation of an abscission layer where the leaf petiole joins the twig. This layer stops the transport of water and compounds, but biochemical processes continue in leaf cells. The breakdown of leaf biochemicals changes the reflective properties of leaves. Senescing leaves change color in part because of reduced reflection of green wavelengths of light, but more because of increased reflectance of yellow and red wavelengths once chlorophyll is gone (Figure 4.1). Yellow colors result from carotenoid compounds which absorb wavelengths other than yellow. Anthocyanins absorb wavelengths other than red, and concentrations of anthocyanins may build up in leaves between the time that the abscission layer shuts off transport out of leaves and the time when cells die. Leaves fall from twigs once the abscission layer is fully formed and cells dry out.

The annual loss of leaves represents a tremendous carbohydrate loss for trees. Some trees avoid this cost by retaining leaves through the winter and rejuvenating them in spring. This ability of conifers entails a set of processes that allow needles to endure severe freezing temperatures (Hari et al. 2013). The seasonal "cold hardening" starts with increasing sugar concentrations in cells (which lowers the freezing point of water). Water solutions can remain liquid below their freezing point, and this "supersaturation" depends on a lack of solid surfaces to begin the process of precipitating ice. Colder temperatures mean that freezing is unavoidable, and ice crystals begin forming between cells. The formation of these crystals pulls water out from within cells, and dehydration further protects cells from damage from expanding ice crystals. The seasonal course of cold hardening follows a time trend in the severity of cold that needles can survive (Figure 4.13). This same challenge applies to buds, twigs, and stems of trees, so even trees with deciduous leaves need to develop frost-hardy tissues. The latitudinal limits of the distribution of some species is limited by their ability to tolerate severe winter temperatures. Frost hardiness is less challenging for tree roots, as snowpacks often keep soil temperatures close to freezing (soils remain warmer than the air).

Most Trees Die Young

Trees produce many more seeds than the number that successfully germinate, and a small proportion of germinants survive long enough to become seedlings. The winnowing of individuals continues with the transitions into sapling sizes and on to the largest classes of trees. Survival at each of these stages depends on a host of mortality factors: microenvironmental stresses (cold, heat, insufficient light, water or nutrients), competition with

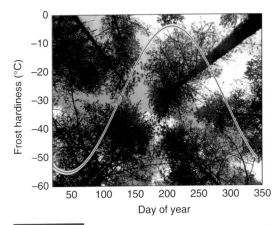

FIGURE 4.13   Scots pine trees in Finland can withstand temperatures of −50 °C (or colder) in midwinter, but temperatures of −20 °C would kill leaf cells in spring and autumn. **Source:** Based on data in Hänninen 2016.

other plants, herbivory, and diseases. Some survivors may be more effective at coping with these stresses, either for genetic reasons or more commonly from advantages that come from differences in plant size (positive feedbacks can lead to greater success for larger individuals). The patterns of survival and mortality for a cohort of trees (or other organisms) can be plotted as a function of age (Figures 4.14 and 4.15). The survivorship curves begin with 100% of the original organisms, and then chart the percentage (or numbers) remaining over time. As with most questions in ecology, the question of survivorship applies to all forests, but the answers depend strongly on local details (contrast Figures 4.14 with 4.15).

The patterns of survival and mortality for a cohort of trees may be only half the story of the composition of a forest. Most forests have trees of multiple ages, so a full accounting of the population dynamics of the forest includes the numbers of trees establishing each year (or each decade) as well as mortality. The combination of new trees (recruits) and loss of already established trees leads to trends in the overall age distribution of trees in forests (Figure 4.16). In planted forests, the age structure is dominated by the single age class of trees, but most forests have large numbers of young trees establishing (and dying) and fewer old trees.

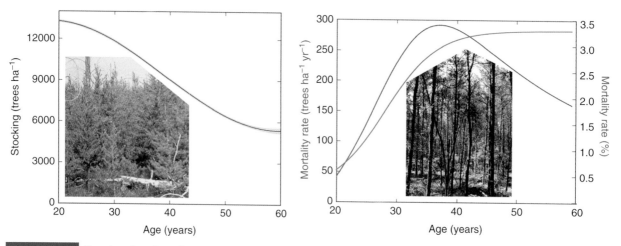

FIGURE 4.14   Two decades after a forest-replacing fire, a forest in northern Ontario had more than 12 000 jack pine trees per hectare. Most of the trees died over the next four decades, as the surviving trees grew larger (left). The mortality rate began low, increased and then declined in terms of tree deaths per year, but in terms of percent of surviving trees mortality rose to a constant of about 3% annually (see Chapter 7 for a discussion of competition-related mortality; **Sources:** Data from of Yarranton and Yarranton 1975, photos by Dan Kashian and Julia Sosin).

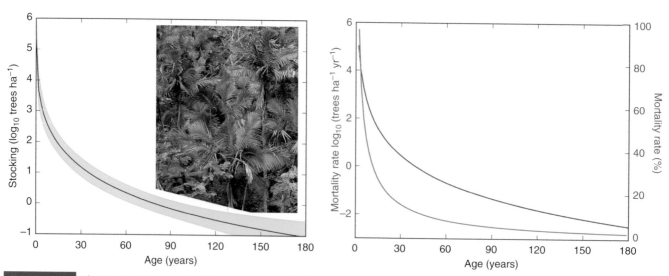

FIGURE 4.15   The survivorship curve (left) for a mountain palm forest in Puerto Rico begins with over 250 000 germinating seeds ha^{-1}, and over 95% fail to survive the first year. A trend across so many orders of magnitudes may be shown on a logarithmic scale to give clearer illustration across the full range. In this palm forest, the rate of mortality declined with age for the number of trees and for the percentage of trees. **Source:** Data from Van Valen 1975, photo by Jose Oquendo).

Reproduction Is the Beginning and the End of Life History Stories

The incredibly complex physiology and growth of plants is only possible because of the success of a prior generation of plants, and the important outcome of the life history of a species is the propagation of genes into the future. One approach to maximize opportunities to send genes into future generations is to live a very long time, and trees are masters of persisting across centuries and even millennia. A thousand years is a long time for a tree to survive the challenges of winds, storms, and fires, so many species evolved the ability to recreate new stems from root systems that survive the death of a stem (vegetative propagation, Figure 4.17). Some species, such as aspen in western North America, use this ability to produce new stems from roots grown by older stems to expand clonally identical forests across more than a hectare, allowing the genotype to persist for centuries longer than the oldest stems.

Most trees around the world each developed from a seed, and the diverse stories of the ecology of seeds and trees could fill an entire library. For any situation that might be imagined, more than a single story has evolved that enables more than one species to colonize an area. Some trees produce billions of tiny seeds that are carried on the wind (as noted above) with a tiny percentage landing on suitable sites for germination and establishment. Larger-seeded species produce fewer seeds that travel only short distances, unless transported by animals. Most seeds in forests have a "shelf-life" of only weeks or a few years before viability is lost.

The probability of a seed encountering favorable environmental conditions can be a matter of time as well as space. Some species produce seeds that remain viable for decades to more than a century, giving the opportunity for germination to occur at a time when favorable environmental conditions develop (Figure 4.18). Forest-replacing fires are common in some landscapes, and seeds can have

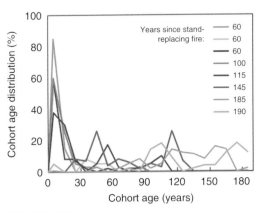

FIGURE 4.16 The combination of the addition of new trees and the death of established trees gives the population structure of a forest. The population structures of eight forests of jack pine in Quebec, Canada showed broad ranges of tree ages; even with forest initiation after severe fires, the forests were not composed of only a single cohort of trees (**Source:** Based on Gauthier et al. 1993). Single cohort forests occur mainly in planted forests.

FIGURE 4.17 Many (most?) broadleaved trees can form new tree stems from root systems and stumps from a previous stem. This ability to sprout new stems allows aspen "clones" to expand with new trees establishing in meadows (left). In other cases, new stems sprout from stumps, like this Fremont cottonwood (sections of stems can be planted in soils, sprouting both roots and shoots; **Source:** photo provided by Scott Abella).

FIGURE 4.18 Most forest trees originate from seeds produced by trees in the current season, or a few seasons in the past, such as this germinating Scots pine (upper left, with the seed coat still sitting on the cotyledons of the germinant). Some forest plants, such as this nitrogen-fixing ceanothus (upper right) develop from seeds that germinate after more than a century, with new flowers restocking a persistent seed bank in the soil. Some pines (such as lodgepole pine in the lower pictures) produce cones which open when ripe, and others on the same tree that remain closed and retained in the canopy for decades until intense fires kill the trees and create favorable seedbed conditions. The cone in the lower right charred when a severe fire consumed the canopy of the lodgepole pine. The cone opened after the fire (the interior is red, the charred scales are black), releasing seeds for the next generation of the forest.

a higher probability of successful establishment and growth after fires have reduced competition between plants, and have altered soils. Seeds with very long-lasting viability may accumulate in the upper soil, with germination triggered by heat, smoke, or other signals indicating a favorable window. Some conifers have developed a system of retaining seeds inside sealed cones that persist on trees for decades. Intense fires kill the trees, but the seeds within cones survive and are released when the serotinous cones open.

Another dimension of time is important for tree reproduction by seeds. Some years are more favorable for successful establishment of seedlings, so it might be logical for trees to produce more seeds in those years and fewer seeds in less favorable years. This logic is not generally applicable to trees, however, because the time course of developing flowers and seeds takes months or more than a year, and trees cannot anticipate when suitable environments for seedlings will come in the future. One opportunity for timing seed production with favorable conditions for seedlings entails developing the primordia (buds) for flowers, and then aborting the full development of seeds later if environmental conditions are poor.

An alternate approach to matching seed production with favorable environmental conditions would be to produce large numbers of seeds each year, ensuring some seeds will be present when conditions are good. Most trees have high variations in seed production across years, so this approach seems like a poor strategy. One drawback would be that seeds are a major food source for many birds and mammals, and consistent, reliable seed crops might sustain large populations of seed eaters that would eat almost all seeds. Producing large seed crops at intervals of several years (called masting) might saturate the ability of smaller populations of seed eaters to consume all the seeds. The second reason that constant production of large number of seeds may be a poor approach is simply that seed production is expensive, consuming a sizable portion of a plant's carbohydrate supply. Wood growth during years of high seed production may drop by 5–50% compared to low-seed years. Not surprisingly, the sacrifice in wood growth varies with environmental conditions; losses of wood growth may be near 0 for big seed years with favorable weather, and quite large during droughts (Figure 4.19).

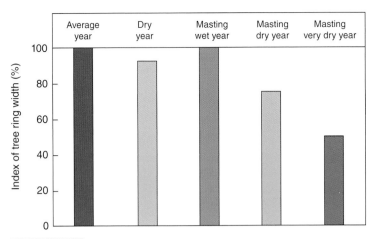

FIGURE 4.19   Beech trees have masting years of high seed production at intervals of five or more years. The reduction in wood growth during masting years varies with environmental conditions. During wet years, wood growth may be unaffected. Masting during a dry year might lower wood growth by 10–15% beyond the reduction directly associated with the dry conditions. Masting during a very dry year can drop wood growth by half. **Source:** Data from Hacket-Pain et al. 2017.

Ecological Afterthoughts: What Benefit Comes from Aspen Having Chlorophyll in Its Bark?

The surface of aspen bark is grayish white, but under the outmost layer is a green layer, as dark green as aspen leaves (Figure 4.20). What benefit would come from chlorophyll in bark? If there is a benefit, why isn't the green layer on the surface rather than beneath another layer? The bark surface is also whiter at higher elevations, and duller/greener at lower elevations. The pieces of this puzzle include many of the processes described in this chapter, from the importance of light intensity, temperature, CO_2 concentrations, and how photosynthesis and respiration are opposite reactions.

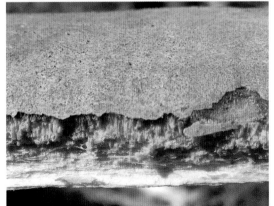

FIGURE 4.20   The bark of aspen is white/gray on the surface, with a layer of green just beneath. The chlorophyll is as rich as in leaves. Thinking about the processes of photosynthesis and respiration leads to insights about how this chlorophyll layer is a useful adaptation.

CHAPTER 5

Ecology of Wildlife in Forests

All the forests of the world evolved with animals shaping them. Animals are major dispersers of seeds, consumers of leaves and branches, and moderators of which tree species last long enough to have a chance to grow from seedlings to trees. Many birds and mammals nest in trees. Beavers use trees to engineer ponds that reduce predation and serve as a place to store branches for a winter food supply beneath the ice. Insects are important pollinators of most broadleaved species, and probably every tree species has one or more specialized insects that feed on leaves, in leaves, under bark, or on roots. Forest ecology focuses on the trees that provide the structure and production that animals depend on, and the animals also shape the forests. This chapter focuses on mammals and a bit on birds; insects are included in Chapter 10. A forest ecology text can barely begin to cover topics of wildlife ecology. This chapter focuses on some of the basics of animal populations, especially on how animals respond to, and shape, forest composition and structure.

Many Species of Trees Coevolved with Animals as Seed Dispersers

What would seed dispersion look like for the forests of the world if there were no animals to help spread seeds? Species with tiny seeds transported by wind might seem like they would be unaffected by the absence of animal seed-dispersers, but in fact they might have more success wherever large-seeded species currently offer competition. The evolution of large seeds provides germinants with stores of energy and nutrients that tiny-seeded competitors lack. The large seeds make such good food that many species of birds and mammals have evolved to rely on seeds as a major seed source.

The islands of Hawaii lie about 3000 km from the nearest continent, and only two large-tree species were dominant on the islands before humans arrived. Ohia is in the same family as eucalyptus (Myrtaceae), with seeds so small that wind velocities of about 10 km hr^{-1} can keep them airborne. Koa is an acacia with large seeds (2000 times larger than Ohia seeds) borne in pods (Figure 5.1). A hard seed coat keeps seeds viable for decades, including as they pass through the digestive systems of birds. The first seeds of the Koa ancestors arrived in Hawaii with birds that crossed 9000 km of ocean about 5 million years ago. About 1.4 million years ago, seeds from Hawaii were dispersed 18 000 km to Reunion Island in the far-off Indian Ocean (near the east coast of Africa) to found a new population (Le Roux et al. 2014). Another 2000 km and the seeds would have been dispersed halfway around the world!

Mammals disperse trees seeds at scale of meters to kilometers, which is a good scale for getting away from a site already dominated by the mother tree, but still within a general environment where conditions might be suitable for a new generation. Seeds and fruits coevolved with dispersing animals, in a tradeoff between costs of producing animal-food-quality fruits and seeds, and benefits of dispersion of seeds to new, suitable sites.

FIGURE 5.1 Koa forests in Hawaii descended from seeds brought from Australia by birds about 5 million years ago. About 1.4 million years ago, a bird started a new population on Reunion Island with seeds from Hawaii, 18 000 km away. **Source:** seed photo by J.B. Friday, University of Hawaii.

Some Animal Species Specialize in Eating Trees

Thousands of species of mammals specialize in eating leaves, twigs, bark and branches of trees, including koalas, sloths, monkeys and apes, elephants, giraffes, deer, and moose. Tree species that coevolved with browsing pressure generally respond productively with regrowing tissues, growing new leaves, new twigs, or entire new stems (Figure 5.2). Even when beavers cut down trees, the stumps may resprout, or send up new shoots from the surviving roots. Severe browsing might prevent the new shoots from transitioning into tree sizes, and in some cases sustained high-intensity browsing can eventually kill resprouting trees.

Tree species that are not favored by browsing mammals may benefit from reduced competition from heavily browsed species. The Białowieża National Parks span the border of Poland and Belarus and the forest has been largely kept as a preserve across centuries. The forest has a long history that includes dynamic populations of browsers (ungulates): bison, moose, red deer (same species as elk in North America) and roe deer. Ungulate populations increased dramatically in the past 80 years, and increased browsing drastically reduced the establishment (recruitment) of new birch trees (favored fodder) and increased the numbers of unpalatable hornbeam (Figure 5.3). Overall, the recruitment of new trees dropped by about half with heavy browsing pressure.

Similar trends happen around the world in response to browsing impacts, including the effects of high populations of whitetail deer in eastern North America and declining recruitment of palatable species, along with strong increases in unbrowsed species. In some cases invasive species can be palatable; high populations of whitetail deer substantially reduced invasive Japanese honeysuckle. Any story of browsing influences on shifting the composition of forests occurs in a milieu of other shifting factors, such as rising CO_2, changing climate, altered fire patterns, and invasive species.

Livestock Grazing and Browsing has been a Core Part of People's Livelihoods Through History

Forests are remarkably poor at producing food for humans. Humans cannot eat the vast majority of plant material produced in forests. Over the past 10 000 years or so, human societies cleared forests for agricultural use, at scales that matched their abilities to work the land. Livestock were used to take advantage of surrounding forests, converting forest biomass into a form that

FIGURE 5.2 Beaver-cutting of aspen stems (top left) typically does not kill the organism. Decades of intense browsing by high populations of elk in Yellowstone National Park did eventually kill this aspen clone (top right). Reintroduction of wolves to the Park may have contributed to the survival of aspen clones by reducing browsing pressures from an overpopulation of elk – or not (see later discussion about Figure 13.4).

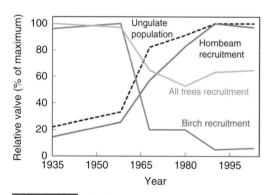

FIGURE 5.3 A fivefold increase in ungulate populations (Y axis is scaled to the maximum value during the period) reduced recruitment of new trees by half, with recruitment of palatable birch falling by 90%, and recruitment of unpalatable hornbeam increasing fivefold (**Source:** Data from Kujiper et al. 2010).

humans could digest. Forest grazing shifted the composition and structure of the forests. The impacts of course depended on situation details, including which livestock species were used (sheep, goats, cattle, reindeer) and forests. Some of the impacts were direct, such as eating palatable species and allowing less palatable species to increase. Indirect effects could include changing the structure of fuels that would carry fires through the forests, and changing the available browse for wildlife (further reducing wildlife populations beyond the effects of human hunting). Livestock use of forests continues in many parts of the world (Figure 5.4), along with dynamics of wildlife (both herbivores and predators).

The impacts of wildlife often relate (directly and indirectly) to the influence of livestock. Livestock may compete with wildlife for food, and may be a source of exotic diseases. Human hunting can further complicate the interactions. Changes in land use can connect wildlife/livestock/forest interactions across long distances. The development of market economies can lower the value of agricultural products, leading to abandonment of crop fields and reductions in livestock populations. Recovering forests provide new opportunities for some wildlife species, at the same time that any food competition with livestock declines. In the early twentieth century Sweden had millions of livestock using forest lands for foraging. Economic development led to dramatic changes in land use, with almost all livestock removed from forests (with the exception of native reindeer, managed as semi-wild livestock in the north). The reduced browsing impacts of livestock on trees and forests might have been moderated by an equally dramatic increase in moose populations (Kardell 2016).

FIGURE 5.4   Domestic livestock are common in many forests, including reindeer in boreal forests, cattle and sheep in temperate forests, and goats in drier forests. **Source:** Finland reindeer photo by Heather Sunderland.

Was Aldo Leopold Right About the Kaibab Deer Herd?

One of the early stories in ecology textbooks was about the effects of predator removal on the Kaibab Plateau on the north side of the Grand Canyon, USA. Aldo Leopold was typically credited with the story, though he was reporting ideas and evidence from other people. The Plateau was a remote area with forests of ponderosa pine, aspen, and a few other conifers. Historically fires would have burned across most of the Plateau every few years. A game preserve was established in the early 1900s, prohibiting hunting by humans. At the same time, predators were greatly reduced to support livestock operations. Leopold related a story that predator removal was followed by an explosion of the deer population, which degraded vegetation and led to a massive die-off of the deer. The story was logical and fascinating, but it began to disappear from ecology textbooks about 50 years ago when critics questioned the legitimacy of the evidence.

A century ago there were no well-developed methods to census populations, and the Kaibab deer story was pieced together from rough estimates (guesses) of deer populations by several visitors to the Plateau. The local forest managers also estimated deer populations every year, but their estimates were far lower than the ones reported by Leopold as a population explosion. Critics suggested the "selective" use of limited data did not provide strong evidence that an explosion of the deer population had occurred, and if the explosion had occurred, it might have resulted from the large reduction in livestock on the Plateau after the game preserve was established.

Stories can be powerful, whether they have a strong base in evidence or not. The idea that predators were important for controlling populations of herbivores led land managers in the US National Park Service to cull herds of elk, replacing the role of Native Americans and extirpated wolves and grizzly bears. But once the Leopold story of the key role of predators was questioned, the National Park Service stopped culling the elk herds. The elk populations increased, with major impacts on forest vegetation. Beliefs in ecological stories lead to real-world impacts, and confidence placed in a poor story can have unintended, undesirable consequences.

Does the classic Kaibab deer story warrant confidence? Are predators important regulators of herbivore populations? Evidence can be assembled from a variety of sources, including historical records, and legacies in the structure of the forests. Aspen suckers (shoots rising from an older root system) are a primary food for deer, and a high deer population that crashed because of overbrowsing should have prevented a cohort of aspen trees from developing during that period. A sampling of the age structure of aspen trees on the Plateau did reveal an absence of new trees from the period when deer populations were supposed to be high (Figure 5.5). Fenced exclosures from the same period showed abundant aspen recruitment, demonstrating that the climate was favorable for aspen. The age structure of aspen also showed a major increase after livestock grazing stopped the frequent fires on the Plateau (fires would burn up suckers), and another period of high deer population that was verified by modern wildlife population estimates.

The information on age cohorts of aspen had the potential to refute the classic story. If a normal-size cohort of aspen had established successfully when the deer were supposed to have been starving, the story would clearly have been wrong. The evidence of the missing cohorts supported the story, though it could not test (or challenge) whether the increase in deer population resulted from the removal of most of the predators or other factors. The actual ecological history on the Plateau at the time scale of a century was complex, with fluctuating climate, livestock populations, fire occurrences and intensities, predator removals, and human hunting. Each piece of this ecological story might be both a response to other pieces, and a cause of changes in other pieces. It's tempting in science to simplify the possible interactions in a system to test and explore possible individual mechanisms. But can

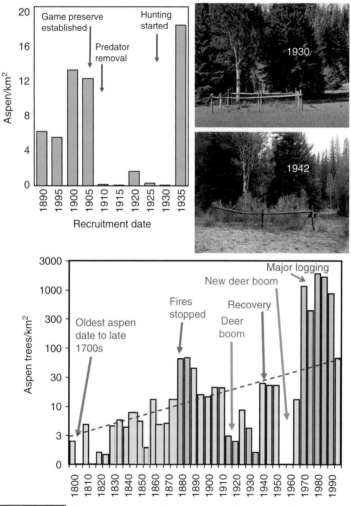

FIGURE 5.5   The age structure of aspen across the Kaibab Plateau supported the classic story. Almost no aspen present on the Plateau in 2000 were recruited during the period when the deer populations was claimed to be high (upper graph). Experimental exclosures showed that climate and other factors were sufficient for aspen regeneration if deer browsing was prevented (photos). Wildlife ecology happens across space and time with many varying factors, and the aspen age structure (lower graph, note Y axis is logarithmic; dashed line is long-term average null expectation for aspen recruitment) reflected the cessation of frequent fires after livestock browsing began, a second boom in deer population in the 1950s/1960s, and a very large increase in aspen after logging of conifers from the 1970s into the 1990s.
Source: Based on Binkley et al. 2006.

those pieces be reassembled into a whole, ecosystem-scale story? The next few pages focus more directly on dynamics of wildlife populations, moving back into forest-scale impacts near the end of the chapter.

Wildlife Population Dynamics Occur Within Complex Ecological Systems

Animal populations must exist in a state of balance for they are otherwise inexplicable.

A.J. Nicholson (1933, cited in Botkin 1990)

The balance of nature does not exist, and perhaps has never existed. The numbers of wild animals are constantly varying to a greater or lesser extent, and the variations are usually irregular in period and always irregular in amplitude. Each variation in the numbers of one species causes direct and indirect repercussions on the numbers of the others, and since many of the latter are themselves independently varying in numbers, the resultant confusion is remarkable.

Charles Elton (1930, cited in Botkin 1990)

Hopes and beliefs in the goodness and balance of Nature may be appealing, and some people may not be able to imagine that reality could be anything other than beautifully balanced. But if hopes and beliefs run counter to both logic and evidence, it might be time to dive into the first step of Science: come up with a clever idea that might account for prior observations.

Ecological systems are too complex to understand in full, so we come up with simple ideas that have a chance of representing the major pieces and interactions with enough detail to be useful, but not too detailed to be understood. One of the simplest ideas would be that the population of a herbivore, let's say a species of rabbit, would be limited by the productivity of vegetation in the ecosystem. Not all the vegetation would be suitable rabbit food, so let's consider only the plant growth available to support rabbits. How many rabbits could the ecosystem support? A century ago this question would begin by trying to gauge the rate at which the rabbit population could grow if food was not limiting. The second step would be to gauge the growth of plant food. Maybe the rate of growth of the rabbit population would slow down as food became scarcer at higher rabbit population level. Putting these ideas together would give an "s" shaped curve for the size of the rabbit population, where growth is slow at low populations because of the small number of breeding rabbits. Population growth would be fastest at a moderate rabbit population level when the food supply was still good. Population growth would slow asymptotically toward zero as the high population of rabbits consumed all the plant food. This maximum, sustainable population level could be given a name such as the carrying capacity of the ecosystem for rabbits (Figure 5.6).

What processes and interactions in a real ecosystem were left out of that rabbit/food story? Maybe the most important missing piece would be the influence of rabbit browsing on plants. When rabbit populations are high, can the vegetation produce as much food every year despite heavy browsing? High rates of browsing might damage plants to the point where next year's growth is less than this year and the subsequent year could be even lower. This would not be a sustainable situation, and the declining food supply would lead to a declining rabbit population. It might seem that some point would be reached where vegetation growth could outpace the consumption by a reduced population of rabbits, but keep in mind the only reason the rabbit population declined was starvation, and a low population of rabbits would not be at the point of starvation if plant food production somehow increased. And what about all the other herbivore species that might eat the same vegetation as the rabbits? The bottom line is that this rabbit/food scenario would lead to extinction of the plants, and then extinction of the rabbits. So this story is too simple to give clear insights into real ecosystems.

How could the rabbits avoid an extinction scenario? One possibility would be that some of the plants could be somehow out of reach of rabbits – too tall, or hiding in the soil in the form of seeds (waiting for reduced rabbit populations to germinate). If some of the vegetation had a refuge from rabbits, then the rabbit population could plunge without first exterminating all the plants. Another complication that could be added would be adding coyotes who eat rabbits. Coyotes do not reproduce as fast as rabbits, so the rabbit population could still increase to the point of damaging the vegetation. But by the time this happened, the coyote population would have been building up to the point that there could be plenty of coyotes to eat all the rabbits that were limited in population by their food supply. This point also would not be sustainable, because the peak of the coyote population would come at a time of declining rabbit populations, and the coyotes would resist dying of starvation until the last rabbits were eaten. This might be prevented by the refuge idea: maybe some rabbits could hide out in protected areas until the coyotes starved, then starting a new increase in rabbit populations.

This rabbit/vegetation/coyote story could be made more complicated, but it already has reached the limits of quantitative abilities in the early twentieth century. The dynamics of this story could be quantified using the Lotka-Volterra equations, with

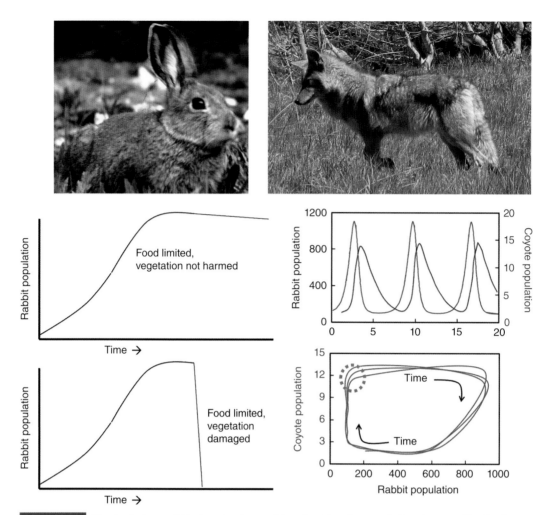

FIGURE 5.6   Simple stories could be imagined as useful explanations for population changes. The simplest might be a world in balance, where a reproductive rate of rabbits slows at high population, asymptoting with vegetation able to cope with browsing (upper left). More realistic would be that the rabbit population would collapse and go extinct when browsing pressures could not be sustained by the vegetation (lower left). A more complex approach that could still be calculated using a slide rule (a mechanical calculator) would lead to fluctuating populations of rabbits and their predators, coyotes, giving a set of stable oscillations (upper right). Including the effects of predation is a good step, but any stability might be easily disrupted by many factors (weather, other predators, other prey, diseases) – especially in the circled domain of the lower right graph, where very low rabbit populations are at risk from very high coyote populations. The value of these stories might be limited to showing why the simplest of stories cannot represent real ecosystems well enough to be useful.

solutions calculated with the help of a mechanical slide rule. There are several key questions about how good an idea this quantification of the story could be. If the concept of a refuge (for plants and for rabbits) actually existed, the quantification might be useful. If refuges aren't realistic, the story might be intriguing but too wrong to be useful (unless the overall trends were realistic, but for reasons unrelated to the characters in the story). What would the carrying capacity of the ecosystem be for rabbits? Would the answer depend on all the other herbivores in the forest, and whether coyotes were present? What if intense herbivory by rabbits shifted the vegetation to species that were unsuitable food for rabbits? What if the spread of diseases among rabbits became severe at high populations? What about the influence of predation by hawks? What if coyotes shifted their predation from rabbits when they were scarce to squirrels and mice, preventing any decline in coyote populations that would ever allow an increase in rabbit populations? What if severe winter weather occasionally led to major mortality of rabbits or coyotes? The potential weak links in the chains of logic in this story could also be expanded to include spatial aspects of landscapes and nearby landscapes.

The good news is that modern quantitative techniques allow simulation models to represent all these factors and all their interactions. Two pieces of bad news are hard to avoid: it's unrealistic to think all these factors could be measured with enough accuracy to be useful, and even if they could be, the complex models would be unlikely to yield any overriding, simple insights about complex ecosystems. A list of morals from this thought-story of rabbits and coyotes might include:

- We can create ideas and give them names (such as carrying capacity), but that doesn't mean we should expect they really exist.
- We might continue to believe that simple ideas really do shape ecosystems, but messy details always interrupt the story before its true outcomes ever appear (like Plato's belief that reality was composed of imperfect representation of ideal essences – interesting philosophy, but poor science).
- We can realize that indeterminate systems, where complex interactions are not required to produce simple, predictable outcomes, can be addressed with questions that suit indeterminate systems. Wishes for simple stories can be left behind.

The rabbit/coyote thought story was not based on any evidence. The following sections cover some of the evidence behind dynamics of actual animal populations in the broader context of forest ecosystems. Two famous case studies give a useful foundation for expectations about the sorts of questions that are likely to be productive in understanding animal populations in complex, indeterminate ecosystems.

Moose and Wolves Established New Populations on Isle Royale in the Early 1900s

Isle Royale is a 15 by 70 km island in Lake Superior, about 25 km from the Upper Peninsula of Michigan, USA (Figure 5.7). A population of moose established on the island in the early 1900s, and some time in the 1940s a pack of wolves crossed the winter ice and founded a population. The island is large enough to support a population of wolves, and small enough for thorough censusing of populations. This case study has been considered a "model ecosystem" for investigating interactions between predators and prey.

The wolves certainly preyed on moose on the Island, but neither population showed any clear trend that might have been predicted by simple stories (Figure 5.8). Wolf populations were somewhat constant for the first 20 years of monitoring, and then doubled. The moose population declined when the wolf population was highest. The moose population showed a second period of major increase, and this time the decline was not associated with an increase in wolf populations. The lowest moose population during the period of record occurred in the early 2000s when wolf populations were intermediate, and then the moose population

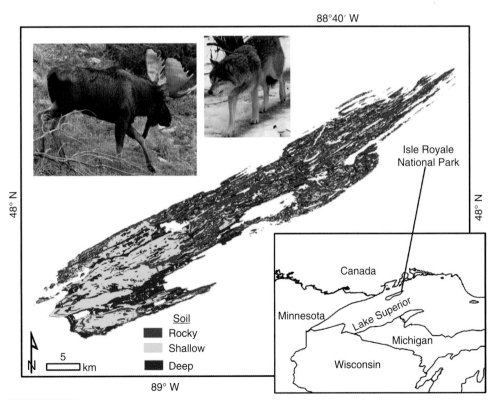

FIGURE 5.7 Map of Isle Royale, where dynamics of populations of moose and wolves have been followed for over 60 years (**Source:** Modified from De Jager et al. 2017).

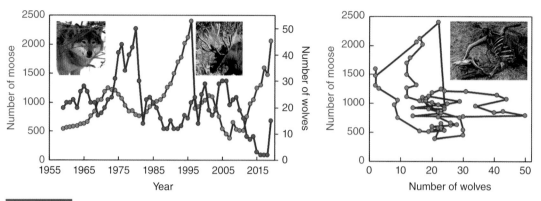

FIGURE 5.8 The dynamics of moose and wolf populations on Isle Royale followed no simple story across 60 years, in relation to trends over time (left), or in terms of coordinated populations of wolves and moose within each year (right; **Source:** Data from Vucetich and Peterson, 2020).

increased dramatically. Meanwhile, the wolf population would have gone extinct on the Island if not for wolves added the Island by managers.

A long list of factors influenced the populations, in addition to the influence of predators on prey and prey on predators. Wolf populations may have responded to varying populations of beavers (another major prey species), an outbreak of canine parvovirus, and inbreeding effects on fitness (countered in part by the immigration of a new male in the 1970s). Moose populations may have been influenced by severe winter weather, and by major changes over time in vegetation. A severe fire burned about 20% of the Island in the 1930s, with young forests providing abundant food for moose. No major fires have occurred since then, and forest vegetation shifted in relation to tree growth and impacts of moose browsing.

Predator–prey interactions can have major consequences for forests. The opportunities for beavers to thrive in a landscape may change if moose reduce the supply of suitable trees for beavers, or if moose-supported wolf populations grow large enough to increase beaver predation. Some tree species are preferred by moose (such as balsam fir) and others are not eaten (such as white spruce). Modeling scenarios indicate that an absence of wolves would have a cascading effect of reducing balsam fir and increasing white spruce across the Island (Figure 5.9).

The current generation of scientists working at Isle Royale concluded the real world does not line up with Nicholson's insistent quotation a few pages above, but does indeed resonate well with Elton's quotation. Their summary thoughts are:

But the more we studied, the more we came to realize how poor our previous explanations had been. The accuracy of our predictions for Isle Royale wolf and moose populations is comparable to those for long-term weather and financial markets. Every five-year period in the Isle Royale history has been different from every other five-year period – even after 50 years of

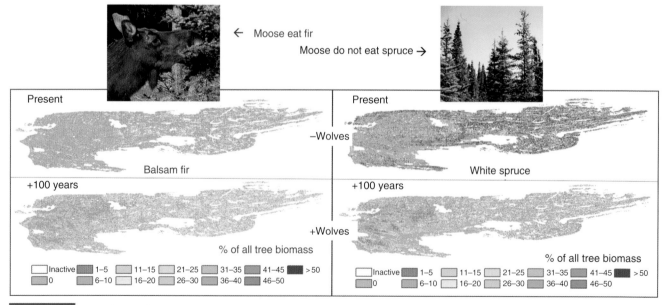

FIGURE 5.9 A simulation of forest vegetation in response to intensity of moose browsing illustrates the size of forest responses that would be expected with and without wolves. **Source:** Modified from De Jager et al. 2017.

close observation. The first 25 years of the chronology were fundamentally different from the second 25 years. And the next five decades will almost certainly be different from the first five decades. And the only way we will know how, is to continue observing. The most important events in the history of Isle Royale wolves and moose have been essentially unpredictable events – disease, tick outbreaks, severe winters, and immigrant wolves.

(Vucetich and Peterson 2020)

One of the major limitations for the Isle Royale case study of populations was that so many factors were varying that teasing out the influence of separate factors was not possible, even in this tractable, "model ecosystem." Stronger confidence in ideas requires experimental manipulation of factors.

The Cycles of Snowshoe Hares and Lynx Repeat, but They Are Far from Simple

Another classic story of animal populations over long periods comes from northern Canada. The Hudson's Bay Company kept records of the number of Canada lynx trapped in the 1800s, and a strong cycle of about 10 years was obvious (Figure 5.10). Snowshoe hares showed a similar cycle, with a peak that commonly preceded the peak in the lynx cycle. This classic case could result from a two-level "tropic cascade," where hare population limits lynx populations, and predation by lynx limits hare populations. The case could also result from a three-level tropic cascade, where the peak hare population is limited by food (and predation by lynx are more important during the hare-decline phase). Or perhaps only the hare population drives all the cycles, with disease or stress determining hare populations (and lynx populations are limited by hare populations). The possible explanations are only part of a long list that has been considered (even including a possible influence of cycles of sunspots). If all explanations are consistent with available data, there is no clear way to see which explanation warrants more confidence. Confidence would need to be based on evidence that supported or refuted some explanations, and experiments might provide that evidence.

Over 40 years of monitoring and experimenting near Kluane Lake in the Yukon Territory of Canada provides a very large base of evidence for testing explanations. C.J. Krebs and colleagues established very large plots where snowshoe hare populations could respond to changes in food supply (with addition of food), to removal of large terrestrial predators (by fencing out lynx, coyotes, and wolves), and to a combination of extra food along with predator reduction. The overall population of hares was clearly limited by food supply, with up to 10-fold increases in population (Figure 5.11). Reducing predation by large terrestrial animals also had some effect. Curiously, hare populations still crashed even with an abundance of food and reduced predation.

These strong population cycles were influenced by many factors. Lynx predation was important, but even more hares were killed by red squirrels (preying on baby hares still in nests) than by lynx: more than half the hares were eaten by squirrels in their first 30 days. The populations of squirrels was not constant, responding in part to timing of cone crops from spruce trees (their major food). Other predators included coyotes and great horned owls, and all these predators also had other sources of food that allowed them to subsist with or without hares. Indeed, red squirrels are normally considered to be herbivores, not predators.

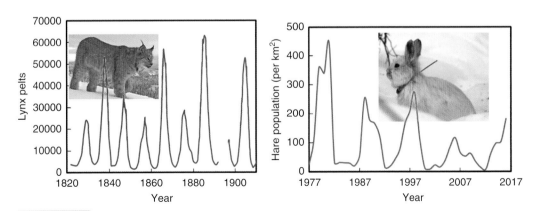

FIGURE 5.10   Some animal populations fluctuate on a semi-regular cycle, though the period and amplitude of the cycles may vary a lot. An early record of a cyclic population comes from the records of lynx pelts collected by the Hudson's Bay Company from boreal forests in Canada (left, **Source:** Data from Elton and Nicholson 1942). The hare in the photos has a GPS collar that gives precise tracking (**Source:** from Oli et al. 2020, used by permission, photos by A.J. Keeney).

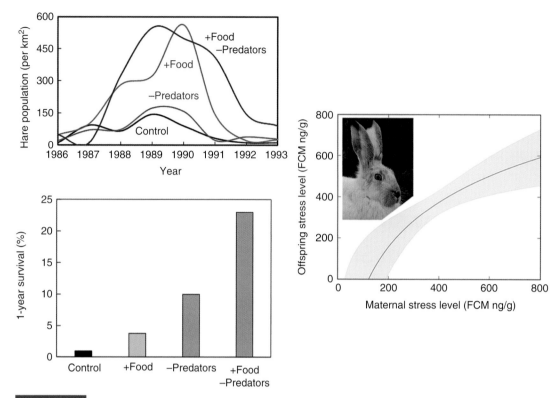

FIGURE 5.11 Populations of snowshoe hares increased with the addition of food, reduction in predators, and especially with the combination of the treatments, but the crash of the hare population was not prevented by any treatment (upper left). Survival of radio-collared hares increased with food addition, but reduction in predators gave an even larger response (lower left). A key feature of the population cycle related to stress experienced by mothers (which was high at high populations, as indexed by cortisol levels measured in hare poop, FCM) and passed on to offspring, leading to low populations for a year or two even when predation was low (**Source:** Data from Krebs et al. 1995, 2017).

Another factor that proved to be important was the stress experienced by mother hares. High stress led not only to lower birth rates, but to a generation of hares that somehow "recalled" the stress level of the mothers. This factor seemed to be particularly important in keeping the population low for two or more years after pressure from predators declined (Figure 5.11).

Four decades of experimentation also echoed the quotation from Elton a few pages above. Predator–prey relationships were clearly important in the Kluane ecosystem, but the interactions were not simple (many species were involved, Figure 5.12) and patterns over time had both some regularity (a 10-year period) and sizable variety (sizes and shapes of peaks and troughs). A concluding comment from Krebs (www.zoology.ubc.ca/~krebs/ecological_rants/the-snowshoe-hare-10-year-cycle-a-cautionary-tale):

> Models of the hare cycle have proliferated over time, and there are far more models of the cycle in existence than there are long-term field studies or field experiments. It is possible to model the hare cycle as a predator–prey oscillation, as a food plant–hare oscillation, as a parasite–hare interaction, as a cosmic particle–hare oscillation, as an intrinsic social–maternal effects interaction, and I have probably missed some other combinations of delayed-density dependent factors that have been discussed. That one can produce a formal mathematical model of the hare cycle does not mean that the chosen factor is the correct one.

The presence of many interacting factors in an ecological system sets the stage for very indeterminate futures. Predictions of the future of indeterminate systems can give artistic insights, but the insights may have little value for quantitative understanding or reliable prediction.

Nature clearly did not agree with Nicholson's belief that it must be in balance, and that's very fortunate. A well-balanced system would be like an airplane: there are not many pieces that can break before the airplane crashes. Indeterminate ecosystems have incredible ability to change and persist. Populations rise and fall, but rarely go extinct. If nature was well balanced and had only one state that was viable, ecosystems would be brittle and require great knowledge, skill, and investment for humans to engage and use. Humans easily alter ecosystems, for better or worse, but ecosystem failure is typically not a risk. The majority of human-driven failures for wildlife species have come from direct overhunting, and intentional and unintentional introduction of novel pests, pathogens, predators and competitors (Chapter 13).

Fortunately, nature does not need to be finely balanced for us to understand and interact with animal populations and ecosystems. We can examine the history of a situation and see how the current condition came to be. We can carefully frame questions so

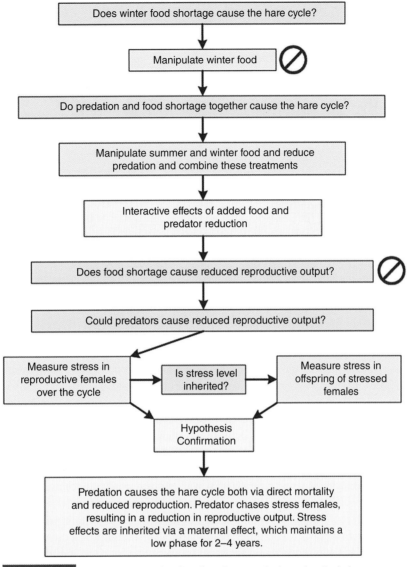

FIGURE 5.12   The food web in the "lynx/hare" system in the Yukon includes strong and weak interactions with a very broad ranges of animals and plants (upper diagram; blue boxes are questions, yellow boxes are topics studied intensively, and pink boxes received less attention). The flow chart illustrates the complexity of experimentation needed to develop strong evidence on factors behind observed population dynamics (negative symbols indicates possible explanations that were ruled out by experimental evidence). **Source:** Krebs et al. 2017 / John Wiley & Sons.

they will be consonant with the indeterminate nature of the systems. We can ask, "What would happen if predators were removed from an ecosystem?" and then develop some useful insights from case studies and models (Stolzenburg 2008). A precise prediction would likely not be possible, but general trends might be informative. Another question might be, "How do populations manage to persist when predator populations are high?" One answer might involve local refuges that prevent the last pair of animals from being eaten. Another might be that local extirpation occurs, and immigration from another area would be vital to continuation of the population.

Patterns and Processes of Wildlife Population Dynamic Shift Across Space and Time

Many censuses of wildlife populations are available from the past century around the world. One example is a large increase in roe deer across Europe in recent decades (Figure 5.13). The documented patterns could result from a variety of driving factors (as described above).

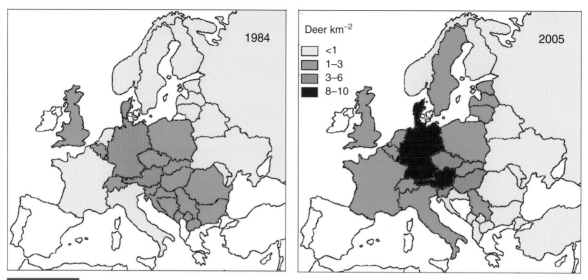

FIGURE 5.13 Roe deer are medium-size (20–30 kg) browsing herbivores, with densities increasing substantially in recent decades across Europe (**Source:** Based on Burbaitėa and Csányi 2009). The size of populations can be tracked over time and space, providing a foundation for investigating the likely impacts on other features of forests.

Another step could be a comparison of two patterns, looking to see if the combined patterns add extra information. The population of elk increased by fourfold in three decades on the Uncompahgre Plateau in Colorado, USA, while mule deer populations dropped by two-thirds. A plot of the two populations shows that the inverse relationship is extremely unlikely to be random (about one chance in 10 million). A very good story could be that an increase in elk populations should be expected to lower deer populations because of competition for food. The correlation pattern is very strong, and clearly does not refute this story. However, even the best correlation patterns do not provide solid evidence about cause and effect. It's possible the elk population simply continued to increase after reintroduction to the area, with no interaction with deer populations. Deer populations could drop over time as a result of severe winters, disease, or vegetation change (either in the summer range on the Plateau, or winter range in the valleys) that harmed deer but not elk. The information in Figure 5.14 is representative of what wildlife managers typically have available when they try to establish management plans and hunting quotas for wildlife. The level of confidence in management schemes should not exceed the quality of the evidence available.

Ideas of cause and effect usually require experimental evidence beyond the documentation of patterns. Red deer (same species as elk in North America) were thought to severely limit tree seedling survival and growth in Scotland. The population of red deer was reduced to fewer than 1 deer km^{-2}, and browsing on shoots of tree seedlings and saplings dropped to near zero (Figure 5.15). Both lines in Figure 5.15 are a function of year, and it would be possible that some other year-related factor actually drove the decline in browsing. However, the convergence of a logical story, an observable process (browsing), and a response to a manipulation (deer removal) provides good evidence for confidence in the story. The observations were spread across time, though, so possible confounding factors (related to year) still could play a role. Repeating the experiment, in part or whole, would give a stronger test of whether the trends resulted from red deer impacts, or some other factor that changed over time.

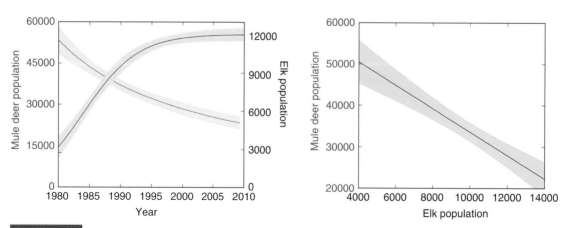

FIGURE 5.14 Over a 30-year period, the population of elk on the Uncompahgre Plateau in Colorado, USA, increased fourfold, as the deer population dropped by two-thirds (left). The inverse relationship between the two was very strong, with each increase in elk associated with a reduction of 2.8 deer **Source:** Data from Colorado Parks and Recreation.

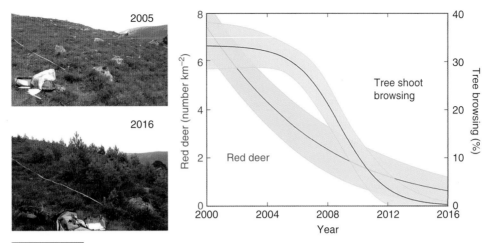

FIGURE 5.15   Reducing a high population of red deer in the Cairngorm mountains in Scotland, UK almost stopped browsing impacts on tree seedlings, allowing reforestation. **Source:** based on Rao 2017.

Good Ideas Without Good Evidence May Be Unreliable, or Wrong

Animals need more than just food to survive. These needs include avoiding predators and diseases, and avoiding too-stressful environments. Production of food for many species of browsing mammals is higher in more open forests, but might forests be too open? Recalling from Chapter 2, open areas may experience colder conditions at night in winter, and it seems reasonable that species such as elk and deer would thrive if they had access to protective thermal cover, reducing the need to burn fat reserves for warmth in winter. It's also possible that thermal cover is not particularly important. Experimentation would be needed to see which of these logically possible ideas would be supported.

An experiment in eastern Oregon, USA followed the loss of body mass of female elk through a winter, comparing elk that were in pens with dense forest cover, with moderate forest cover, and in the open with no cover. The elk were fed all winter, but the food supply was limited to match what might be available to free-ranging animals. The elk did best in the open, no-cover treatment, refuting the idea that thermal cover was important (Figure 5.16). But would the same pattern apply in another winter, with different weather conditions? The experiment was repeated with different individuals another year, when the weather was warmer and rainier. The loss of body weight over winter was less than during the colder winter, but the benefit of having no cover was consistent.

If elk do not benefit from thermal cover in winter, it would still be possible that other species would. A similar experiment during a severely cold winter in Colorado, USA with mule deer found there was essentially no difference between deer held in pens with cover and pens in the open with no cover. Of course it's still possible there are situations or species where thermal cover in winter is helpful, but the studies in Figure 5.16 would support an expectation that thermal cover would not be expected to be important, unless contrary evidence arose from other experiments.

Strong Evidence Comes from Comparisons of Treatments at the Same Point in Time

The possibility of confounding factors covarying in time can be ruled out when factors are tested at the same time. Wildlife impacts are commonly examined with fenced plots (exclosures) where one or more species are not allowed to influence vegetation (as in the Kluane experiments). The influence of browsing animals is often very apparent with comparisons inside and outside exclosures (Figure 5.17). The influence of animal species in forests often happen at scales that are not represented by exclosures. For example, the differences in aspen and grasses on the two sides of the fence in Figure 5.17 might have implications for the ignition and propagation of fires, but fires often occur at landscape scales that are not captured at the scale of a plot.

Sometimes larger scale experimental evidence can be developed. An archipelago of islands occurs between the North American continent and Vancouver Island, Canada. For various reasons, some of the islands have no deer, and others have low or high populations. The scale of the effects of deer covers enough area that impacts could also be examined on populations of birds (Figure 5.17), providing strong evidence that too many deer leads to very few bird species.

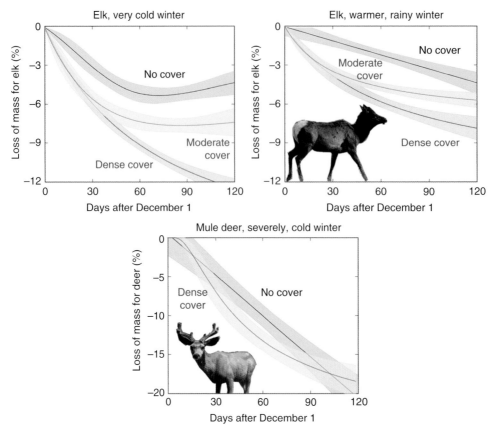

FIGURE 5.16   The hypothesis that elk benefit from thermal cover in winter was tested with yearling female elk in western Oregon, USA with pens under dense forest cover, moderate forest cover, and in the open. Across two winters of contrasting weather, elk did better (lost less weight) in the absence of cover, refuting the hypothesis. A similar study in Colorado, USA with mule deer of various ages concluded that cover had no effect on loss of mass over the winter, also refuting the hypothesis (**Source:** Data from Cook et al. 2004).

FIGURE 5.17   Fencing can be used to test for the effect of animals in a single location over a single period of time (**Source:** upper photo, fence to exclude elk in the Jemez Mountains of New Mexico, USA). A comparison across locations can provide insights where locations differ in animal populations, but are otherwise similar (**Source:** lower photos from islands between Vancouver Island, Canada and the North American continent; **Source:** from Martin et al. 2011, used by permission).

Ecological Afterthoughts

The interactions between wildlife species, populations of species, and forest composition and structure are so large that many possible outcomes over time and space are possible. The systems are too complex for tightly constrained interactions and outcomes. What powerful insights can be developed about such indeterminate systems? Some factors might be strong enough that their presence or absence could determine the overall state of the forest. The presence or absence of a top carnivore might have such strong influence in a foodweb that broad trends in a forest could be expected. A story like this was told for two canyons in Zion National Park, USA (Figure 5.18). One canyon had a road with millions of tourists each year, very few cougars, many deer, few regenerating cottonwood trees, severely eroded riverbanks, and low fish populations. The other was visited only by occasional hikers, had a strong population of cougars, fewer deer, more cottonwood, stable riverbanks, and higher fish populations. The story is logically possible (always the first important point in scientific explanations), but how much confidence is warranted in the overall story of trophic cascades? Could cougars actually influence the geomorphology of rivers and the quality of habitat for fish? How might evidence be developed to challenge (or support) this story?

Many People → few cougars →
many deer → few cottonwoods
→ wide, shallow river → few fish

Few People → many cougars →
few deer → many cottonwoods →
narrow, variable river → more fish

FIGURE 5.18   Two canyons in Zion National Park, Utah, US, look very different. The one on the left has a road with millions of tourists visiting each year, and the one on the right is managed as wilderness with no trail and few hikers. Ripple and Beschta (2006) expected the two canyons would have been quite similar before European settlement, and that present differences resulted from a trophic cascade of changes in predators, herbivores, and geomorphology. **Source:** photos from Ripple and Beschta 2006, used by permission.

CHAPTER 6

Forest Soils, Nutrient Cycling, and Hydrology

Forests are intimately dependent on soils, and soils are fundamentally shaped by the forests growing on and within them. Soils strongly influence water availability to plants, and the movement of water into groundwater, springs, and streams. The majority of plant biomass is comprised of carbon, oxygen and hydrogen which are abundantly available in air and water. More than a dozen other elements (nutrients) are essential for plant nutrition, and the supplies of these all depend on soils. This chapter has an overview of basic features of soils, nutrients and water, including how they change across landscapes over time.

Forests Need Soils for Physical Support

A core characteristic of a tree is a crown elevated high above the ground, with the inevitable consequence that the crown is exposed to strong winds. The taller the tree, the greater the leverage that is transmitted to the base of the tree. Trees can anchor strongly into soils by sending roots deep into cracks in rocks (Figure 6.1), or deep into the soil (increasing the mass of soil that would need to be moved for the tree to topple). Trees can also gain support with broadly spreading root systems, though if the soil is too thin, the shallow roots may tip up if there is too little mass of soil to weigh the tree down.

Soils Here Are Different from Over There, and Soils Now Are Different from Soils Then

Forest soils vary at every scale of space, from the nanometer scale of bacterial cells to the kilometer scales of landscapes of forests. Most of the chemistry that is so fundamental to soils happens at the nano end of the spectrum, at interfaces between mineral particles and organic compounds, between surfaces and microbes, and between air and water-filled pores. Soils often show pronounced differences with depth below the surface, from an organic horizon (O horizon) of decaying plant parts to a mineral-dominated A horizon strongly influenced by plants, down to B and C horizons (and subsoil) with weaker signatures of plant influences. Moving across landscapes changes the parent materials within which soils form, along with differences in topography and other factors that lead to very large differences among soils (Figure 6.2). The differences in soils, from biology to physics to chemistry, provide broad ranges of conditions for plant growth. Tree productivity commonly differs by a factor of two (or more) within local areas as a result of differences in soils (Figure 6.3).

Life in soils can be precarious, as conditions change drastically over time periods of hours, seasons, years, and millennia. Water enters and leaves soils over time frames of hours, with the two-edge effect of providing good hydration for biological processes, and constraining the supply of oxygen and limiting biological activity. Freezing of soils can happen overnight, shutting down biological

FIGURE 6.1   Individual trees can find sufficient support, water, and nutrients from cracks in rocks, such as these Masson pines in Zhangjiajie National Forest Park, Hunan, China (left). A centuries-old Engelmann spruce tree that established in a very thin soil layer atop a rock in Colorado, USA did not have enough anchorage to keep the tree from toppling over to the right in a very strong wind (right).

FIGURE 6.2   These profiles come from sites just a few km apart in Brazil. They experience the same climate and both are supporting eucalyptus plantations, but they differ in three major ways: the Entisol on the left formed in deposits of sand from old beaches, and the Ultisol on the right formed from granite rocks. The Entisol has been developing only for a few tens of thousands of years (in relation to sea level and continental uplift cycles), while the Ultisol has been forming (and dissolving away) in place for millions of years. A key distinction for plants is the Entisol has very little clay, and very low water holding capacity. The Ultisol stores more plant-available water between rainstorms.

processes and rupturing cells. Temperature fluctuations can be very large in the upper parts of soils. Over longer time periods, very slow annual changes accumulate to create very large differences in soil structures and processes. Even though soils may look constant (almost inert) with a casual glance, they're actually as dynamic and changing as the forests they support (which are also the forests that help shape them in return).

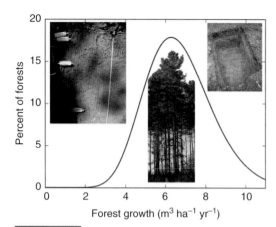

FIGURE 6.3 Forest growth differs by more than a factor of two across a 100 000 ha landscape in the Coastal Plain of South Carolina, USA. The climate is the same across the landscape, and the distribution of forest productivity results from differences in soils. **Source:** Based on data from the USDA Soil Survey, https://websoilsurvey.nrcs.usda.gov/app/WebSoilSurvey.aspx.

Organic Matter is the Top Feature of Soils

Soils are a relatively recent creation on Earth. For almost 90% of the Earth's existence, the absence of life on land meant that only geological processes were occurring across landscapes (as noted in Chapter 3). Rocks weathered in relation to exposure to oxygen and carbon dioxide, freezing and thawing, and other non-biological processes. The evolution of plants that could cope with living out of water provided the novelty of organic matter to interact with minerals, bringing new potential for both energy flow through the soil, and more elaborate soil structures.

Forest soils are most commonly characterized by a layer of organic matter on top, called the O horizon (for more background on forest soils, see Binkley and Fisher 2020). The O horizon forms as aboveground litter falls and gradually decomposes. Many forests have roots that grow into the O horizon, so dead roots also contribute new substrates for the development of the O horizon. Some of the O horizon is undecomposed materials from plants, and other parts result from the life and death of small animals and microorganisms that contribute novel compounds to the complex horizon. Material from the O horizon can be mixed into the upper mineral soil (A horizon) by worms and other animals. Some organic matter is soluble in water, and leaching can move substantial amounts from the O to the A horizon. The solubility of organic compounds depends in part on the acidity of the solutions, so molecules may dissolve in one location and be precipitated in another, less-acidic location (forming an organic-enriched layer in the mineral soil). Mineral soils also receive inputs of organic matter in the form of roots and mycorrhizae that grow and die within mineral horizons.

Organic matter is the most important part of a forest soil for three reasons. Soils contain more than 99% of the biotic diversity found in forests, and the vast network of life is fueled by the energy that comes from oxidizing organic matter to form water and carbon dioxide. Soil organic matter contains large amounts of essential plant nutrients, and the annual decomposition of organic matter provides the largest portion of nutrients used by plants. The third reason is structural: organic molecules are great at forming complex aggregates (often in association with mineral particles) that enhance soil porosity, aeration, and water holding capacity.

A fascinating aspect of organic matter is that a carbon molecule would typically reside in the soil for only a few years to a few decades before being returned to the atmosphere. This means the average "turnover time" for soil C is about the same as the average turnover time for C in trees. Most of the C generated as tree biomass is lost from the tree as litterfall, root death, and respiration, so even though the inner stem may have centuries-old C, the average duration of C within a tree is much shorter. The same is true for soils. Many soils contain large amounts of C that entered the soil centuries (or longer) in the past, even though the turnover time is quite high and the average C atom sits in the soil only a few decades.

Clay Content Comes in Second to Organic Matter

Mineral soil particles are often lumped into three size classes. Clay particles are the smallest, less than 2 μm in size. Silt comes next, with sizes between 2 and 50 μm in size. Sand particles are the largest (not counting rocks), and range in size from 50 μm to 2 mm. These size classes are extremely important because of water movement, water storage, and water access for plants. If a 2-mm diameter grain of sand was the size of a soccer ball, a grain of silt would be the size of a grain of rice, and a clay particle would be 1/10 the size of an actual grain of sand. Size matters because of the relationship between mass and area. One gram of sand has about 25 cm^2 of surface area, compared with 450 cm^2 for silt. The tiny size of clay particles translates into a surface area of more than 10 000 cm^2 for 1 g. The vast surface area of clays means that a large amount of water can be held in the soil. High-clay soils may actually hold some of the water so tightly (with a very low potential) that it is difficult for plants to extract all of it. Water moves easily through sandy soils, draining away within a few hours or days after storms. Soils with finer texture have a moderate or low rate of water movement, but retain more water for longer periods than sandy soils.

The chemistry and structures of clay-sized particles vary, with important consequences for soils. Some clay-sized particles are unstructured groups of iron oxide and aluminum oxide particles, and these amorphous (without form) clays accumulate on the surfaces of larger soil particles. Clays also come with very regular three-dimensional structures involving multiple layers of silica tetrahedra layers and aluminum-rich layers. The layers may also include other cations such as magnesium and calcium in place of aluminum. The charge on calcium and magnesium (2^+) is less than that of aluminum (3^+), so clay particles with substitutions like these end up with a net negative charge. The net negative charge on the clays contributes (along with organic matter) to a soil's capacity to bind cation (positively charged) nutrients in readily available pools (termed cation exchange capacity). The structure and charges of clays also lead to differences among clay and soil types for shrinking and swelling with drying and wetting cycles, and for binding clays and organic molecules together to enhance the three-dimensional structure of soils.

Soils Breathe

Walking through a forest, about half of the volume of the soil below is actually empty, with no rocks, sand, clay or organic matter. The physical components in soils don't fit together perfectly, leaving spaces between particles. Particles also bind together to form small-to-large aggregates, again leaving spaces between aggregates. A typical value for a forest soil would be that about 40% of the volume is comprised of mineral particles, 10% would be organic matter, and the largest volume (50%) would be pore space. Some pores are so small that their volume is dominated by nearby surfaces, and water would be held very tightly (a very negative potential). Larger pores have spaces that are largely unrelated to any nearby surface (at molecular scales), so water drains away easily and air diffuses rapidly. Plants depend on aerobic metabolism, so healthy roots require supplies of oxygen to support cellular biochemistry. The ability of soils to provide air (and oxygen) is generally not a problem, except where saturation by water drastically reduces oxygen diffusion (oxygen diffusion in air is about 10 000 times faster than in water). Some soils may be too wet for good tree growth, though the direct problem is not too much water, but too little oxygen. Water-saturated soils may have such slow decomposition that the addition of new detritus is faster than decomposition, leading to the development of organic soils in bogs, mires, and fens. Vast areas of the world's forests occur on organic soils, across all latitudes where forests are found.

The Variety of Soils Is Parsed into Soil Taxonomic Groups

The whole idea of a "soil" existing as an entity is a tough problem, because changes are so very large across small distances and over moderate spans of time. Nothing in soils is consistent for any distance or time. Nonetheless, the characteristics of a soil in one location typically will have more in common with those of another nearby location than with a very distant location. There are many reasons why soils at two locations would have similar characteristics, and many reasons why soils differ.

Soils at two locations maybe more similar if they both developed from granitic rocks. Sharing a common "parent material" means initial soil development likely had some similar features, including the ratios of elements (such as potassium:calcium, or calcium:iron), the weatherability of the minerals, and the sizes of particles that remain as the rocks weather. Soils that develop at locations from granitic parent materials still differ for a number of reasons. The group of rocks that fit into the "granite" class in geology has substantial variation in minerals and chemistry, and these variations remain important in soils even after a million years of processes have worked on the soil. Some granite-derived soils occur on flat areas or the shoulders of slopes, and others occur near the bottom of slopes. These differences in topographic position influence what happens with water, with follow-on implications for plant growth, soil aeration, erosion, and rates of exposure of unweathered rocks to the environments at Earth's surface. Another reason for soil differences is climate: soil formation tends to go faster (and in somewhat different directions) under warmer and wetter environments than cooler and drier environments. Soil differences also result from the influences of plants and animals, which are not the same everywhere. Tree species substantially alter soil chemistry, biology and structure, as do animals that burrow through soils. Tractor plows allow people to be included in the group of animals that burrow through soils.

Soils Differ in Age, Even if Most Don't Have Birthdays

The idea of soils having different ages is easy to picture for some situations. Floods lead to creations of new terraces, where soils begin to form in the rubble of earlier soils that were washed down from upstream. Volcanoes deposit fresh layers of rock and ash, and soils begin to develop as soon as the materials cool. The retreat of glaciers leaves behind deposits that vary in time since exposure, giving a chance to examine how soils might change over time. However, most forest soils occur on flat lands, slopes, and mountains that lack exciting recent geological events, and it might seem like those soils would be very old, having begun development after some very distant, exciting event. Soils are not organisms though, and they continually form and fall apart. Components of soils fall apart when erosion physically removes particles, and when biological and chemical processes break down soil particles that are then lost as water moves through. If overall development of a soil could be considered to have a rate of formation, a common value might be about 1 mm every century. Such a soil would also have a rate of falling apart, and that rate would also be about 1 mm every century (though the actual removal of materials would be a combination of chronic annual losses and big losses after major events). The soil might be about 1 m deep, which would mean that no part of the soil could be older than about 100 000 years. Many parts of the Earth's surface are relatively stable at a time scale of 100 000 years, so the soils found on the landscapes today would not be the first generation to develop, and the on-going processes of formation and falling apart would not fit any idea of a given birthdate or age for a soil.

How could soils without birthdays differ in age? Again, it's important to think of soils as being quite different from organisms. Some rock types dissolve more readily than others, and the rate of soil formation (and falling apart) will be faster than on harder rock types. The soil from one rock type might take twice as long to form (and fall apart), making it twice as old on average, even though both soils are continually changing rather than being created fresh over the eons. As a final note, very stable landforms show major changes in rocks (parent material) far below the surficial meter or two that we might characterize as currently being soil. The transition from original geology to fertile soil is a story that includes a large amount of geochemical changes as well as life processes.

Trees Affect Soils

Moving across landscapes, the dominant tree species typically shift in response to changing water supplies, soils, and other factors. Very favorable sites usually support a different set of tree species than nearby poor sites. It might be tempting to think that good-site species created the fertile soils beneath them, and the poor-site species fostered the development of infertile soils. This logic would have an assumption that soils differ primarily in response to the trees that occupy them, ignoring the very important influence of topography (and its effects on water supplies), parent materials and other soil-forming factors. An informative test of the effects of tree species on soils would require a comparison of species at a single site, where other soil factors do not vary. Many experiments have used this approach, and common-garden comparisons have revealed that soils do change to reflect the influence of tree species (Figure 6.4). The evidence indicates that tree species do contribute to the shaping of soils, along with a suite of other important factors.

Biological activity is highest in the upper soil, and the largest effects of tree species are commonly on the characteristics of the O horizon. Some species produce dead leaves and roots that decompose slowly, often as a result of being poor-quality food for soil animals. Other species develop only seasonal O horizons that accumulate after autumn leaf fall, and disappear by late summer. These species produce dead material that is good food for earthworms and other animals that mix the fresh litter into the mineral soil. It might seem obvious that a large accumulation of an O horizon means a species leads to higher total carbon storage in a soil. However, species that develop thin O horizons as a result of animal mixing tend to have more C accumulated in the mineral soil, and the total C stored in the O horizon + A horizon may be about the same.

FIGURE 6.4 A common-garden experiment in Toronto, Ontario, Canada found a large O horizon developed under white pine, compared with only a thin O horizon under red oak. The organic matter accumulation in the A horizon was the reverse, so the total contents of organic matter in the O + A horizons were similar.

By definition, all stories in ecology are about interactions, among organisms and environmental factors. The influence of tree species on soils is also a story of interactions. The organic chemistry of leaves, for example, can include differences in the acidity of dead leaves. Acidity is measured on a logarithmic scale, with each unit represent a 10-fold increase in acidity as pH numbers go down. Acidity of litter influences the suitability of leaf litter as food for earthworms. The suitability of litter for earthworms affects the degree of mixing of O horizon material with underlying mineral soil. The degree of mixing of the organic matter and mineral material also influences acidity, with follow-on implications for earthworms. A common-garden experiment in Poland illustrated the overall outcome of these interactions for the mass of the accumulated O horizon (Figure 6.5). Acidic conditions (low pH) had higher O horizon mass, and low biomass of earthworms. The graph makes it look like O horizon mass is a dependent variable, passively responding to the driving variables of acidity and worm biomass. As noted above, the interactions are not simple, and do not run in only a single direction over time. A graph can be a good way to illustrate overall, simple outcomes of ecological stories, but simple patterns should not be mistaken for actual representation of the fascinating details of interactions in ecological systems.

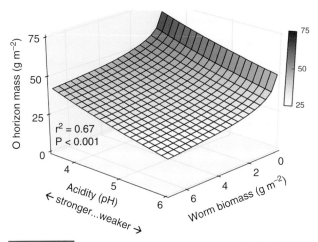

FIGURE 6.5 A common-garden experiment in Poland included plots with linden, sycamore maple, Norway maple, European beech, silver birch, silver fir, red oak, Douglas-fir, hornbeam, European larch, Scots pine, and Austrian pine. The mass of the O horizons ranged from 10 to 80 g m⁻², and about 55% of the variation in O horizon related to the number of earthworms present and the acidity of the O horizon. **Source:** Data from Reich et al. 2005.

Decomposition Reverses Photosynthesis and Nutrient Uptake

Photosynthesis uses solar energy to split water molecules and add hydrogen onto carbon (derived from carbon dioxide) to form sugars (carbohydrates) and release oxygen. Decomposition reverses all of this, with oxygen joining with hydrogen (forming water) and releasing carbon dioxide and energy. The processes are the opposite of each other, but some differences are important. Photosynthesis produces simple sugars, and the simple sugars could be rapidly decomposed by almost all organisms on the planet. Biochemical processes take sugars from photosynthesis to produce thousands of more complex organic compounds, and these organic compounds have so many types of chemical bonds and three-dimensional shapes that breaking them down is far from simple.

The heart of decomposition of organic detritus in soils is a story of enzymes. When animals consume plant materials, enzymes in their digestive systems break down large molecules into small ones that can be absorbed and transported in the bloodstream. Breakdown of organic detritus in soils is also a story of enzymes, but these enzymes are produced by soil fungi, bacteria, and Archaea that excrete the enzymes out into the soil. An enzyme that digests material inside an animal is guaranteed to benefit the animal, but an enzyme excreted out into the soil may or may not result in benefits to the microorganism: there's no guarantee that the resulting small molecules will diffuse directly back to the enzyme-producing microbe. This seemingly unreliable approach to making a living in the soil is successful enough that it sustains the intricate soil ecosystems that sustain all forests.

The reliance on external enzymes to breakdown organic material means that microbial ecology depends very much on access to surfaces. The molecules inside a solid piece of wood are inaccessible to microbes and their enzymes, but if a beetle burrows into the wood, microbial decay can follow more rapidly. Similarly, an earthworm may eat a dead maple leaf, greatly increasing the surface area of material passing through its gut. Earthworm poop has great surface area for microbes, and indeed microbes may colonize the material while it's passing through the worms.

Microbes need nutrients to form new cells, just like plants and animals. They obtain these nutrients from decaying organic matter. But if the microbes need nutrients and they are at the exact scene of decomposition, how do plants ever obtain nutrients? There are four key features that ensure both microbes and plants get a sizable portion of the nutrients released from decaying organic matter. Microbes depend (mostly) on organic molecules in litter for both energy and nutrients, and when they've used up the energy portion, they have little demand for nutrients. This idea is sometimes described as a carbon-to-nitrogen (C:N) story. If the C:N is high, microbes have a good supply of energy and may retain most of the N released in decomposition. If C:N is low, microbes have too little access to energy to compete strongly for N, leaving more available for trees.

The second feature that keeps both microbes and trees supplied with nutrients from decomposing litter comes from the short lifespan of soil microbes. Some microbial cells last only for a few days in soils, and cell death is followed by leakage of cytoplasm that might be scavenged by roots. Death of microbial cells can be substantial after episodes of extreme temperature (especially freezing) and droughts. The life expectancy for a microbial cell can also be determined by the rate at which thousands of species of minute soil animals graze on them.

The third mechanism that helps plants compete with microbes for nutrients is water uptake. Plants and microbes both absorb nutrients that diffuse into their vicinity, but plants have the additional double benefit of mass flow. Absorption of water brings along nutrients that enter roots along with the water, and this flow of water also enhances the diffusion supply of nutrients by replenishing the local microenvironment around roots (rhizosphere). Microbial cells cannot pull in vast quantities of water that would bring along nutrients, and once their local microenvironments have been depleted of nutrients, they have no way to foster replenishment (except through additional decomposition).

The final mechanism that ensures plants can compete with microbes is the symbioses with mycorrhizal fungi. These fungi subsist on carbohydrates supplied by the plants (though some can get a bonus from their own decomposition skills), and their tiny size lets the fungal hyphae exploit the same micro-environments at the same microscopic scales as free-living microbes. As a side note, the mycorrhizal fungi have the same needs for energy and nutrients as other microbes, and the balance between sending nutrients to the plants versus retaining them for themselves is another complex story of costs, benefits, and shifting balances (Chapter 4).

Almost all Forest Biodiversity Is Found in the Soil

A diverse temperate forest might have a dozen tree species in a hectare, and a tropical forest could have 10 times as many. The diversity of trees is commonly exceeded by that of understory and canopy (epiphyte) species. Forests also have diverse assemblages of large and small animals. The total number of species found aboveground adds up to less than 5% of the diversity of a forest. There are too many fascinating aspects of soil biology to fit into a few paragraphs of a forest ecology textbook. Perhaps the most useful presentation here would be a selection of pictures of the sorts of organisms that exist in soils (Figure 6.6). Time would be well spent looking through a marvelous website and free atlas of soil biodiversity (Orgiazzi et al. 2016).

Leonardo da Vinci Couldn't Figure out How Water Got to the Top of Mountains

The history of ecology, along with the history of most domains of science, included attempts to understand problems by using analogies. Five centuries ago in Leonardo's time, people were unclear on how water got into high-elevation springs and streams:

> It must be that the cause which keeps blood at the top of man's head is the same as that which keeps water at the top of mountains. . . There are veins which thread throughout the body of the earth. The heat of the earth, distributed throughout this continuous body, keeps the water raised in these veins even at the highest summits.
>
> (Leonardo da Vinci, cited in Duffy 2017)

This might be a good hypothesis to consider, but an analogy that begins "it must be. . ." sounds risky. Analogies that are deemed to be irresistible or captivating are bound to influence thought and action (Neustadt and May 1986), so it's always wise to list similarities and differences when engaging in analogies. Of course, in Leonardo's time no one understood how blood circulates in bodies either, so it was more than a century too soon to list similarities and differences between blood circulation and the Earth's hydrology. The idea that rainfall and snow are the sources of water in streams, springs, and rivers was closely examined by Pierre Perrault more than a century after Leonardo (Duffy 2017). Two centuries after rain and snow were used to explain mountain streams, other very basic features of hydrology still remained unknown. George Perkins Marsh's pioneering book on environmental science (Marsh 1864) tallied all the evidence about the effects of forest removal on the flow of springs and streams. He started his summary:

> Almost every treatise on the economy of the forest adduces facts in support of the doctrine that the clearing of the woods tends to diminish the flow of springs and the humidity of the soil, and it might seem unnecessary to bring forward further evidence on this point.

Well, it turns out that all the perceived evidence up to that time had the story exactly backwards. Clearing forests increases soil moisture and the amount of water flowing in streams (described in Chapter 2, and later in this chapter). One of the pillars of science is challenging conventionally accepted ideas with better evidence.

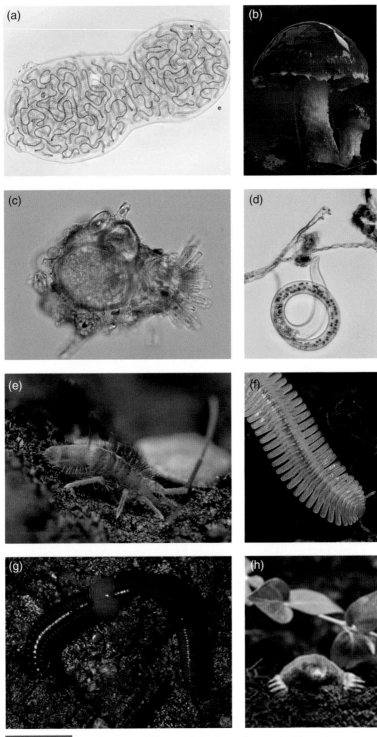

FIGURE 6.6 The diversity of life in soils is almost inconceivable, so here are a few samples, ranging from microscopic to large enough to hold in hand: (a) bacteria, (b) fungi (fruiting body pictured; the hyphae are of course tiny), (c) protists, (d) nematodes, (e) collembolas, (f) myriapods, (g) earthworms, and (h) moles (**Source:** from Orgiazzi et al. 2016, photos by W. van der Putten, E. Mitchell, S. Axford, D. Robson, A. Murray, M. Hedin, D. Heard, and Serena).

The Atmosphere Holds Only a Few Days of Precipitation

The twentieth century included massive advances in understanding Earth's hydrology, from small scale forest water budgets to the globe. One of the largest insights is that the atmosphere holds an average of just 18–19 mm of water vapor (Figure 6.7), yet this small concentration of water molecules provides enough greenhouse effect to keep Earth from freezing (CO_2 is only the second most

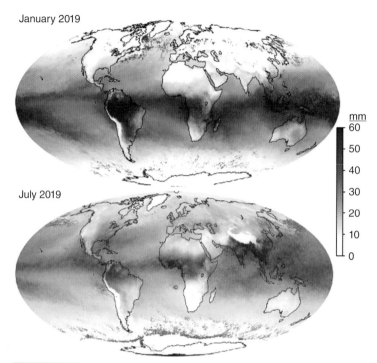

January 2019

July 2019

FIGURE 6.7   The water content of the atmosphere varies with latitude (high in the warm tropics) and through seasons (higher in summers). These global images show the dryness of India in winter compared to the monsoons of summer, and the shift in moisture from southern to central Africa. **Source:** NASA Earth Observation System.

important greenhouse gas). About 2.7 mm of rain falls from the sky each day, matched by an equivalent evaporation that restocks the atmosphere. The entire water content of the planet's atmosphere recycles every week or 10 days.

Forest Water Budgets Begin with Precipitation

Water enters forests as rain, snow, and sometimes fog and dew. Water leaves as gas from evaporation and plant transpiration, and as a liquid in springs, streams, and deep seepage into groundwater. All of these interacting processes vary across time and a space (a regular theme in this book), and intensive measurements are needed to understand patterns and the processes that influence those patterns.

The quickest way for water to leave a forest is to evaporate from the surfaces of canopies before ever reaching the soil. This evaporation from the outside surfaces of canopies is termed canopy interception loss, or simply interception. Forest canopies display about two to eight layers of leaf surfaces (or double that, if counting the undersides of leaves), which means water falling onto a canopy is likely to stay in the canopy, until so much water has fallen that leaf surfaces are fully wetted and any excess has to fall off (Figure 6.8). Interception can also be important in snowy environments, where snow may sit in canopies for days, exposed to dry air for evapotranspiration. Typically between 5 and 15% of the precipitation falling onto forest canopies is lost back to the atmosphere as interception.

Most of the precipitation falling on forests does make it to the soil. A small fraction reaches the soil by falling through gaps in the canopies, but most falls after a brief time on leaf surfaces (termed throughfall). Another small amount of precipitation flows down the stems of trees (stemflow). The amount of water added to the soil in stemflow is only a few percent of total precipitation, but this can still be a large amount of extra water in the local vicinity of stems.

The soils of almost all forests begin with an organic layer, and the O horizons can almost always absorb water faster than it can fall in precipitation. The rate at which water enters a soil is called the infiltration capacity, and the infiltration capacity of O horizons is high enough that it typically isn't even measured. When the input of water to an O horizon exceeds the ability of the horizon to hold water (the water holding capacity), water moves into the mineral soils. The ability of mineral soils to absorb water may be high or low, depending on soil texture (the amount of clay) and soil structures (degree of aggregation, and macropores from dead roots that can channel water).

FIGURE 6.8   Precipitation falls mostly on canopies in forests, and high surface areas can result in enough water sitting long enough on surfaces to evaporate between 5 and 15% of incoming precipitation.

A rainstorm of moderate intensity might drop 2 cm of rain in an hour, and almost all forest soils have infiltration capacities that would let the water move into the soil. A very intense storm might drop the same amount in 10 minutes, or double that amount over longer periods. Intense storms rarely exceed the infiltration capacity of soils with O horizons. In other cases, O horizons may be absent because of recent fires, and intense storms can exceed the infiltration capacities of mineral soils leading to overland flow. Storms that come after fires sometimes generate tremendous erosion, moving massive amounts of soil and debris down to areas where deposition occurs on gentler slopes. These deposits become new parent material for soil development in riparian settings, and indeed without periodic inputs of new sediments the fertility of riparian ecosystem would decline.

Once water has entered a forest soil, it's unlikely to evaporate directly back to the atmosphere. The surface of a forest O horizon dries out between storms, and very little water will be conducted upward from moist mineral soil below even when the air is very dry. Most water that enters forest soils leaves through the continuum of mycorrhizal fungi, roots, stems, branches, and leaves into the atmosphere.

A sizable portion of precipitation falling on forests percolates deeply, ending up in springs and streams, especially during periods when water inputs exceed the ability of plants to use water. Intense storms sometimes exceed the water holding capacities of soils, and the springtime melting of snowpacks also can add more water to soils than they can retain. The flows of springs and headwater streams often rise and fall in response to these events that exceed the soil's water holding capacity. Not all water stored in soils between storms is destined to wait for use by plants. Gravity is also a good competitor for water, and small amounts of water are always moving downward in moist soils. This slow, background movement of water sustains the base flow of springs and streams, between storms and snowmelt events.

Some movement of water in soils goes against gravity, mediated by plant roots. Deep soils may have water at times when shallow regions are dry. Most of the water taken up by roots at depth moves into stems and crowns, but some leaks out into the dry upper horizons of the soil during transport. During dry periods, this upward redistribution (and leakage) of water may equal about 20% of daily water use by trees (Bleby et al. 2010).

The simple image of water leaking from a forest soil into a stream actually has fascinating (and important) details. All points along a 100-m slope would receive the same average precipitation, but soils near ridges are often thinner and have less clay, limiting their ability to hold on to water after a storm (Figure 6.9). Water not retained by the ridgeline forest seeps downslope, where it may be used by trees, stored in the soil (later used by trees), or move on down the slope. Forest growth is often faster at the foot of slopes, as a result of this increase in annual water supply, and also as a legacy of historical movements of nutrients and sediments from upper slopes. When storms (or snowmelt episodes) lead to substantial flow down hillslopes, water percolating from high on the slope would take longer to reach a stream then water that fell lower on the slope. The short-term rising and falling in the flow of water in springs and streams are influenced most strongly by water coming from close to the bottoms of slopes. The patterns of flow over time include rapid increases in flows after storms, more gradual and sustained increases when snowpacks melt, and then long periods of base flow until another event triggers resurgent flows. These time patterns of flow in springs, streams and rivers are often represented in hydrographs of flow quantities across time.

Another set of important details come together close to streams. Soils, and the subsoil material beneath them, can be saturated with water at lower-slope positions. Water tables may be close enough to the surface to be reached by roots (which can reach more than 10 m below the soil surface). Springs and streams appear when the saturated region reaches the surface. We may picture that water flows across landscapes only in streams, but it also flows below the surface through the sediments that fill riparian zones. The water-saturated zone around streams is called the hyporheic zone (Figure 6.10), and the zone is a bustling ecosystem of microbes and small animals. Water flowing in streams also mixes with water flowing in the hyporheic zone, influencing water temperature, oxygenation, and other traits.

FIGURE 6.9   The growth of trees often increases going down slopes. The two black arrows show the height of trees at the lower slope is double the tree height (and four times the productivity) at the ridge. Hillslope hydrology can be pictured as the amount of water falling as precipitation (the blue arrows, same for all locations), the water moving downslope (yellow arrows), and the total amount of water available for trees represented by the size of the buckets, influenced by soil depth and clay content (**Source:** photo by J.L. Stape).

 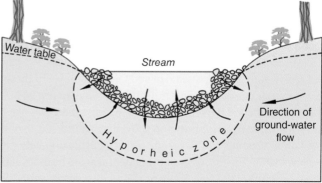

FIGURE 6.10   Streams flow with water down a valley, but water also moves through sediments beside and below the bed of the stream. The predominant direction of flow is of course downstream, but mixing occurs between the stream and the hyporheic zone. The zone is ecologically very active, supporting diverse biotic communities among the sediments, and influencing the temperature and chemistry of the exposed surface water. **Source:** Based on Winter et al. 1998.

Water Use by Forests Can Be Measured Across a Range of Scales

Chapter 1 mentioned that a tulip poplar tree might transpire 70 liters of water in a day, with the dryness of the air providing the energy to lift the water from the soil upward more than 30 m into the tree crown. At the scale of a single tree, water use can be assessed by applying heat from a probe inserted into the stem (Figure 6.11). If a large volume of water moves up the stem, the applied heat will be dissipated more than if water is not moving. Evapotranspiration at a scale of few hectares can be assessed by a technique called eddy covariance (or eddy flux). This method uses very rapid measurements of the water concentration in the air, coupled with equally precise measurements of air movement (up and down). Putting together patterns of concentration and air movement allows a calculation of water moving out of the forest. The same technique is used for measuring the net balance of C across small landscapes. A watershed approach is used to examine hydrology at the scale of landscapes. If the bedrock of a

FIGURE 6.11 Evapotranspiration can be measured at varying scales in forests. Heat-dissipation methods are used to trace water flow up tree stems, with one probe providing heat and another determining how fast the heat disappears as water moves up the stem (upper row). Eddy flux techniques measure water loss from forests at scales of hectares by very high-frequency measurements of water concentrations in the air, and three-dimensional wind speeds (middle row). Evapotranspiration can be estimated for catchments with water-tight bedrock as the difference between incoming precipitation and outflowing stream water (lower row). Adding in measurements of changes in soil water content (over time, at various depths) can provide additional insights into forest hydrology (right).

watershed allows no seepage into deep groundwater, then water use can be estimated simply by measuring incoming rain and snowfall, and then subtracting the amount of water flowing out in a stream. Streamflow is gauged with a weir that is calibrated to estimate flow based on height of the water going over the notch of the weir. The insights gained at any of these scales may be enhanced by tracing changes in the storage of water in the soil, across time and depth.

Trees Use Most (or All) of the Water

A great deal of information has accumulated over the decades on the water budgets of forests. The most important conclusion is that transpiration by trees, along with interception losses from canopy surfaces, removes most of the water available within ecosystems. In drier locations that support trees, the amount of water use by forests is limited by the water available (Figure 6.12). Sometimes forests use more water than comes in as rain, as trees in riparian settings are very good at accessing water supplied from

FIGURE 6.12 The amount of water returned to the atmosphere by forests in Australia matches rainfall in dry sites. Wetter conditions lead to higher evapotranspiration, but the rates taper off as rainfall exceeds the amount of energy available to evaporate water (**Source:** based on Zhang et al. 2001). Compare to the patterns summarized for conifers and hardwoods in Figure 2.18.

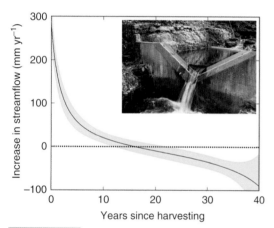

FIGURE 6.13 Streamflow increased after clearcutting a watershed in the Coweeta Basin, but after a dozen years the new forest had the same streamflow as the unharvested control watershed. The young forest then increased water use relative to the control, either because of species changes in the young forest, or in the old forest. **Source:** Data from Jackson et al. 2018.

upstream locations. Some forests receive more rain than there would be energy in the atmosphere to evaporate from leaves, and then large proportions of precipitation are channeled into streams and groundwater.

George Perkins Marsh (and Everyone Else) Was Wrong About the Effect of Forest Cutting on Water

Marsh was quoted a few pages earlier as saying everyone knows that soils, springs and streams become drier after forests are cut. He discussed a wide range of anecdotes that all pointed in this direction. However, no actual measurements had ever been made to test this idea. The twentieth century brought an explosion of measurements that supported the logical conclusions that (i) trees use water; and (ii) without trees, more water remains behind to show up as streamflow (not less). One exception to this generalization may be important. Some desert savannas have soils with very low infiltration rates (especially if grazed heavily), and in these cases the soil infiltration capacity may be higher under trees, leading to higher moisture under trees than between trees.

The evidence of increased flow after forest cutting is extensive enough to provide insights into the magnitudes, timing, and important driving factors. One of the watersheds at Coweeta (Chapter 1) was clearcut in the 1970s to find out what happens to streamflow. Two watersheds were monitored for 11 years before the planned harvest. Monitoring entailed building a "cutoff wall" at the bottom of the watershed so than any water leaving the watershed must go via the stream (not as hyporheic flow). The bedrock was expected to be mostly watertight, minimizing losses of water to groundwater. The calibration period allowed the flow of one watershed to be predicted based on the flow of the other. This relationship was applied after cutting to predict likely streamflow if the watershed had not been cut. For a few years, the stream in the cut watershed had about 200 mm yr^{-1} greater flow than predicted from the calibration phase (Figure 6.13). Water use by the regrowing forest increased until it matched the control forest's level after about 15 years. The use of water by the regrowing forest continued to increase over the next 25 years. Transpiration rates vary among the tree species in these forests, and the 30-year-old forest might have had species that simply used more water. It's also important to realize that the "control" watershed was also getting older during this period, subject to all the changes that accumulate over decades in forests (including mortality of large trees, and shifts in species composition). As watersheds diverge after a calibration period, it becomes harder to identify the actual drivers of the differences in streamflow.

Reliable Generalizations Require Evidence from More than One Case

The case study in Figure 6.13 examined the effect of forest cutting on streamflow by characterizing a single pair of watersheds. If streamflow was tightly controlled by only one or a few factors, a single case study might provide solid evidence for generalization (as a model ecosystem). Most questions in ecology and forestry vary across landscapes (and over time), so the insights from one case gives limited (or no) insight for anywhere else. This is why research sites such as Coweeta often include more than a single pair of small watersheds, and why hydrologists are so interested in looking at the evidence from many studies as a basis for generalizations. One such assessment of many sites found two broad patterns of responses of streamflow to forest harvest (Figure 6.14). The post-harvest increases in flows were higher in wetter locations, and increases were generally larger for conifer forests than broadleaved forests. The confidence bands around the two types of forests overlap at both ends of the precipitation gradient, but not in the mid-range. Even so, some individual conifer sites might have lower responses than some broadleaved sites; clear average differences between groups does not mean there is no overlap among their populations.

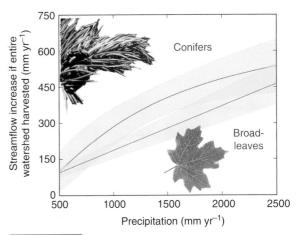

FIGURE 6.14   Streamflow increases when trees are removed, as a result of lower interception loss of water that evaporates from the outer surfaces of canopies, and lower transpiration of water from within leaves. Removing conifers generally results in greater increases in streamflow because interception losses are often higher from the high leaf area of conifers compared to broadleaved species. **Source:** Based on Brown et al. 2005.

Nutrients Make Life Possible

For historical (and fascinating) reasons, the chemistry of carbon compounds in plants and animals is called organic chemistry. Carbon by itself would have no chance of sustaining the massive numbers of chemical reactions that are necessary for cells to function. Carbon atoms join with oxygen (O) and hydrogen (H) to form small molecules such as sugar, and in other cases vast molecules such as cellulose. Life depends on reactions and structures that just aren't possible with only the potential bonding of C, O, and H. Nitrogen (N) is next to C in the periodic chart of the elements, and it brings important opportunities for variation to C-based molecules. Carbon atoms can form bonds by sharing up to four electrons with other atoms, either gaining or losing electrons in the bonds. Nitrogen brings more variation, by being able to gain three electrons, or lose five. Adding N to a long molecule of C backbones allows new shapes for molecules, and new possibilities for energy storage and release.

Another dozen elements are needed for the full suite of biochemical processes in cells. Phosphorus is a key part of lipid (fat) molecules in cells, as well as the key energy compounds of adenosine diphosphate (ADP) and adenosine triphosphate (ATP). Calcium is crucial for forming cell walls, and magnesium is important in enzyme-mediated processes (including photosynthesis). Potassium is especially important as an activator of enzyme reactions, and also as a salt ion that determines osmotic potentials in plants (including closing and opening of leaf stomata). Most of the other nutrient elements are needed in smaller amounts, and they serve specialized purposes in enzymes and some other compounds.

What happens when plants have an insufficient supply of nutrient elements? Almost all forests have a limited supply of one nutrient or another, and yet the trees are very healthy. A limited supply is not the same thing as a severe deficiency. The spectrum from an abundant supply to a limited supply to a severely deficient supply occurs in various soils around the world. Supplies of nitrogen and phosphorus are the most common growth-limiting nutrients in forests (Figure 6.15), but limitations are also common for potassium and even nutrients such as boron (B) which are required in only miniscule amounts.

Forest fertilization is a basic component of management of fast-growing forest plantations. How do foresters determine which elements limit growth? It would be very handy if the foliage of trees could simply be analyzed for concentrations of nutrients, showing which ones are low enough to constrain growth. This idea works in extreme cases, where concentrations are really low or really high, but most forests have foliar chemistry that does not indicate clearly which nutrients limit growth. It might be that looking at ratios of nutrients would make the limiting nutrient more apparent; perhaps a ratio of P:N would indicate which is more likely to be limiting than simply looking at the concentration of each element individually. This hypothesis has not proven to be more useful than simple concentrations alone, so forest managers simply test for nutrient limitations with fertilization trials, and then extrapolate across landscapes based on soil types (Binkley and Fisher 2020).

Nutrients Come From the Atmosphere and From Rocks

The biogeochemistry of each nutrient element is unique. Nitrogen enters the forest from two-and-a-half major sources. The atmosphere is mostly N_2 gas, but no plants have evolved the chemical possibility of using this source. A wide diversity of microbes, however, evolved a nitrogen-fixation process more than a billion years ago, converting the almost-inert N_2 gas into ammonia for constructing amino acids and proteins. The process takes a great deal of energy, and the enzyme system is very sensitive to high

FIGURE 6.15 Slash pine trees develop sparse crowns and grow poorly on this sandy, wet soil in Florida, USA (left side of picture). A single application of phosphorus led to full canopy development and very rapid (and profitable) growth (right). Adding nitrogen fertilizer is also profitable in many forests.

concentrations of oxygen. The process of nitrogen fixation can be very important to the microbes themselves, but the rates of addition are so low (often about 1 kg N ha^{-1} yr^{-1}) that many years of accumulation would be needed to affect a forest. Some higher plants have developed a symbiosis with nodules on roots that protect the microbes from damaging concentrations of oxygen and supply them with sugar, leading to very high rates of nitrogen inputs to forests dominated by these species (rates of 10–100 kg N ha^{-1} yr^{-1} are common).

The second source of N for forests is deposition from the atmosphere. The air contains some traces of ammonium-N (from biological processes), and some of nitrate-N (formed by lightning or other high-energy phenomena). In the modern world, the air contains far more of these N compounds than historically, with ammonia emissions from agriculture (especially livestock) and nitrate from combustion in car engines and industrial plants. The rates of N input for forests now depend strongly on geographic proximity to these sources. The rates of N input were more than 10 kg N ha^{-1} yr^{-1} for many forests in the northeastern North America and central Europe in the late 1900s (Figure 6.16). These rates were probably high enough to increase forest growth across the regions, and perhaps tighter air pollution regulations that reduced these sources of nitrate will reduce forest growth. In the USA the reductions in nitrate have been offset by huge increases in ammonia deposition in agricultural areas. The major story of N deposition at a global scale is massive rise in N deposition (mostly nitrate) with industrialization in China and India; deposition in excess of 20 kg N ha^{-1} yr^{-1} is now widespread.

The final source of N for forests may be unexpected. Rocks are the major sources of most of the other elements, but typically not for N. Actually, a few types of rocks (such as micas) can have enough N that mineral weathering can provide notable N inputs in areas with these rock types.

The inputs of other nutrient elements, such as P, calcium (Ca), and potassium (K) are largely from weathering of rocks, though deposition from the atmosphere is also important. Clouds moving onto shore from the ocean carry trace amounts of sea salt, and some of the salt ions are nutrients. Very long-range transport of dust high in the atmosphere is probably one key for sustaining fertility of the really old forest soils around the planet. The P originally contained in rocks may be leached away over hundreds of thousands of years, but dust storms in the world's deserts can lead to high enough rates of P input to sustain soil fertility far across the globe.

Biogeochemical Cycles Are Complex

The complexities of the biogeochemical cycles of each nutrient are beyond the domain of forest ecology, but a few points give some general perspective. The cycle of N is one of the most complex, because N has major phases of gases (N_2, N_2O, NO, and others), of salts (ammonium, NH_4^+; nitrite, NO_2^-; and nitrate, NO_3^-), and very diverse organic-N compounds (Figure 6.17). The transformations and

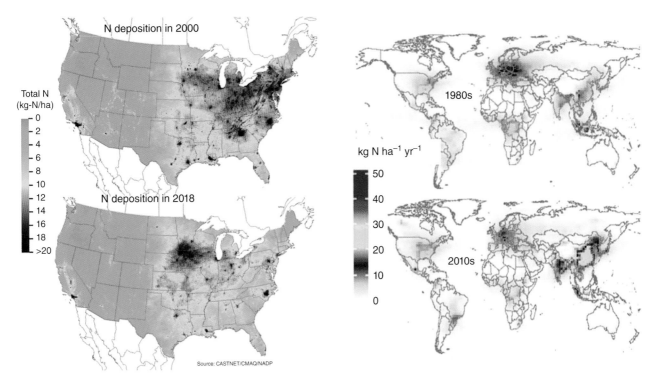

Source: CASTNET/CMAQ/NADP

FIGURE 6.16 The addition of nitrogen from the atmosphere to forests and other areas in the USA in the late 1900s was dominated by nitric acid, formed during combustion in car engines and industrial processes (top left). Tighter air pollution standards led to major reductions from car and industrial sources, but the inputs of ammonium-N (unregulated in the USA) dramatically increased in agricultural areas (lower left). Across the globe, air-quality regulations led to a substantial decline in nitric acid input in Europe (right upper and lower), but massive industrialization in China and India resulted in the highest levels of N input in the world. (**Sources:** USA maps from NADP 2020, world maps adapted from Ackerman et al. 2018 / John Wiley & Sons).

movement of N through the ecosystem have names like mineralization (which is not about rocks), and nitrification and denitrification (which are not quite the opposite of each other). Trees take up ammonium, nitrate, and soluble organic compounds, each of which is dominant in different sorts of forests. An atom of N inside a tree leaf contributes to capturing more than 100 atoms of C through photosynthesis.

Decomposition is the Centerpiece of Nutrient Cycling in Forests

The annual addition of nutrients into forests could provide only a few percent of the nutrients taken up by trees and other plants in a forest. The majority come from recycling of nutrients used in prior years, as decomposition releases energy for use by microbes, CO_2 to the atmosphere, and nutrients into the soil solution for use by microbes and plants. Rates of decomposition of organic materials of course varies, ranging from a few months for some highly decomposable leaves to decades or even centuries for woody material (Figure 6.18).

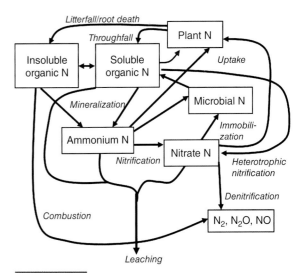

FIGURE 6.17 The forms, transformations, and movement of N in forests are complex, with oxidations, reductions, ionic interactions, and organic bonds. **Source:** Adapted from Binkley and Fisher 2020.

Many studies have investigated rates of decomposition, and how it varies across species and types of materials. The implications of rates of decomposition and nutrient release are not as straightforward as might be expected. Consider two tree species, one with easily decomposed litter and the other with very recalcitrant litter. Both might drop about 30 kg N ha^{-1} yr^{-1} in dead leaves (litterfall) each year. The dead leaves of one species might decompose in a single year, and the other over the course of five years. What are the implications? In the first year, the forest trees could access the full 30 kg N ha^{-1} from the very decomposable litter,

FIGURE 6.18   The log in the upper left decomposed over about a century, and the one in the upper right was consumed in a fire in just a few hours. It might seem that the carbon that used to be in the log on the left would have moved down into the soil, but most of the carbon from both logs ended up as CO_2 in the atmosphere. Rates of log decomposition differ across species, as a result of differences in chemistry and suitability for insects and fungi. An experiment in Germany investigates these trends by placing logs of similar sizes, but different species, in several forests with different environmental conditions. Insights from the experiment will take decades to accumulate. Answers to many of the important questions in ecology depend on evidence from long-term experiments.

but perhaps only $6\,kg\,N\,ha^{-1}$ from the recalcitrant litter. But what would the budget be over 5 or 10 years? The rapid decomposer species would provide $30\,kg\,N\,ha^{-1}\,yr^{-1}$, obtained from the decomposition of the previous year's litterfall. The recalcitrant decomposer would have five cohorts of litter present in this example, each releasing $6\,kg\,N\,ha^{-1}$, for a total of $30\,kg\,N\,ha^{-1}\,yr^{-1}$. The different rates of decomposition would only be important if the litterfall input rates differed, or if some of the litter never decomposed.

Another key point about the larger picture of forest nutrient cycling is that nutrients accumulate over decades in the accumulating tree biomass. The nutrients must have come from somewhere, and unless inputs are unusually high, the accumulation in tree biomass comes from depletion of pools of nutrients that were formally in the O horizon and mineral soil. If the O horizon does not decline in nutrient content over time (which is often the case), then the net accumulation of nutrients in tree biomass must come from depletion of nutrient pools in the mineral soil.

Nutrient Losses Are Chronic and Episodic

Water leaching out of forests contains nutrients, and nutrient losses in streamwater are commonly the largest annual pathway of loss. What determines the rates of annual loss of nutrients in stream water? Early speculations assumed that high rates of forest growth would provide a pool (a sink) for nutrients, and slow-growth rates would lead to higher rates of nutrient leaching. The evidence from nutrient cycling studies showed this idea wasn't useful; as forests across the northeastern USA got older, there was a drastic and unexplained decline in leaching losses of N. The best current prediction for the rates of N loss from forests is the ratio of C:N in soils: high C:N sites tend to retain N better than sites with low C:N (perhaps reflecting the importance of the microbial community in retaining N). Nitrogen losses in stream water sometimes increase after forest harvest, as forest cutting reduces both plant uptake and the C supply to microbes. Annual loss rates are commonly on the order of 1–5 kg N ha^{-1} yr^{-1}, mostly as soluble organic-N compounds. Sites with high inputs from N fixation or pollution have much higher rates, with nitrate-N comprising the majority of the losses.

Episodic losses of nutrients can overshadow annual losses in streamwater. Fires burn (oxidize) the N in organic matter to form N_2O, NO, N_2, just as the C is burned to form CO_2. Burning organic matter leads to a loss of about 5 kg N for each 1000 kg of biomass, summing to 10–25 kg N ha^{-1} for low intensity surface fires, to more than 100 kg N ha^{-1} for more intense fires (Chapter 11).

Another major episodic loss of nutrients is associated with removal of forest products. The N concentration in wood is relatively low, and removal of wood entails a moderate loss of nutrients (Chapter 12). More intensive harvesting that includes bark, branches, twigs and leaves can double or triple the nutrient loss (even though the total harvest mass increases by less than 50%).

Ecological Afterthoughts: Consequences of a Warmer World for Snow, Streams, and Forests

The total precipitation around the globe will be higher in a warmer future, but warmer mountains may have shorter seasons where snow will persist in the forests. The amount of water flowing annually in streams is low where seasonal snowpacks are brief in the Rocky Mountains and Sierra Nevada of the western USA, increasing with increasing length of snowcover (Figure 6.19). This graph represents a spatial pattern, where shorter periods of snow cover come from sites that are warmer than those with longer snow cover. Should the same trend, or even a similar trend, be expected to happen within individual sites as the climate warms? How does the period of snowpack manage to influence total streamflow for an entire year (how, exactly, does it matter if precipitation comes as rain, as snow that melts quickly, or as snow that accumulates for longer periods?). What are the implications for other pieces of the hydrology of a forest, for growth of a forest, and for the occurrence and impacts of major events?

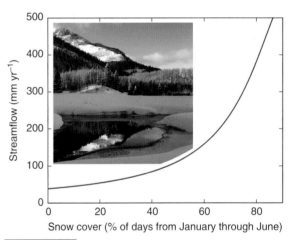

FIGURE 6.19   The annual flow of streamwater increases across watersheds of the Rocky Mountains and Sierra Nevada forests in the western USA as the period of snow cover increases. Watersheds with short-durations of snow cover have low streamflow, compared to those with long-lasting snow cover. The overall annual precipitation would be relatively similar across these 71 watersheds, so the difference in streamflow does not reflect a simple difference in total precipitation. **Source:** based on Hammond et al. 2018, photo by Cary Atwood.

CHAPTER 7

Ecology of Growth of Trees and Forests

Forests change over time as individual organisms arrive, mature, reproduce, and die. Populations of individuals of a species can show similar types of changes: appearing in an ecosystem, increasing in numbers, perhaps sending off individuals that colonize new areas, and in some cases becoming locally rare or extinct.

The growth of an individual can be considered from various perspectives. In a chemical sense, organisms gain energy, assimilate more than a dozen elements from the environment, and split cells to produce new biomass. The formation of new cells ranges from bacteria and Archaea, where daughter cells are identical to the original cell, to fungi, higher plants and animals where cells differentiate for specialized roles. The genes of an individual never change, with two key exceptions. Genes can flow from one type of bacterium to another, and the concept of "species" doesn't apply well. In more complex organisms like plants and animals, the genes don't change during a single lifetime, but genes may be activated or deactivated. Not all the DNA on chromosomes is "read" in the protein-making systems of cells, and some of these regions can be turned on or turned off, sometimes in response to environmental signals (this is the domain of epigenetics).

A population is a group of individuals of a species that occur close enough to each other for regular interbreeding. Populations of species can increase over time, and an analogy with the growth of individual organisms could be developed. A population could be "born" in a sense, when it arrives in a new area. The population grows, perhaps sending off new populations that colonize new areas (reproducing the population), and might die if no individuals manage to reproduce. The power of this analogy between organisms and populations depends on a list of the similarities and differences. Perhaps the most important difference entails the dynamics of genes in populations. Genes change in frequency within populations over time, with some genes showing up in larger (or smaller) proportions of a population in response to environmental "selection" or just random drifting. New genes also can arrive in a population when individuals from other populations arrive, and of course they can disappear if the subset of all members of a population with a particular gene die. (This shifting gene pool might be an analog for the turning on and off of genes in epigenetics, another analogy with similarities and differences that might be considered carefully.) The most important way that growth of organisms differs from growth of populations is that organisms are mostly on a one-way trend to increasing (or constant) size, whereas populations are almost always increasing or decreasing (often by orders of magnitudes).

These two perspectives on growth, of individuals and populations, can be joined to consider the "growth" or development of an ecosystem. After an intense fire, a rapid response of surviving plants, germinating buried seeds, and newly arriving seeds leads to a new assemblage of plants. The assemblage changes over time, as individuals increase in size, new individuals arrive (of the same species, or novel species), and some individuals die and perhaps even all members of some species die. The total biomass of all plants increases over time for some span of years, and may then decline, or oscillate, or remain almost constant. What similarities and differences does this sort of ecosystem change present compared to growth of a single organism or a population of a single species? The biggest difference again deals with genes: ecosystems have entire "packages" of genes come and go as species arrive and depart. The changes in an ecosystem that result from the arrival of a new species can cascade with major impacts on ecosystem composition and functions, greatly exceeding the typical magnitude of change that follows the activation of a single gene in an organism, or the arrival of a new gene in a population of one species.

The key point is that opportunities for change over time are much more limited for an organism than for a population, and for a population than for an ecosystem. An acorn can only develop into an oak, with the environment influence the size and rate of change (but not outcome). A population can shift from one end of a gene pool to another (such as shifting dominance of coloration of moths in response to predation). An ecosystem can change in almost every imaginable way, with the coming and going of species. The key implication of this greater range of change is the truly huge difference in degree of flexibility. The future is largely written for an acorn, which must develop into an oak tree or die. The future is almost unlimited for an ecosystem, with nearly complete flexibility in number and sizes of trees, and especially in species composition with genes coming and going over time and space. If early ecologists had listed (and appreciated) the vast potential of genetic changes in determining ecosystem changes, they might never have fallen prey to the seductive (but weak and misleading) analogy of relating organism growth to ecosystem change (Chapter 9). An analogy between populations (or evolution by natural selection) would have been a notch more powerful, with much less risk of confusion and misdirection.

This chapter explores the growth of organisms, primarily focusing on insights from trees. The changes in ecosystems over time are so large and varied that separate chapters are needed (Chapters 9 through 14).

Forests Are Small and Large, and Growth Is the Key Process Driving Increases

The average biomass of forests around the world varies with climate, soils, and frequencies of severe events that kill trees. The forests with the highest rates of growth may not show the highest accumulations of biomass, because longevity of C atoms in trees depends in part on the longevity of trees. Not surprisingly, forests with high biomass have the combination of relatively high growth rates, and very long lifespans. The old-growth forest types with the highest biomass accumulation all come from the western USA, from areas of moderate temperatures and high precipitation (Figure 7.1). The top-four types from the western USA are closely followed by ancient kauri forests in New Zealand, and then by massive eucalyptus forests in southeastern Australia. Some species of eucalyptus can approach 100 m in height on fertile sites, but few live more than 500 years, failing to reach the extreme biomass accumulations of longer-lived conifers. On a global average basis the accumulation of biomass is far below these old-growth values. Any forest with more than 500 Mg ha^{-1} of biomass would be quite large (and uncommon). The forests with the fastest growth rates are intensively managed plantations, such as eucalyptus plantations is subtropical areas of Brazil. These forests have the potential to accumulate large amounts of biomass, but harvesting at young ages ensure the C atoms do not have a high longevity in stemwood.

Growth is Examined in a Variety of Ways

Growth at the level of a whole organism is commonly examined by dimensions such as diameter, height, volume and mass. Each of these scales could be used for examining part of an organism (such as a branch of a tree), or the entire organisms. The numbers that characterize growth of single organisms can be summed up to represent groups, and growth in forests is often examined at the scale of whole trees, and also as growth of whole forests (sometimes called "stands" in forestry).

One additional way to characterize growth is from the perspective of energy. The organic matter that makes up animals and plants contains about 16–20 MJ kg^{-1}, and so the mass of a tree or a whole forest

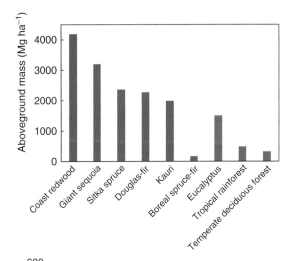

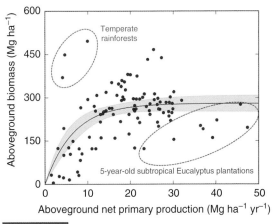

FIGURE 7.1   The biomass of old growth forests differs among forest types around the world, with five major conifer types exceeding all broadleaved types (upper bar graph; **Source:** Based on Sillett et al. 2020). On average, forests with higher growth rates accumulate more biomass (lower graph; **Source:** Modified from Pan et al. 2013), but as with all overall trends, it's important to consider variations around the trend, and what factors might relate to cases that fall above (such as temperate rainforests) and below (such as eucalyptus plantations).

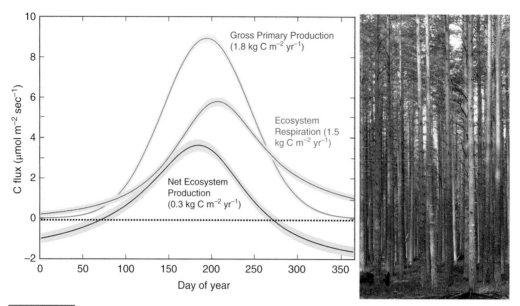

FIGURE 7.2 During the summer, total photosynthesis is quite high for a 60-year-old Scots pine forest in southern Finland. Much of the carbon added by photosynthesis (GPP) is matched by the release of carbon (and energy) from respiration of plants, animals, and decomposition processes (ecosystem respiration), and less than 20% of the gross energy (and carbon) fixed by the forest remains after a year. **Source:** Data from Heinonsalo et al. 2015.

can be considered as mass (g, kg, Mg) or as energy (kJ, MJ, GJ). The mass might be counted as total dry mass of organisms, or just as the mass of C atoms (which is handy for comparison with atmospheric pools of C). An accounting scheme is used for examining the components of energy flow through organisms in forests (Figure 7.2):

- Gross Primary Production (GPP) is the total photosynthesis from all the plants. Technically, some of the energy obtained through photosynthesis could be used within leaves before the actual measurement of what is considered to be GPP.
- Net Primary Production (NPP) is GPP reduced by the amount of respiration conducted by plants. This would be the net increase in plant biomass over a given period, as plant respiration consumes biomass and produces CO_2, H_2O, and heat.
- Net Ecosystem Production (NEP) is NPP minus energy (or matter) released as biomass is digested and decomposed by heterotrophs (browsers and decomposers). The net increase, or decrease, in the total biomass of a system from one period to the next would be NEP. (Atmospheric scientists call this same accounting term Net Ecosystem Exchange, NEE, and put a minus sign in front of it as it represents net C removal from the atmosphere.)

This accounting scheme can be elaborated with net secondary production being the energy (or biomass) content of all herbivores in a forest, and net tertiary production representing the carnivores (though more useful accounting terms might be defined).

The diameter of a tree may be measured with a tape, and the height by an angle gauge (with some geometry) or laser. How is the volume or mass of a tree measured? In rare cases, entire trees may be weighed for total mass. The more common approach is to cut trees into pieces, and use a subsampling scheme to measure diameters along the stem, the density of wood along the stem, and perhaps a subsample of leaves and roots (Figure 7.3). The volumes and densities can be calculated as sums of various parts. This is of course an expensive undertaking, and it might entail killing the tree. Fortunately the sizes of whole trees (which are determined by this destructive sampling) is tightly related to tree diameters and heights, and equations are developed (called allometric equations) that extrapolate the measurements of a few destructively sampled trees into estimates of many trees and whole forests. Once an equation is available for estimating tree sizes, the change in diameter or height can be used to estimate the biomass at a later point in time, and subtraction of the two biomass values gives the growth of the tree.

In addition to characterizing growth by the change in diameter, volume, height or mass of a tree, forestry has a long tradition of focusing on basal area as an index of the size and growth of trees. The basal area of a tree is the cross-sectional area of its stem, at breast height (1.3 m above the ground). The basal area of a forest is the sum of the basal areas of all the trees in the forest. This has been a useful approach because it combines a measure of the sizes of trees and a measure of the number of trees in a forest. Two forests would have the same basal area of 28.3 m² ha⁻¹ if one had 400 trees with a diameter of 30 cm, and the other had 144 trees with a diameter of 50 cm. If the growth of forests of the same basal area would be roughly similar, then basal area could be a useful index for characterizing forest. A related approach, stand density index, uses the average diameter of trees and the density per ha.

Yield Tables Were an Early Example of Parsing Variation in Forests Across Landscapes

Every forest landscape has a current average amount of wood in living trees per hectare, and this average could be determined by measuring many forests and averaging the values. The central tendency or trend would also have a measurable amount of variability around it, with some sites supporting much more tree wood than others. The next logical step is to see if some factors could be identified that account for a useful amount of the variation. Classic forestry used two factors to account for variation: forest age, and site index. Older forests typically have more wood than younger forests, though this idea may not work very well for forests that do not have a particular birthday (sometimes called uneven-aged forests). Site index was an approach to gauging the basic productivity of a site, combining the effects of soils, local topography and any other factor influencing tree growth. Site index was defined as the average height of dominant (or nearly dominant) trees at a given reference age (such as 50 or 100 years in temperate and boreal forests). Height was chosen because it is less sensitive to the density of trees per hectare than diameter. A forest inventory would measure the wood accumulated in many forests, and then the mass of information would be parsed by ages and site indexes to produce yield tables (and graphs, Figure 7.4). Not all of the variation across a landscape would be accounted for by age and site index, and other factors might be the number of trees per hectare, the influence of diseases or pests in some sites, and difference in uniformity (or heterogeneity) among tree sizes within forests.

This classic (and powerful) statistical approach to accounting for forest growth was modified over the years, as quantitative analysis techniques improved, computer simulations became easy, and better information came available from remote sensing approaches and geographical information systems. The accumulation of wood and rates of growth could be related to location on landscapes, with lower-slope positions typically showing higher rates of growth than ridge tops. The primary limitation of a statistical approach to understanding patterns is that currently observable (or historical) patterns resulted from processes that might not be destined to repeat in the future. Climate and weather are not constant across decades and centuries, with implications for both average growth rates and the influences of pests, diseases, and fires. Silviculture also changes over time, leading to very large increases in rates of wood production that can limit the value of yield tables from one rotation for predicting growth in the next rotation.

FIGURE 7.3   The mass of a eucalyptus tree was estimated by cutting down the tree, taking measurements of diameters with height above the ground, taking wood cross sections for later determination of mass and density (visible in the upper left of the upper picture, stacked on the black box), and weighing all branches, twigs and leaves. The mass of the root system was determined with a similar approach (lower photo). Subsets of each tissue type were taken to the lab to determine oven-dry mass, allowing calculation of the dry mass of all the pieces of the tree. **Source:** photo by Otavio Campoe.

Patterns in Yield Tables Were Explained Based on "Growing Space"

Why does forest growth differ from one forest to another? And why does growth change as forests age? A yield table could give very powerful estimates for patterns of forest biomass and growth, but the statistical approach was not aimed at providing insights about processes that generate the patterns. An attempt to provide explanations (especially in North America) was the development

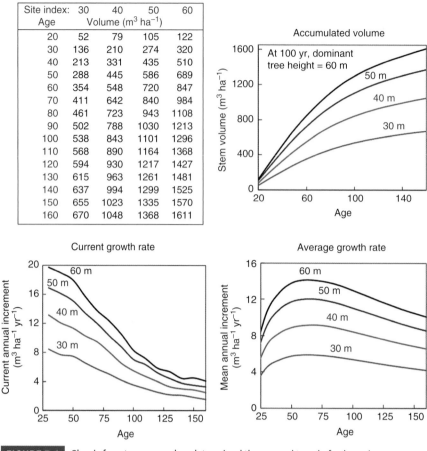

Site index:	30	40	50	60
Age	Volume (m³ ha⁻¹)			
20	52	79	105	122
30	136	210	274	320
40	213	331	435	510
50	288	445	586	689
60	354	548	720	847
70	411	642	840	984
80	461	723	943	1108
90	502	788	1030	1213
100	538	843	1101	1296
110	568	890	1164	1368
120	594	930	1217	1427
130	615	963	1261	1481
140	637	994	1299	1525
150	655	1023	1335	1570
160	670	1048	1368	1611

FIGURE 7.4 Classic forestry approaches determined the general trend of volume increases with forest age, and then accounted for the variation around the general trend with an index of site productivity (based on heights of dominant trees, in this case ranging from 30 to 60 m at 100 years). Yield tables were constructed that listed expected traits for stands as a function of age and site index, and the traits might include number of trees per ha, average tree size, or total forest volume (as in the top left example). The trends over time could be plotted for biomass, as well as the current growth rate (lower left), and the average growth rate for the life of the stand up to a given age (lower right; **Source:** Modified from Douglas-fir in McArdle 1930). The US Forest Service maintains a registry of classic yield tables (https://esi.sc.egov.usda.gov/html/fsnatreg.htm).

of the idea of "growing space." Growing space was not the area of ground occupied by a tree or a forest, nor the three-dimensional volume that would also include height. It was described by Oliver and Larson (1990) as:

> *It is often convenient to consider that a site contains a certain amount of intangible growing space, or capacity for plants to grow until a factor necessary for growth becomes limiting… The amount of growing space varies spatially and temporally… When plants have filled all available growing space, they begin competing with other plants to obtain and maintain growing space.*

This approach may have seemed like a good idea, but explaining a pattern (such as growth of a tree or forest) based on something intangible, which varies in space and time, was only a case of circular logic. When growth was higher, it must be because of more growing space, and the evidence for more growing space was simply that growth was higher.

Production Ecology Parses Growth into Ecophysiological Factors Constrained by Mass Balance

This section heading packs so much into a short phrase that it needs to be unpacked to be useful. The growth or production of a forest results from a suite of ecological and physiological processes and interactions. These processes can be measured (and modeled) in ways that are limited or constrained. Estimating growth based on forest age and site index is an unconstrained approach: any value that comes out of an analysis is possible (though some values would match evidence better than others). Production ecology is constrained by mass balance, which means that if GPP is estimated for a forest, the production of wood must be a number smaller than GPP. Similarly, if the amount of wood grown per unit of water transpired is known, then the total wood production is constrained by the amount of water used by the forest. This mass balance approach doesn't automatically provide more accurate estimates of forest growth, but it provides a framework where any model used to explain growth has to be consistent from start to finish. Aside from the technical points, production ecology provides a clear framework for understanding the processes that result in the observable, measurable patterns of forest growth.

Forest Growth Is a Function of Resources in the Environment, Resources Acquired, and Efficiency of Resource Use

This section heading is only a little less tightly packed than the previous one. The basic idea is that the growth of a forest depends on the amount of water available in the environment, and on the proportion of that water actually taken up by trees. Growth further depends on how efficiently the trees use water in photosynthesis and growth. The beauty of this production ecology approach is that all pieces of the puzzles are tangible, and directly measurable with known precision.

An example of production ecology based on light as a resource would start with the amount of photosynthetically active radiation reaching a forest during the growing season (a good number might be $2\,GJ\,m^{-2}$ for a temperate forest). Forest canopies often intercept 80–95% of incoming radiation; so 90% of $2\,GJ\,m^{-2}$ would be $1.8\,GJ\,m^{-2}$ of light actually used. Light is a "raw material" used in photosynthesis, and a typical efficiency of light use might be 1.5 g of GPP per GJ of light used, giving a GPP for this case of $2.7\,g\,m^{-2}\,yr^{-1}$ (= $27\,Mg\,ha^{-1}\,yr^{-1}$). If wood growth is of interest, it would be important to factor in the partitioning of GPP to wood. If 20% of GPP ended up as wood, then wood production would be $0.54\,kg\,m^{-2}\,yr^{-1}$ (= $5.4\,Mg\,ha^{-1}\,yr^{-1}$).

Production across landscapes varies because of differences in supplies of resources in the environment. Some locations receive more light because of aspect and slope, and water supplies are typically greater at lower slope positions than near ridges. Soil nutrient supplies also differ substantially across locations, and production depends on the efficiency of using resources. A watershed in Idaho was modeled in a geographical information system, and incoming sunlight was related to modeled rates of production. Production increased with increasing sunlight, but the high variation around the average trend revealed that differences in light use efficiency was also very important (Figure 7.5).

The production ecology approach provides a functional (process-based) explanation for *why* patterns of rates of growth develop. This is not a guarantee that a production ecology approach gives more accurate predictions of growth than non-process-based statistical approaches, and both approaches may have value for some situations. A process-based approach may have high value for estimating how growth will respond to changing conditions (environmental or silvicultural), or when growth is estimated for an area without solid historical information for statistical approaches.

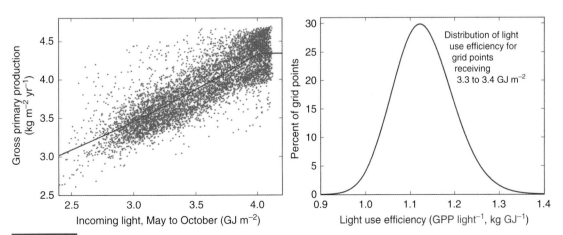

FIGURE 7.5   The production of forests across a watershed in Idaho, USA (Figure 2.7) depended on the light received across slopes and aspects (left). An increase of 15% in incoming light was associated with about a 15% increase in gross primary production (left). This central tendency had a large variation of points above and below, reflecting the importance of light use efficiency (20% differences among points was common; right, **Source:** from data of Wei et al. 2018).

The Growth of a Forest is the Sum of the Growth of All the Trees

This point is obvious, as the total for a group has to be the sum of all the members of a group. The reason this point is interesting for forests is that trees interact with each other, so explanations for the growth of whole forest increases with understanding interactions among individual trees. The growth of a forest with 1000 trees might result from roughly equal rates of growth from each tree, or from very rapid growth of 100 trees coupled with slow growth for 900 trees. Another example would be how growth of individual trees changes over time to account for changes in forest-level growth. A forest-level increase in growth could result from all trees increasing in growth rate, or from great acceleration in growth of some trees more than offsetting declining growth in other trees.

Two perspectives are commonly used to examine growth of individual trees within forests. The first is a production ecology perspective, where the forest-level approach is applied to individual trees, explaining why smaller trees grow differently than medium or large trees (or trees of one species grow differently from those of another species). The second perspective is a statistical neighborhood approach that asks, "How is the growth of this tree affected by nearby trees?"

Large Trees Usually Grow Faster than Small Trees in the Same Forest

This point is also obvious because differences in growth rates among trees accumulate across years, with faster-growing trees ending up larger than slower-growing trees. A production ecology perspective would examine what factors explain the differences in growth rates among trees. For example, larger trees likely capture more light with their bigger crowns. But do larger trees use that light less efficiently for photosynthesis and wood production because they have an abundant supply? Or are smaller trees with low capture of light less vigorous overall, further limiting their ability to use light efficiently? Or do all trees use light with equal efficiency, and bigger trees grow faster only because they catch more light? The production ecology analysis of individual tree growth has been done for only a few dozen cases, but there is a strong general trend that large trees within a forest grow faster than smaller trees, capture more resources than smaller trees, and use the resources more efficiently to produce wood (Type III response, Figure 7.6). A few examples have found that larger trees are not always more efficient at using resources; some cases of a Type I response have been documented. Overall, this is a situation where there is a general tendency (Type III) that applies across most forests, but not to all forests.

Dense Forests Have the Highest Growth Rates

It's obvious that one tree growing in a meadow cannot match the growth rate of 100 trees growing together in a forest. But would 100 trees growing in a forest have less total growth than 200 trees occupying the same amount of land area? Across all forest types around the world, the general trend is that growth per unit of land area increases with the number of trees packed into the area (tree density, or stocking), though the rate of increase is asymptotic and may reach a plateau (Figure 7.7). High-density forests have smaller trees, but the sum of growth across all trees is higher at higher densities. A forester interested in growing the maximum amount of wood would always choose a denser forest over a less dense forest. However, a forester interested in maximizing profits from wood growth would need to balance the unavoidable trade-offs between high growth per hectare, and smaller individual tree sizes.

The density of trees in forests comes from the opposing contributions of establishment of new trees (often called recruitment) and the death of trees (mortality). In managed forests the establishment of new trees is often a matter of planting of seedlings or plantlets (developed by cloning techniques), and mortality comes with chainsaws and large harvesting machines. Managed forests may have some intermediate harvestings before the final harvest, sometimes with a goal of improving individual-tree spacing for faster individual-tree growth (precommercial thinning), or to obtain income (commercial thinning). Reducing the density of a forest always reduces total growth, even if the removed wood is added to the total harvest at the end of a rotation (Figure 7.8). Some of the trees removed in thinnings may not have survived to the final harvest, so thinning can "recoup" some wood that would otherwise become dead wood in the forest.

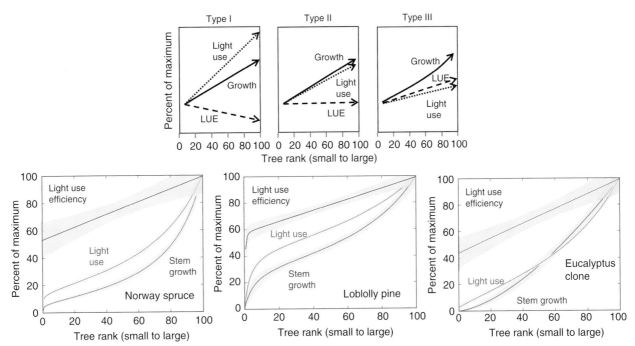

FIGURE 7.6   The production ecology of individual trees can be used to understand the contribution of large and small trees in forests (sometimes called dominant and suppressed). Three types of responses are possible when trees are lined up from the smallest to the largest within a forest (top row). Growth would almost always be higher for larger trees (that's how they got to be larger), and larger trees would almost always capture more resources. The efficiency of using those resources to produce wood might decline when big trees are not very efficient (Type I), or the efficiency could be constant across tree sizes (Type II), or the big trees might actually use the resources more efficiently than smaller trees (Type III). The three cases in the lower row show that differences in growth rates from small to large trees did indeed depend on larger trees capturing more light. Each case also showed a Type III response, where the bigger trees were up to twice as efficient in growing wood per unit light intercepted (**Sources:** from data of Gspatl et al. 2018, Campoe et al. 2013, Binkley 2010).

Forest Growth Peaks at a Young Age and Then Declines, but Not the Growth of the Biggest Trees in the Forest

All forests around the world seem to share a common pattern of growth: growth increases in young forests as crowns and root systems develop, and then growth declines substantially (often by half or more, Figure 7.9). One might expect that a strong universal pattern would have a clear explanation based on some universal process, but the explanation for age-related decline in forest growth remains elusive. A simple idea might be that respiration costs of keeping trees alive increase as forest biomass accumulates, but most of the tree mass is deadwood (with no respiration cost). Other possible explanations include shifting partitioning to belowground in response to declining soil nutrient supply, reduced photosynthesis because of hydraulic limitations in getting water to the tops of tall trees, and physical interactions such as crowns swaying in the wind and damaging neighboring trees. These possible explanations are logically possible, but research seems to have refuted them all, at least for some cases.

Another pattern that appears to be nearly universal is that growth of dominant trees accelerates or plateaus, but doesn't decline (unless the tree is damaged; Stephenson et al. 2014). This suggests that dominant trees continue to expand their consumption of a site's resources, at the expense of smaller trees. Perhaps smaller trees become resource-starved to the point of being very inefficient at using resources to grow, lowering the total growth of the forest even as large trees continue to do well. Or perhaps growth of individual trees is not a simple outcome of the tree's ability to acquire C from the atmosphere; perhaps growth relates to long-term aspects of reproductive success that would not necessarily relate to maximum growth rates for trees that are not dominant in the forest (Prescott et al. 2020).

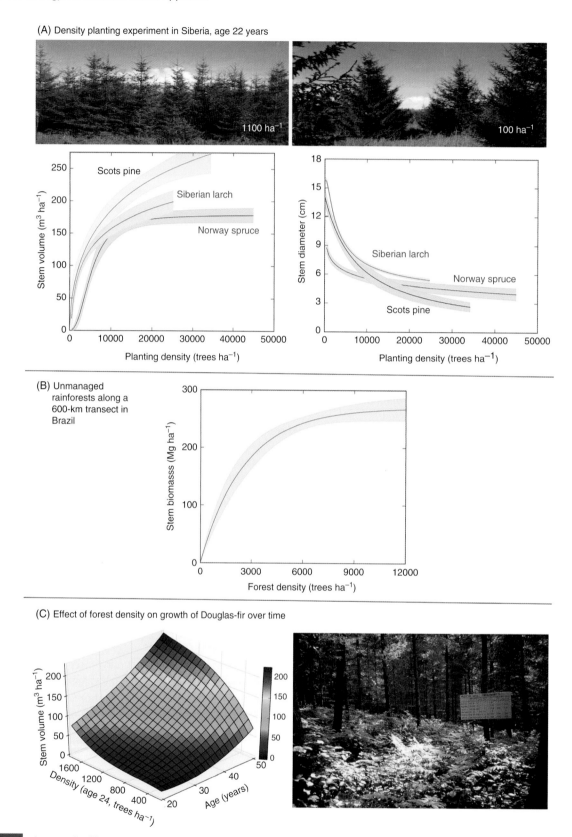

FIGURE 7.7 The growth of forests increases (or plateaus) with increasing density of trees, and the size of individual trees declines. A density experiment in Siberia planted trees across a large range of densities (A, upper pictures and graph). After 22 years, the pattern of surviving trees showed strong increases in plot-level volume up to 10 000–20 000 trees ha⁻¹, and tree diameters decreased with increasing density (**Source:** Data from Sobachkin et al. 2005). At high densities, competition was primarily between trees. At low densities, the trees competed with understory grasses and forbs. The same pattern of increasing accumulated biomass with increasing density of trees was evident in a 600-km transect in the Amazon region of Brazil (B, middle graph; **Source:** Data from Schietti et al. 2016). The influence of density on accumulation of wood shifts across time (C, lower graph and picture), as illustrated for a Douglas-fir plantation that was thinned to varying densities at age 24, on a low-productivity site in Washington, USA (**Source:** Data from Harrington and Reukema 1983).

A production ecology approach can explain how components of production change over time, whether resource supply and use decline, or efficiency of resource use declines, or partitioning shifts. These changes can be examined at the whole-forest scale, or at the scale of individual trees. This approach has been applied only for a few forests, so any generalization at this point might not be supported as research continues to provide more evidence. The decline in forest-level wood growth in older forests seems to relate more to the efficiency of resource use than to reduced resource use (so use of light, water, and nutrients does not decline much if at all). Intriguingly, it seems the drop in resource use efficiency is larger for non-dominant trees: the largest trees may sustain a high efficiency across the years, but forest growth declines from reduced efficiency of resource use by the smaller 90% of trees in the forest. Another decade or two of research should give a basis for much more confidence in whatever general trend is closest to the real story of forest growth, as well as how commonly forests show patterns that are different from the overall average.

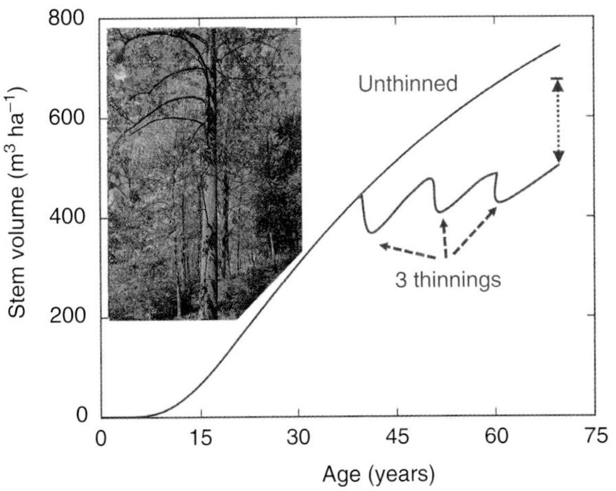

FIGURE 7.8 The growth of an unthinned forest of tulip poplar (on a site where dominant trees would be 33 m at 50 yr) would be greater than a thinned forest, even if the wood removed in thinnings were summed with the accumulated wood at age 70 years (dotted arrow with two heads). Despite lower total yield of wood, the economic value might be higher for the thinned forest because of the accelerated growth of large, more-valuable trees after thinning. **Source:** from data of Beck and Della-Bianca 1972.

Forest Growth Changes over Time, Not Just with Age

As forests age, two things are happening: surviving trees get older, and time passes. Those may seem to be two sides of the same coin, but it's possible that a 50-year-old forest in 1970 would be experiencing different silviculture and environmental conditions than a 50-year-old forest in 2010. In the 1980s, there was widespread concern across Europe (and eastern North America) that forest growth rates were declining because of increased air pollution. Air pollution may affect tree growth, but in fact the growth of forests was accelerating during that time (the acid rain claims from the 1980s provide interesting insights on the use, and lack of use, of evidence). Forests across central Europe steadily increased in growth rates through the twentieth century. Growth of beech and Norway spruce forests increased by about 10–30% between 1960 and 2010 (Pretzsch et al. 2014). Forest management practices did not change much during that time, so the increased growth likely resulted from climate (somewhat warmer and wetter), from deposition of nitrogen, and perhaps from direct benefits of increasing CO_2 on photosynthesis and water use efficiency. Increases in growth of forests in Scandinavia were even larger (Figure 9.14), likely including responses to major improvements in silviculture (including higher forest densities).

Neighbors Influence the Growth of Trees

The individual-tree examples for production ecology considered differences in sizes of trees, but no advantage was taken of the information that can be found in knowing the locations of each tree in a forest (Figure 7.10). Including information on tree location allows assessment of the effects of neighbors (both species and size) on the growth of individual trees (Pommerening and Grabarnik 2019). The resource capture, efficiency of resource use, and growth of a medium-sized tree may depend heavily on the sizes and proximity of neighboring trees. A focal tree (the tree being analyzed) in a location with few neighbors, small neighbors, and distant neighbors likely captures more resources and grows better than the same sized tree would grow in the presence of many neighbors that are close and large (Figure 7.9). These tree-to-tree interactions depend on spatial locations of trees, and change over time as tree sizes change. In most cases, a neighborhood crowded with large trees reduces the growth of a focal tree, but it's also possible that a tree of one species might benefit from neighbors of another species (Pommerening and Sánchez Meador 2018; Figure 7.11).

The statistical analysis of neighborhood influences can be combined with a production ecology assessment of resource supplies, usage, and efficiency of use. Such a complex undertaking would need to be motivated by a fascinating question or opportunity, because the outcomes would likely be very specific for each location and point in time. This level of insight could

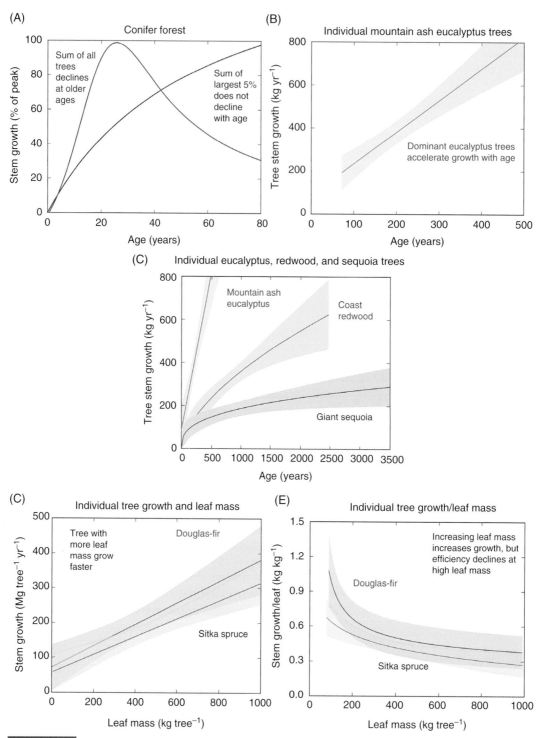

FIGURE 7.9   A long-term experimental plot near the coast of Oregon tracked the growth (and mortality) of all trees in the plot (A). The plot-level growth peaked at about 25 years of age, and then declined by more than half over the next 50 years. The 5% of trees that were largest early in the plot's development continued to accelerate in growth for the whole period (and likely beyond; **Source:** Based on data from Binkley 2003). Growth rates of individual dominant eucalyptus trees in Australia and New Zealand continued to increase for (at least) five centuries (B, **Source:** Data from Sillett et al. 2015a), and for more than 1000 years for redwoods and sequoia (C; **Source:** Data from Sillett et al. 2015b). Accelerating growth of dominant trees results in part from gaining more resources (at the expense of smaller trees), as demonstrated by the relationship between leaf mass and growth (D). Trees with more leaves may not grow as much wood per unit of leaf mass, as a result of differences in either photosynthesis or partitioning to stems versus other parts of trees (E; **Source:** Data from Sillett et al. 2021).

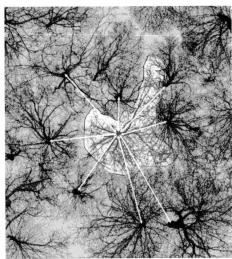

FIGURE 7.10   Growth of a focal tree (circled in red in the left picture) may be influenced by neighboring trees (blue arrows), with the influence scaling with distance between trees and relative sizes of trees. A more complex approach is also possible, where laser-derived images of tree crowns allows for tree influences to be based on crown traits rather than simply on tree diameters (blue area is crown of a focal tree, green crowns are neighbors; **Source:** image from Seidel et al. 2011, used by permission).

lead to virtual forests that live in the joint dimensions of simulation models and geographic information systems, and the mapping of virtual forests onto real forests might lead to unexpected, novel insights. On the other hand, this could be a complex situation reminiscent of the saying, "The most useless scale for a map is 1:1."

How Might a Mixed-Species Forest Grow Faster than a Single-Species Forest?

The production ecology framework lets this question be answered with measurable factors. The possible factors would be that the mixed forest:

A. has access to a larger supply of resources (light, water, nutrients);
B. uses a larger proportion of the environment's resources;
C. has a higher efficiency for using resources to drive photosynthesis; or
D. partitions a larger fraction of GPP to wood (and less to belowground or other sink).

The example in Figure 7.12 comes from a comparison on the same experiment as in Figure 7.11. The supply of incoming light was

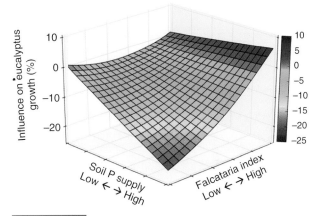

FIGURE 7.11   The size and distance of nearby N-fixing falcataria trees influenced the growth of individual eucalyptus trees (for more on the same plantations see Figures 7.12 and 7.15). The effect depended strongly on the soil supply of phosphorus (falcataria have a high demand for P). Where P supply was low (back, left side of figure) eucalyptus showed little difference between neighborhoods with low or high influences of falcataria (a high index means large, nearby trees). Where P supply was high (front, right-side of figure), eucalyptus trees grew better under strong influences of falcataria. The complex interplay between tree species, sizes, and locations depended on the P and N nutrition of each species. **Source:** Based on Boyden et al. 2005.

identical across the experimental plots, so the mixture effect did not relate to light supply. Light interception also did not differ much between the levels of species mixture, but the mixed plots produced more aboveground biomass (aboveground net primary production [ANPP]) per unit of light intercepted, so factor C would be the key. How did the mixed plot achieve a higher efficiency? There are two possibilities. The first would be the mixed plot had higher GPP per unit of light used, a higher efficiency of photosynthesis. The second possibility is factor D: the photosynthetic efficiency could be the same between plantation, but the mixed plantation partitioned a higher fraction to wood and a lower fraction belowground than the single-species plantation.

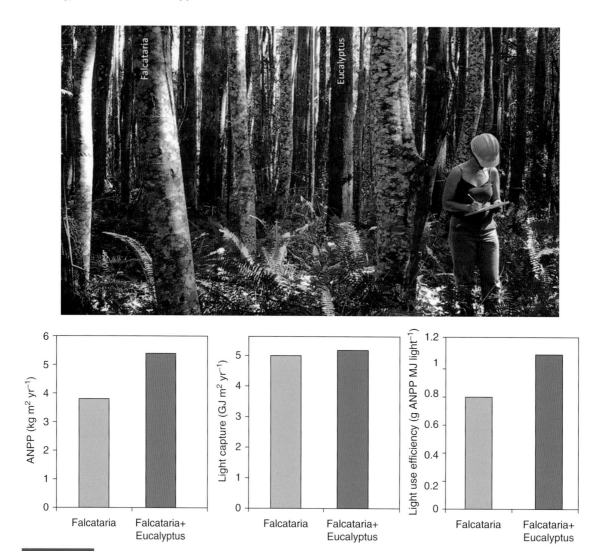

FIGURE 7.12   A mixed plantation of falcataria and eucalyptus grew 40% more aboveground biomass than a pure plantation of falcataria (left). Light capture was similar between the plantations (center), so the greater growth resulted primarily from the increased efficiency of producing biomass with light (right; **Source:** Data from Binkley et al. 1992).

Mixed-Species Forests Usually Cannot Match the Growth of Fast-Growing Monocultures

Trees of different species grow at different rates within a site, and the production ecology approach could explain this as a result of different rates of resource use, the efficiency of using resources, and the partitioning of carbohydrates. Would it be possible that total forest growth might be increased with a wider array of production ecology responses than would be found for a forest with just one species of tree? One approach to answering this question would be to survey growth across landscapes where some sites have forests with more species than others, looking to see if productivity increases with the number of species (Figure 7.13). However, this approach would be inadequate if sites with higher resource supplies (richer soil, or lower-slope sites with higher water supply) tended to be occupied by larger numbers of species. A more challenging approach to testing the possibility is with experimental plantations that compare growth within individual sites (Pretzsch et al. 2017; Ammer 2018).

A comparison of growth rates for pure plots of two species would show that one species grew faster than the other, sometimes with a consistent species difference even across a range of site conditions. In Europe, Norway spruce usually grows faster than beech. This would imply that spruce must use more resources, use resources more efficiently, or partition more to wood growth (and less to other sinks). Would a 50:50 mixed plot of both species lead to growth rates halfway between the rates for monocultures of spruce and beech? Or might the growth be somewhat better than the halfway point, indicating that the mixed forest did not match the high growth rate of pure-spruce, but at least mixing in beech did not lower growth as much as might be expected (based on the average of pure spruce and pure beech)? A comparison across 23 sites in Europe found on average that mixed forests of Norway spruce and beech had 254 Mg ha^{-1} of stem mass, compared to an average from single species forests of 255 Mg ha^{-1} (Pretzsch et al. 2010).

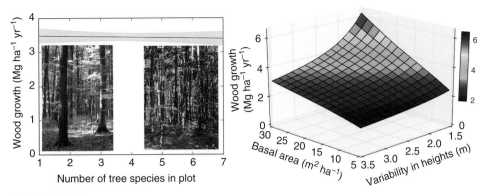

FIGURE 7.13   An analysis of 5000 inventory plots in German forests showed no trend in wood growth in relation to the number of tree species present (left). The plots did show that denser forests (higher basal area) grew faster, and that forests with low variation in tree sizes grew faster than those with heterogenous tree sizes. **Source:** Based on Bohn and Huth 2017.

Given that pure Norway spruce plots grew more wood than pure beech plots, the fact that mixed plots matched the growth of spruce-only plots means the effect of mixing species turned out better than a simple linear mixing of the single-species plots would have suggested. This pattern is sometimes called "overyielding," not because the mixture grew better than the faster monoculture, but because the loss in growth of the mixture wasn't as bad as it might have been (sometimes terms can be awkward). An overyielding mixture means that moving a forest away from a monoculture of the faster-growing species by mixing in other species does not have as large a penalty as it might have.

When a Species Increases Resource Supplies, Mixtures May (or May Not) Outperform Single-Species Forests

The production ecology framework would lead to higher growth rates for a species that somehow increased the supply of resources available to trees. Trees cannot increase the amount of sunlight falling on a forest, but some possibilities exist for one species to root more deeply in soil, accessing supplies of water that may not be available for another species. Tree species also change the rates of decomposition and nutrient turnover in soils, so one species might develop a higher supply of a nutrient. But could access to a higher supply of resources lead to higher growth in mixtures than in single-species forests? A single-species forest of the species that had higher access to resources would lead to higher growth than a mixed forest, unless some other part of the production story differed. Imagine a species that increased soil N supply, but the species had a low efficiency of using light. Mixing in a species with high light use efficiency could lead to higher overall growth, as the benefits of higher N supply could be amplified by a species with high light use efficiency. This logical possibility does not appear to be common.

   In almost all cases, the fastest growth is found in single-species forests of the fastest-growing species. The only notable exception to the central tendency involves mixtures that pair N-fixing species with fast-growing, high-efficiency species (Figure 7.14). The N-fixing species may not use light and water as efficiently as the fast-growing species, but the N-fixer could boost the N supply to the neighboring species that is more efficient at using light and water, resulting in high overall growth of the mixture. The situation where a mixed forest actually out-performs both single-species forests has the awkward name of transgressive overyielding. The mixed-species bonus with adding an N-fixing species of course would be likely only on sites where the N supply in not already high.

The Growth of Mixed-Species Forests Changes over Time

The patterns shown in Figure 7.14 were for a single point in the development of the experiments. What would happen if the experiments examined the effects of mixtures for two or three times as long? One case study is available that followed the mixed species effects for two decades for a highly productive tropical location (Figure 7.15). In this case, the extra growth in mixed-species plots increased over time, from not-very-impressive at six years to about a doubling of total stem biomass after two decades. The take-home story from an experiment always needs to keep details in mind, and in this case the long-term performance of the mixtures

FIGURE 7.14 Experiments on mixed-species forests compare the growth in monocultures of two species with growth in mixed plots. In an experiment in Brazil, plots were planted at a constant total density, with ratios of 100% eucalyptus, 100% acacia, or 50% of each. Eucalyptus grew more wood in a six-year rotation than N-fixing acacia (**Source:** photos from plots at age three years). If the species had a neutral interaction, the mixed plots would have wood growth at the blue dot. In this case, the soil was rich enough in N that the eucalyptus in the mixed plot did not benefit from N-fixation by acacia, and competition between species led to a negative interaction (blue bar below blue dot). A similar experiment on a poorer soil in the Congo showed the eucalyptus indeed benefited more from improved nitrogen nutrition, compensating for any competition with acacia (blue bar above blue dot, a case of transgressive overyielding; **Source:** from data in Epron et al. 2015).

might be overestimated. The original experiment was designed to give insights into growth over six to eight years, and the size of the plots was chosen with that timespan in mind. By the time 20 years had passed, the trees were much taller than the width of the plots, so any plot that managed to have taller trees might capture extra light, shading a neighboring plot. The real benefits of N-fixation in mixed plots may have been overstated in later years if those plots also captured extra light at the expense of neighboring plots.

Mixed-Species Forests Are not Only About Growth Interactions Between Species

Well-designed studies that compare mixtures of species may start with a constant total number of trees per hectare, with varying proportions of species. Many things shift over time in forests, so an experiment initially designed to test species effects may begin to show other effects. For example, some trees may die over time, shifting the ratio of species away from the initial design. Differences in the number of (surviving) trees per hectare will lead to changes in total forest growth, aside from any influence of species. Forest growth also depends in part on differences in sizes of trees within neighborhoods. Uniformity of tree sizes leads to higher growth rates in forests (Figures 7.13 and 8.24), though the penalty may be less severe if the difference in tree sizes includes trees of different species (Torresan et al. 2020). Perhaps the most important point about forest mixtures is that decisions about species in forest management are rarely based only on growth. Long-term forest management considers risks (of single-species and mixed-species forests), non-wood resource values (including wildlife habitat), and other goals of landowners.

FIGURE 7.15   The effects of mixture changes over time, as shown in this experiment in Hawaii, USA that followed constant-density plantings of eucalyptus and N-fixing falcataria, with seven levels of mixtures (on a poor soil). At age six, the interactions between species were largely neutral, but in later years eucalyptus benefited from increasing N supply, and its higher resource use efficiencies led to greater growth than monocultures of either species, a case of transcendent overyielding. **Source:** Data from Binkley et al. 2003.

Understory Vegetation is Important in Most Forests

Understory vegetation (sometimes called ground flora) is an important part of most forests, and importance can be described in many ways. Sometimes trees that will form part of the forest canopy in the future spend many decades as small plants in the understory. Understory vegetation (whether comprised of grasses, forbs, shrubs or small trees) can also substantially reduce growth of overstory trees. Trees have the advantage for capturing light, but other plants can compete effectively for soil water and nutrients. Elimination of understory vegetation can increase growth of overstory trees, and of course removing overstory trees can benefit the understory.

Understory vegetation shapes much of the environment for insects, amphibians, reptiles, mammals and some birds, including serving as habitat and sources of food. Humans may also place high values on understory plants; the berries produced in a very slow-growing pine forest may have a higher economic value across a rotation than the production of wood.

The composition, structure, and dynamics of understory vegetation is strongly influenced by the overstory. Uniform overstory canopies lead to different understories than patchy canopies, even if the total number of trees is similar. Tree canopies that intercept almost all the light (along with resource acquisition by roots) provide little opportunity for understory plants; the paucity of

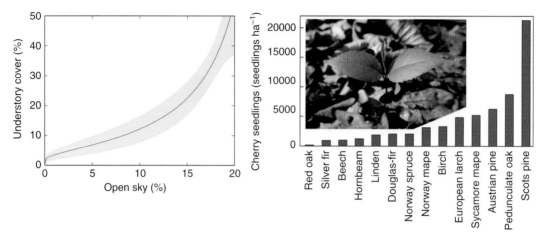

FIGURE 7.16 A common garden experiment in Poland included replicated plots of 14 tree species. Many things differed under the influence of the overstory trees. The cover of understory plants showed a typical trend of increasing in locations where the overstory canopy allowed more light to reach the ground (left). Interestingly, the number of understory plants did not relate to incoming light, highlighting the importance of resource supply for broad features of forest (such as understory cover), but not for details such as species diversity. The establishment of cherry seedlings (an invasive exotic species from North America) did not relate to light passing through the canopy, but it differed by more than an order of magnitude among overstory tree species (right). The differences in cherry establishment could include covarying factors of light, water, and nutrients, but also factors such as habitat for birds that are a prime vector for dispersing cherry seeds (**Source:** Data from Knight et al. 2008, photo by Krzysztof Ziarnek).

understory plants has been a complaint against monoculture plantations that are targeted for maximum wood production. Forest thinning operations increase understory production as the tree-dominance of resource use is reduced.

A side effect of increasing growth of forests, whether from silviculture or climate, may be a reduction in understory vegetation. For example, increased densities of trees in Sweden appears to have lowered understory biomass by about 12% from 1999 to 2011 (Fridman and Nilsson 2017). If the increase in growth of overstory trees was overlooked, it might be tempting to look for a different explanation for changes in the understory (such as enrichment of the soil from N deposition).

Understory vegetation is also influenced by the species of trees comprising the overstory. Some of the effects may result from simple competition for resources (such as usurpation of soil water and sunlight) by the overstory. The identity of species is usually very important too (Figure 7.16).

Clearly the composition and growth of understory vegetation responds to the composition and resource use of the overstory, but the relationships do not fit clear patterns that allow confident generalizations across forests and time. A prime time for understory vegetation might be when forests are young, when trees are too small to compete strongly for soil nutrients, water, and light. This might happen in some cases, but that pattern does not happen generally. For example, a chronosequence study of boreal forests found that understory growth peaked in forests that were more than a century old, either plateauing or declining the second and third centuries (Figure 7.17). Chronosequence studies try to piece together stories of forest change over time by examining forests of different ages at a single time, and time effects may be confounded by space effects (as described in Chapter 9). The precise trajectories of understory growth over time may deviate from those graphed in Figure 7.17, but the general story of relatively high growth of understory vegetation in older forests is probably robust (see Chapter 12 for more examples from managed forests).

The growth of understory plants might be expected to decline as the density of overstory trees increases, but again this would probably not be a very useful generalization. The widespread fires in 1988 in Yellowstone National Park, USA led to intensive research that characterized patterns in regrowing forests across the landscapes. Some areas had little tree establishment, while others had more than 100 000 trees ha^{-1}, two-and-a-half decades after the fire. The density of trees had almost no effect on understory production (Figure 7.17), from a tree density of 0 to over 3000 ha^{-1} (an average spacing between trees of 1.8 m, less than the height of trees). At the highest densities (with more than 10 trees crowded into each m^2), understory production was only reduced by about half compared to having no trees at all.

Why aren't patterns of understory growth more consistent relative to forest age and to tree overstory density? Understory communities are typically comprised of dozens (or many dozens) of species, and these species have varying tolerances of low resource supplies, and varying efficiencies of using the resources acquired. Perhaps increasing competition with overstory trees shifts the composition toward understory species with higher efficiencies of using water, nutrients and lights, partially compensating for reduced resource supplies. Dense tree cover clearly impedes the supply of light to understory plants, but any influence on supplies of water and nutrients would not be so clear. Overstory trees might have advantages of larger carbon supplies to grow roots, but at the scale of a single understory plant, the potential for investments in roots and mycorrhizal fungi might not be so heavily in favor of trees. A final point would be that even though understory plants receive less light than overstory trees, the competition among

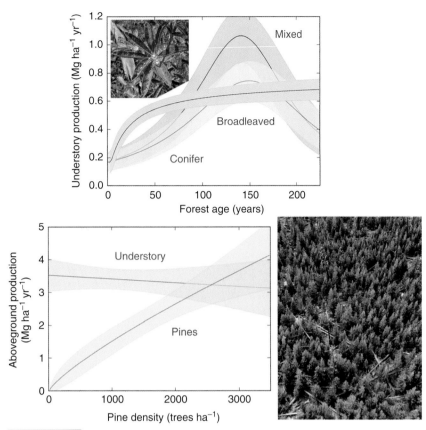

FIGURE 7.17   Forests in Ontario, Canada were sampled for understory biomass in relation to forest age, separated into types dominated by conifers, by broadleaved species, and mixtures. The growth of understory plants was relatively low in the first few decades of forest development, and it peaked or plateaued during the second century (upper, **Source:** Based on data in Kumar et al. 2018). In Yellowstone National Park, USA, the growth of understory vegetation two and a half decades after fire was similar across a wide range of densities of trees (lower). Even at an extreme density of 10 trees in every m^2 (100 000 trees ha^{-1}), understory production remained about half the level for sites with no trees (**Source:** Based on data from Turner et al. 2016).

all plants for water and nutrients may result in understory plants restricting the growth of large trees (see Chapter 12). Overall, this is a case where the current and future composition and function of overstory and understory plants is simply not constrained tightly enough for simple, general patterns to have high value.

Mortality Gets the Final Word on Forest Production

The death of one large tree in a hectare of a forest could remove more biomass from the hectare than is added by all the other trees in a year. Does that mean the forest didn't grow that year? This question highlights the distinction between forest productivity and the rate of accumulation of forest biomass. If all trees survive during a given period, the "gross increment" is the same as the "net increment." If biomass of trees that die is large, the net increment for the period is lower than the gross increment, and might even be negative. However, gross increments are always positive. The long-term development of a forest always includes the deaths of many trees, so the accumulation of biomass always results from the opposing contributions of tree growth and mortality.

The long-term changes in a forest can be characterized in many ways, including changes in overall forest total values, or parsed by species. Growth rates may be examined as a function of the gross rate of growth, or net rates after the subtraction of trees that died during a given period. The death of trees is a core part of normal, long-term changes in forests.

These features are illustrated in Figure 7.18 for a mixed-species forest in Oregon. The biomass of all species increased over time, but the rate of growth of alder trees declined as growth of Douglas-fir increased. The net changes in living biomass depended on biomass removed from the tally when trees died. The mortality loss of alder increased substantially over time, and Douglas-fir

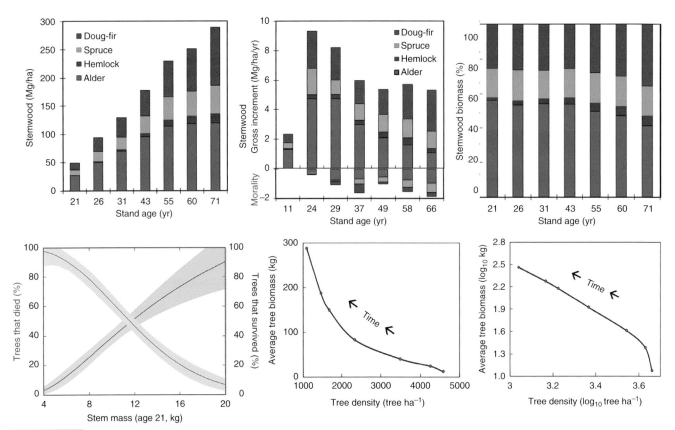

FIGURE 7.18 The 50-year changes in a forest in coastal Oregon (same forest as in Figure 7.8) included an almost sixfold increase in biomass. The increase resulted from opposing influences of growth and mortality, including shifting dominance among trees and species (upper row of graphs). Mortality was concentrated in smaller trees, which had more than a 90% probability of not surviving five decades (lower left). As the average size of trees increased (moving up the Y axis), forest density declined (moving to the left on the X axis of the middle graph in the lower row). The average tree size in the forest increased by 25-fold, total forest biomass increased sixfold, and 75% of the trees died. The self-thinning trajectories are sometimes plotted on logarithmic axes (bottom right).

trees accounted for increasing percentages of total biomass for the forest. Mortality can also be examined in other ways that highlight some of the factors behind the observed rates. Trees that were small early in the forest development were unlikely to survive over the decades, whereas initially large trees were very unlikely to die.

Why exactly do trees die? During periods between major events (such as harvesting or windstorms), mortality may seem to be driven largely by competition among trees. This is sometimes called self-thinning mortality, and many studies have characterized these trends and the slopes of the trend between increasing average tree size and decreasing number of stems per hectare. However, the direct cause of death is usually not simply related to competition. Long-term studies have documented causes of tree death within forests, and shorter-term studies have examined causes of death across regions. The rate of tree death in an old-growth forest of Douglas-fir and western hemlock in Washington, USA averaged 0.75% annually over 36 years. The cause of death could be identified for 70% of the dead trees, with wind being the most important (Figure 7.19). More small trees died than large trees, but that was simply because the forest had many more small trees: mortality did not show a relationship with tree size (unlike the forest in Figure 7.18). Some new trees were establishing as others died, and the overall population structure did not change. Across the forests of Washington and Oregon, USA, almost half experienced mortality rates of <1%, and about 20% of the forests had mortality greater than 2.5% annually. High rates of mortality tended to be associated with fires, and low rates with diseases and insects.

Death is Not the End of the Story for Trees

Trees discard large amounts of dead leaves, twigs, and roots each year, while accumulating wood in stems and large roots. These annual inputs of detrital organic matter fuel the life in the soil leading to long-term changes in chemistry, biology, and physical structure of the soil. Trees don't live forever of course, and the organic matter in a dead stem may equal almost as much as all the annual litter produced in the tree's lifetime. Dead stems take decades or centuries to decompose (unless consumed in a fire),

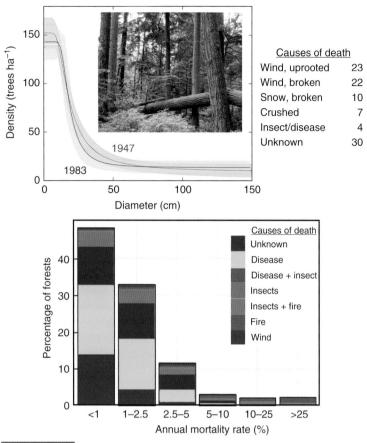

FIGURE 7.19 The number of trees of various sizes did not change much in an old-growth forest of Douglas-fir and western hemlock over 36 years, even though 22% of all trees died (upper). Wind was the primary cause of death (**Source:** Based on Franklin and DeBell 1988). Monitoring of 290 000 trees across 3673 forests in the same region showed a wide range of mortality rates, with about 20% of forests losing 2.5% or more of the trees each year (lower; **Source:** Reilly and Spies 2016 / Elsevier).

providing structure and unique microenvironments in the forest (Figure 7.20). The ecological stories of dead wood vary tremendously among forests, in the amounts of material, the time for decomposition, and residual influences on soils. The decaying logs have thriving communities of arthropods and fungi, and in some cases provide good locations for the next generation of seedlings to establish.

Of course much of the wood accumulated in a forest may be removed for wood products, leading to much less dead woody material in the forest. Does a reduction in the dead woody material in a forest matter? It certainly matters to the species that use the material as habitat or substrate for energy. It might seem that anything so large and pervasive in forests must be crucial for the overall functioning of the ecosystem, but evidence from experiments would be needed to develop confidence (or support skepticism) about ecological consequences of reduced dead wood in forests. Abundant evidence shows that generations of trees continue to grow well where little dead wood remains, but other ecological consequences might be worth considering.

Ecological Afterthoughts: Is it Better to Remove Small Trees or Large Trees When Thinning a Forest?

Foresters have many reasons for cutting some trees before the end of a complete rotation. Thinning out some trees allows the remaining trees to grow faster, and may provide some income from selling some wood. A forester might choose to thin mostly smaller trees, allowing the larger trees to remain until the end of a rotation. Small trees may have low market value, so thinning prescriptions that remove larger trees may be financially favorable in the short run. How would the overall yield of wood through

FIGURE 7.20 Forests may accumulate decaying woody material for decades or centuries. This spruce/fir forest in Colorado, USA has at least three generations of fallen woody material, spanning from recently dead stems to small pieces of decayed wood from trees that died two centuries ago (left). Intensively managed plantations of trees have most of the woody material removed from the forest, resulting in very little dead woody material accumulating on site (eucalyptus plantation in Brazil).

FIGURE 7.21 At the end of a century-long rotation, a Norway spruce forest in Germany is likely to have about 25% more volume if thinning operations during the rotation removed large trees rather than small trees.

a rotation depend on the details of thinning operations? Thinning always lowers total growth (as a result of lower density), but the outcomes may be different for thinning small trees ("thinning from below") versus thinning large trees ("thinning from above"). A compilation of results from over 20 long-term experiments with Norway spruce in Germany (Pretzsch 2020; Figure 7.21) revealed that standing volume of a forest on a good-quality site might be about 1500 m³ ha⁻¹ if thinned from below during the rotation, or 2000 m³ ha⁻¹ if thinned from above. What would be the implications of this overall outcome for the production ecology of the forest during the rotation? Would the outcome likely be the same for other forests at other times, or are there reasons to expect the production ecology would not have a consistent general story?

CHAPTER 8

Forests Across Space

The Three Most Important Things for a Tree Are Location, Location, and Location

This old joke about the most important things in real estate applies to forest ecology, in ways that go beyond humor. The idea of location has several meanings in forests. The obvious one is location in space: the spatial location of a tree or a forest sets the stage for a suite of environmental factors (temperatures and supplies of resources). Location can also refer to a point in time, when all the interacting pieces of the forest come together and interact. The third ecological meaning might seem to be a repetition of the other two, but location can also be used to focus on what's nearby in space (other trees, slopes that influence flow of water and air) and in time (past events with current legacies). This chapter focus on spatial aspects of locations, and Chapter 9 delves into changes over time. These topics are developed insightfully in classic books such as Gergel and Turner (2017) and Turner and Gardner (2015).

How Small Can a Forest Be?

To earn the name of "forest," an area would of course have to have at least one tree, and the smallest area that might warrant the term would need to be large enough to contain a tree. A typical tree might be 10–30 m tall, with a spreading root system that reaches outward 10 m or so, and perhaps 2–4 m downward (Figure 8.1). A volume this large would typically have a few (or many) understory plant species, perhaps some epiphytes growing in the tree, and thousands (or more) species of organisms in the soil. Clearly even such a minimal forest is comprised of a vast set of interacting species.

Forests are much more than just trees, and when trees are spaced far apart, the overall character of the vegetation and ecosystem may be influenced more strongly by smaller plants (Figure 8.2). Trees that punctuate grasslands might be considered to be in a savanna, or if trees are roughly equally as dominant as smaller plants the spatial arrangement might be considered a woodland. Denser aggregations of trees severely restrict the opportunities for smaller vegetation, clearly meeting any definition of a forest. These points are worth mentioning because the two-way interactions between trees and smaller plants scale up to very large influences on spatial and temporal patterns. As an example, trees in a grassy meadow matrix likely experience more frequent, but less severe, fires than trees in denser forests (see Chapter 11).

Forests May Be Divided Into Stands, But Not All Forests Are Structured As Distinct Stands

Recalling the Gregory Bateson quotation near the beginning of Chapter 3, forested landscapes are commonly divided into parts and wholes, though Nature may not provide clear boundaries to dictate how it should be done. Perhaps the most commonly used division of landscapes in forestry deals with the idea of stands. A stand is conceived as an area occupied by a relatively uniform set

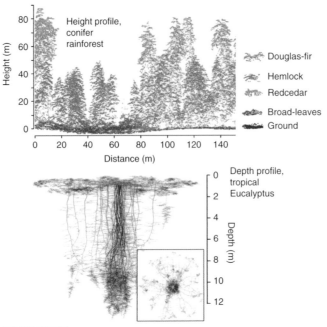

FIGURE 8.1 The spatial dimensions of forests include the land area beneath trees, but also the vertical space of the tree crowns and forest canopy, and the soil and roots. The upper image was developed with ground-based lidar imaging [radar], which reveals not only tree heights but also the intricate geometry of the tree crowns (**Source:** Sillett et al. 2018 / Elsevier). The lower image is a digital simulation of the root system of a single eucalyptus tree in a plantation in the Congo; the inset is a vertical view showing the radial pattern (the scale of the inset box is 6 × 6 m; **Source:** Barczi et al. 2018 / Oxford University Press).

FIGURE 8.2 The minimal size of a forest might be the domain of a single tree, including understory plants, epiphytes, animals, and soil microbes (left). A single tree might occur within a matrix dominated by grasses or shrubs (left and middle images), and such areas might be thought of by other terms (such as savannas and woodlands). Higher densities of trees, especially large trees, clearly qualify as forests, and the dominant influence of trees limits opportunities for grasses, shrubs, and other vegetation (right).

of trees: similar in distributions of species, sizes, and ages. The boundary between one stand and an adjacent stand might be clear, especially where land management leads to units with sharp boundaries of species composition and ages (Figure 8.3). Boundaries may also be diffuse (or non-existent), and stands could be defined based on operationally useful traits, such as distance to a road or a change in topography that would affect management options. The conceptualization of forests being comprised of discrete stands was important when maps were two-dimensional, and each location fell into one unit or into a different unit. For the past few decades, geographical information systems have taken away this binary view of boundaries, and forests can be routinely mapped with gradients of values, and multiple layers of information, removing the classic need for demarcating stands.

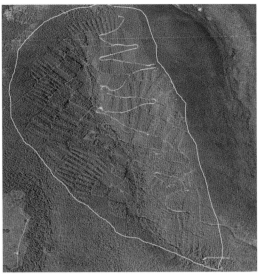

FIGURE 8.3   The delineation of an operationally useful unit, a strand, is clear in the left photo, where a young stand of lodgepole pine is bounded by a road along a ridgetop (upper side to the left) and by a creek (lower side to the right). The unit qualifies as a stand because of sharp boundaries that separate it from adjacent areas of older forests, as well as by topography. The young stand appears greener because the older forests experienced high mortality of large trees in a recent bark beetle event. A topographically delineated watershed identifies the area that contributes water to a given stream (right). The forest in this watershed was harvested in a variety of patterns as part of a hydrologic experiment on forest water use. **Source:** from Google Earth.

The second most common division of land areas in forests would be watersheds. Watersheds are topographic divisions, where ridgelines determine that water flows to different streams (Figure 8.3). Watersheds are nested across landscapes, with first-order watersheds comprising the highest points of the landscape that sustain perennial streams. When two (or more) first-order watersheds join farther down slopes, the combined area is a second-order watershed. This hierarchical (and elevational) nesting of watersheds continues upward to major rivers that get designations beyond sixth- (or so) order. Watersheds are a good example of Nature providing a possible division into real parts and wholes, but watersheds do not necessarily provide units that are useful for dividing up forest landscapes. The value of identifying differences in forests across landscapes of course depends on the purposes underlying the project.

Beyond the level of watersheds, forested landscapes can be divided into any number of units, and some units may be more similar than other units. Similar units might be grouped together and given names, such as spruce-fir or oak-hickory. Groupings could also be based on ages or forest structure, or on productivity for wood (such as site index, Chapter 7). Forests are always changing, and some groupings of units might be based on ideas of what the units would be like after hundreds of years of unmanaged changes. For example, a "habitat type" scheme for forests in northern Idaho, USA, was based on speculating what trees would be dominant after centuries passed without major events (fires, storms, logging), and what dominant species would comprise the understory (Cooper et al. 1991). Ten different habitat types were defined with subalpine fir dominating the canopy, if enough time had passed since major events. Such schemes can be devised from armchairs, or with extensive sampling in the field, but given that forests show great variation across time and space (rarely repeating with high precision), the classifications are only as useful as evidence-based outcomes would support.

Groupings can also be developed to include similar forests that occur in different places. The Society of American Foresters amalgamates the forests of the USA and Canada into about 140 types, such as "#601 Swamp chestnut oak/cherrybark oak" and "#520 Mixed upland hardwoods" (Eyre 2017). This may be too many divisions to be useful for some purposes, so the types can be amalgamated into groups or zones (such as 43 zones, Ruefenacht et al. 2008). Sometimes the nested levels of groupings are arranged as hierarchies, such as an ecoregion approach in the USA (Bailey 1980) that has very broad domains (based on climate) subdivided into divisions (also based on climate), further subdivided into provinces (where vegetation differences are pronounced), and even down to sections (differentiated by topography). It might be appealing to believe these hierarchies actually exist in Nature, similar to the taxonomy of plants. However, that would be a bad use of analogies. The taxonomy of species is powerful because of shared genes across generations, with descent from common ancestors. This fundamental and important process is completely absent from groupings of forest types. The groupings of forest into types, and hierarchies of types, could be useful for communicating about forests, but as Bateson would stress, the groupings are not necessarily provided by the universe.

People Engage with Forests by Defining Areas of Interest

The notion of a "forest" usually comes with ideas about sizes of areas, and of boundaries. The interactions of a single tree and the soil and organisms associated with it might fit into a scale of 10 × 10 m, but typically a forest would be envisioned as something on the order of a hectare (100 × 100 m) or larger (Figure 8.4). As noted above, a forested area may be rather uniform or encompass differences in species, ages, structure and topography. At scales of 100 ha or more, most forests would have quite broad ranges of these features. As the defined area of a forest increases in size, some features change in a predictable way.

Larger Plots Contain More Species

A species might be present in a general area, but absent from a small plot (such as 1 m²). Increasing the size of a plot increases the probability of the species occurring within the plot. This simple idea of randomness leads to "species-area" curves that tally the number of species found in plots as the size of plots increases (Figure 8.5). This is just one important consideration among many when an investigation is designed to answer questions about vegetation patterns (Stohlgren 2006).

Animals also show patterns in the number of species encountered as sampling area increases. An example comes from Hawaii, where lava flows that cover old forests incompletely leave behind forest patches of varying sizes (Figure 8.5). The number of species of birds in the forests increases with increasing size of the forest patch.

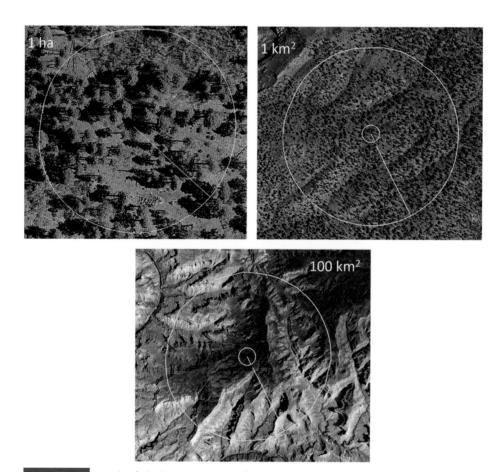

FIGURE 8.4   A scale of 1 ha (0.01 km², upper left) typically includes dozens to hundreds of trees, and is a large enough scale for examining aspects such as competition between trees, and tree/soil interactions. A scale of 1 km² (upper right, showing 1-ha plot in the center) encompasses more variation in soils and slopes, and often a greater variety of vegetation. Increasing by another factor of 100 (lower image, with 1 km² plot in the center) may increase the elevational range and types of vegetation (grassland, shrubland, forests). **Source:** Images of the Kaibab Plateau and Grand Canyon, USA, from Google Earth.

Vegetation Differs Between Locations

This statement would seem to be obvious when locations differ substantially in climate, soils, or other factors that influence plants and animals. Some of these features are discussed below. Less obvious is that the local assemblage of species (or any other ecological feature) is practically never identical between any two locations, or within one location over time. A memorable example of this is the story of "the embarrassing third plot" from Stohlgren et al. (1999). Hundreds of studies have examined how factors such as grazing by wildlife or livestock affect vegetation, using comparisons of two plots that differ in the factor of interest. A fenced exclosure is commonly used to test for the responses to grazing versus no grazing (Figure 8.6). In many cases, general conclusions are based on compilations of responses from many studies that that used this paired-plot design at single locations. Stohlgren et al. (1999) pointed out that what appears to be a result of grazing (or some other treatment) might be strongly influenced by the simple fact that any two plots will not be identical. Could simple differences that result from space be misinterpreted as the effect of grazing? They tested this with a simple idea: they added a third plot to the paired-plot design. One plot was inside a fenced area, one plot was outside the fence, and the third plot was also outside the fence (and a bit farther away). They found that about 60% of the plant species occurred in both fenced plots (excluded from grazing) and unfenced adjacent plots. Forty percent of plant species occurred on just one side of the fence. This would seem to indicate that grazing changed the species composition by about 40%. However, a list of species found in the second unfenced plot found about a 40% difference in species composition with the first unfenced plot! The apparent effect of grazing was not larger than the null expectation of differences that would arise from space alone. Failure to understand the importance of space would lead to an embarrassing overestimation of the effect of grazing.

Space Has Another Dimension for Animals

Forest plants of course are rooted into a single location in space, and the spatial aspects for questions about forest ecology are relatively straightforward. The importance of space for animals in forests has another dimension, as animals move across landscapes and cross into different types of vegetation. The success of an animal in a forest landscape is affected by the spatial extent and arrangement of vegetation across space, in ways that are too complex to fit within a forest ecology text. Two examples illustrate some of this complexity. Elk move across landscapes in Rocky Mountain National Park, USA spending different amounts of time in different vegetation types (Figure 8.7). The elk eat a variety of plants, including grasses, willows and other shrubs, and small aspen trees. Their allocation of time includes both location and activity, especially eating and resting. The use of a landscape would also be affected by responses to the presence or absence of predators; open vegetation types may produce greater amounts of good forage, but might come with high predation risks.

The use of landscapes by roe deer in Bavarian Forest National Park, Germany changed with time of day, and through the year (Figure 8.7). Old forests were used any time of day in winter, and only lightly used in summer. Young forests were used in summer, spreading throughout the entire day in midsummer. Meadows were used in spring through fall, primarily in mornings and afternoons. These time-of-use patterns relate to both the availability of food, and risk of predation.

If a forest manager wanted to provide optimal habitat for elk or for roe deer, how much area should she allocate to young forests, old forests, and meadows? This easy question has no clear answer, for a long list of ecological reasons. A wiser approach might be to ask what pieces in a vegetation mosaic might be undesirably too small, guiding where some marginal changes might be helpful to elk or roe deer. This "undesirable" approach to defining goals in forest ecology comes back around in Chapter 14.

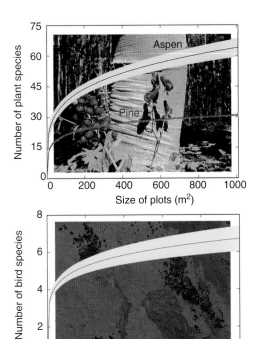

FIGURE 8.5 The number of plant species found in a forest dominated by aspen trees in Colorado, USA increases as the size of the sampling plot increases (upper). The rate of increase is steep as sampling goes from very small plots to medium-size plots, and then slows with further increases in the size of area sampled. These species-area curves differ for different forests; the red line represents the same trend for lodgepole pine forests in the same landscape (**Source:** Data from Chong et al. 2001). In Hawaii, lava flows sometimes flow around patches of forest, leaving patches of varying sizes (lower). The number of bird species found in a forest patch also increases with increasing size of the patches (**Source:** Data from Flaspohler et al. 2010, image from Google Earth).

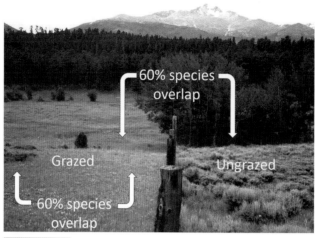

FIGURE 8.6   Grazing impacts have been studied in thousands of experiments where adjacent plots are compared, one inside a fenced area and one outside. The list of species found in each plot could be compared to indicate the effect of grazing, but the plots differ in space as well as in grazing. In this case, an additional plot outside the fence gave a null expectation for the effect of space, and indeed it was as large as the difference between the inside/outside adjacent plots (**Source:** Based on Stohlgren et al. 1999). Note that grazing greatly influenced vegetation structure, even if the lists of species did not show larger differences than would be expected across space with no difference in grazing (photo from Rocky Mountain National Park, USA, after three decades of exclusion of elk and deer).

Differences in Forests Usually Increase with Distance, But Not Always

It seems logical that the species found in one forest will be more similar to the list of species in a nearby forest, and less similar to the list from a distant forest. Evidence generally supports this expectation, especially where distance includes gradients in elevation or other site features (Figure 8.8). The suite of species occupying a particular plot includes some random factors of environment and history.

Would ecosystem-scale features such as tree biomass show greater similarity with distance between plots than species composition? This is a good question, but one that has great variability around whatever general trend might apply. A research project in the Harvard Forest in the northeastern USA examined tree biomass along transects in the 1930s, before a major hurricane decimated the forest. There was no pattern of biomass similarity in adjacent $10 \times 10 \, \text{m}^2$ plots along the transects (Figure 8.9): high-biomass plots were as likely to be adjacent to low-biomass plots as to high-biomass plots. The project also included a resampling of the transects 70 years after the hurricane. The lack of similarity in biomass between nearby plots remained, though the average differences among plots declined substantially. The pre-hurricane forest included massive white pine trees that established a few centuries earlier, probably after a severe fire. White pines did not regenerate after the hurricane, and the dominance of broadleaved trees was more pronounced. Patterns in similarity of species composition, biomass, or other features of forests can show a wide variation across both space and time, pointing to the wisdom of recognizing that details always matter and that generalizing is most successful with large amounts of evidence and humility.

Location Matters Both Locally and Regionally

The diagram of forests along the Front Range in Colorado, USA in Figure 2.1 considered broad patterns that develop as a result of elevation-related differences in precipitation, temperature, and major events (such as fires). The elevational location of a forest is of course important, but more-local aspects of location also matter. Two forests may occur near each other at

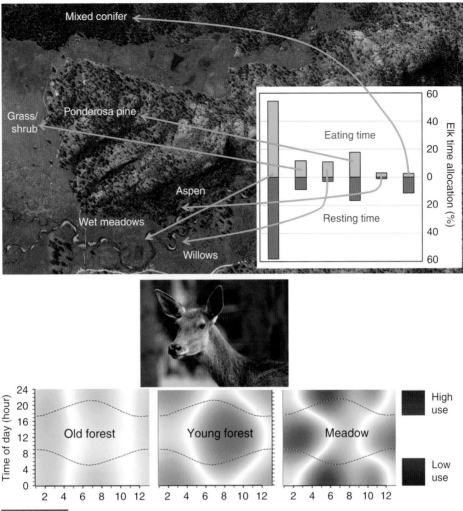

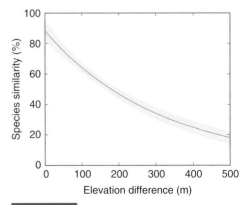

FIGURE 8.7 Elk allocate time across landscapes in the summer in Rocky Mountain National Park, USA, spending most of their time in wet meadows and ponderosa pine woodlands (upper). The time spent in each vegetation type might be split evenly between eating and resting (such as in the wet meadows), or might be primarily resting (as in mixed conifer forest; **Source:** Data from Schoenecker et al. 2004, photo from Google Earth). The use of landscapes in Bavarian Forest National Park, Germany are used by female roe deer at different times of day, shifting through the year (**Source:** Based on Dupke et al. 2017 / John Wiley & Sons; roe deer photo by Max Pixel).

the same elevation, but one occupying a ridgeline will be very different from another at the bottom of a slope. Maps of forests can reflect these aspects, relating forest properties to overall locations in elevation, and locations relative to local topography (Figure 8.10). Many techniques are available for quantifying patterns across landscape. Figure 8.10 also shows the semivariance pattern, which indexes the loss of similarity with distance away from an initial starting point.

Locations on ridges or on convex portions of landscapes tend to lose water, nutrients, and soils to the benefit of soils and forests lower on slopes and in concave positions. The changes in productivity along a topographic gradient were apparent in Figure 6.9 for a eucalyptus plantation in Brazil, and similar patterns occur for most slopes. A broadleaved forest in Japan accumulated twice the biomass at the bottom of a slope than at the ridge, in part from increases in canopy leaf mass (Figure 8.11). Along the transect, higher leaf mass was associated with greater supplies of N in the soil. Wood growth also was higher where high soil N led to a shift in partitioning of carbon from belowground into stems.

FIGURE 8.8 The species composition of plots in an evergreen, broadleaved forest in Japan overlapped by more than 75% when plots were within 75 m of elevation, but similarity declined to less than 20% when elevations differed by more than 500 m. **Source:** Data from Itow 1991.

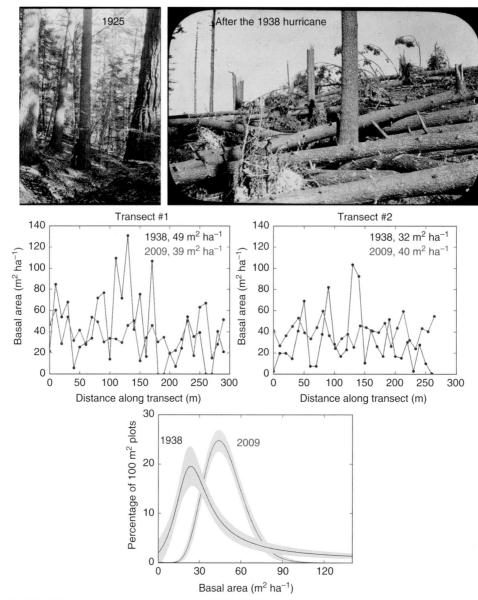

FIGURE 8.9   The basal area (and biomass) of a three-century-old forest in Massachusetts, USA did not show a pattern of high similarity along two transects in 1938 (upper graphs). A hurricane toppled most of the large trees, including the white pines that dominated the forests. Post-hurricane conditions did not favor white pine regeneration, and the subsequent broadleaved-dominated forest also show little similarity among plots along the resampled transects. The plots were more uniform in basal area in 2009 than before the hurricane in 1938 (lower graph; **Sources:** Based on D'Amato et al. 2017; photos from Harvard Forest Archives).

Resource Use Varies Across Landscape Gradients

Trees growing in a flat landscape have access to the amount of water falling in precipitation, minus any water that seeps below the depth of roots. Many trees grow on areas with varying topography, and trees have access to less water on ridges and steep slopes (where percolation downward is rapid), and more water on lower slopes and near streams. The overall supply of water to trees was shown for the Coweeta basin in Figure 1.6; Figure 8.12 shows the estimated use of water across another basin in northern Idaho, USA. It might seem that water use would be highest where trees have the most access to water, but transpiration also depends on the energy available to evaporate water from trees. Low elevations have lower transpiration owing to lower precipitation. Mid-elevations have higher precipitation and are warm enough for the highest rates of transpiration. High-elevation sites have plenty of water available, but low temperatures reduce transpiration rates. The importance of the energy available to evaporate water is illustrated by the differences in transpiration between slopes that face south (with more incoming sunlight) and those that face north. As with difference in elevation, this effect of aspect results from the interaction of available water and energy to evaporate water. These topographic patterns are of course reversed in the Southern Hemisphere.

Mind the Gap: Spatial Patterns of Trees Within Forests Modify Resource Supplies

Gapminder is the name of an organization dedicated to providing the best available information about the state of the world, in the hope that being well-informed about the recent past and the present will lead to wiser choices about the future (www.gapminder.org). In this case the "gap" is between the reality of population, economics, education, and health, relative to people's beliefs. The idea of a gap is that a large difference between adjacent things is a prime situation for changes to occur. A gap in a forest is a place of change in a forest, where spatial patterns of resource availability change.

The spatial arrangement of trees leads to variability in capture of light. The arrangement of trees depends on the outcome of many factors: the original locations of seeds, variations in soil resources, and the locations and rates of resource use by other vegetation. These factors change over time, especially when dominant trees are damaged or die. The removal of part or all of a dominant tree leads to a gap in the canopy, and opportunities for other species to increase access to resources. The formation of a canopy gap gives a localized increase in available light, but it may or may not have a similar spatial effect in soils. The rooting systems of trees intertwine through the soils much more than do the branches of crowns, so roots of surviving trees and understory plants may already be in position to absorb water and nutrients that would have been used by the missing tree.

Forest ecology focuses so strongly on trees that the only words we have available to describe gaps in forest refer to the fact that a location is missing a tree. Available words include gaps, openings, and breaks in the canopy. This perspective on "what's missing" may overlook the ecologically important story of "what's present." A gap in a forest is actually a location where small grasslands, meadows, and shrublands have the opportunity to be present. This distinction between looking at what's missing (trees) and what is present (vegetation communities of other sorts) may seem trivial, but the balance between dominance of trees and shorter plants in drier landscapes has tremendous implications for fires and forests (Chapters 11 and 13).

Remote sensing techniques allow the two- and three-dimensional structure of forests to be determined across large landscapes (and indeed the whole planet). A frequent pattern throughout ecology is that a few things are very common (or dominant), and many things are relatively uncommon (or rare). This pattern (sometimes called a power-law, though patterns are too variable to call it a "law") also applies to the distributions of numbers of gaps and their sizes in forests. A 370-km^2 forest in western Washington, USA contained both

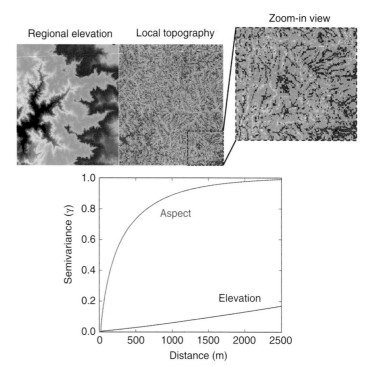

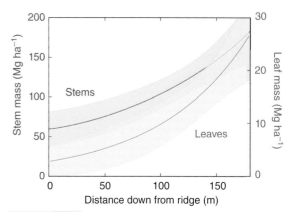

FIGURE 8.10 A 90 000-ha landscape in Sequoia National Park in California, USA (upper left figure) encompassed an elevation gradient from 700 m (black/red colors) up to 3500 m (blue colors). The same area was plotted in relation to local topography. Blue colors are along local ridges, and yellows and reds are lower slope positions. Both scales (elevation and landscape position) are important factors in forests. The similarity of locations across distances can be examined with a semivariogram, which shows how similar characteristics are between a starting point and various distances. In this landscape, locations were likely to have similar aspects within few hundred meters, but beyond 1 km aspects tend to be unrelated. Elevation changes more slowly with distance across this landscape, and distances of more than 2 km still have similar elevations (**Source:** graph based on Urban et al. 2000, images courtesy of D.L. Urban).

FIGURE 8.11 A broadleaved forest in Japan showed increasing stem biomass and leaf mass with distance from a ridge top to lower slopes. The supply of soil N increased down the slope, supporting the higher leaf mass and shifting partitioning of carbon away from belowground and into stem growth. **Source:** Based on data from Tateno et al. 2004.

managed and unmanaged forests. Forest structure was mapped for 95 9-ha plots, across an elevation gradient of 165–1655 m. Most of the gaps were less than 10 m^2 in size, but a few were several hundred m^2 (Figure 8.13). Large gaps were rare, so the combination of gap frequency and gap size showed that two-thirds of the total gap area across the landscape occurred in patches between 20 and 250 m^2.

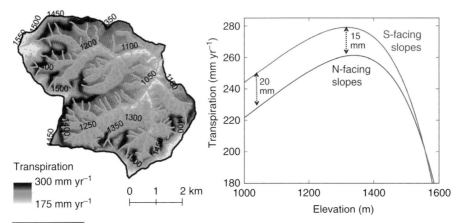

FIGURE 8.12   Transpiration varies from less than 200 mm yr^{-1} at low elevations in a watershed in northern Idaho, USA, to about 280 mm yr^{-1} at mid-elevations where higher rates are possible from higher precipitation, coupled with enough energy to evaporate water. Trees at higher elevations transpire less water despite higher precipitation, because lower temperatures provide less energy. All aspects receive the same precipitation, but south-facing aspects have more energy (see Figure 2.7) to evaporate water. **Source:** Based on Wei et al. 2018 / Elsevier.

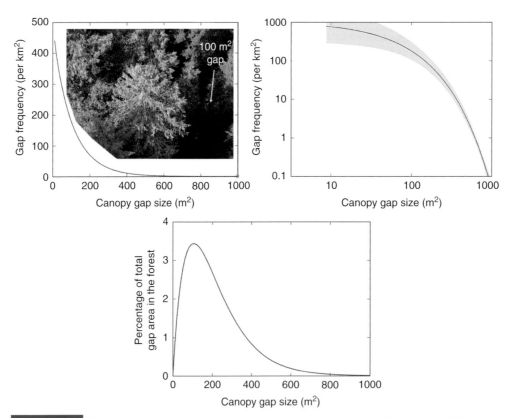

FIGURE 8.13   The pattern of gaps across the Cedar Creek watershed in western Washington, USA showed a very high frequency of small gaps, and very few large gaps. The patterns can be viewed in various ways, depending on the feature of interest. The upper left figure charts the central tendency, and the sample sizes were so large and the pattern so consistent that the confidence intervals around the central tendency are not discernible in the graph. The range in gap size covered two orders of magnitude, and the number of gaps km^{-2} spanned more than three orders of magnitude, so a plot of the relationship on logarithmic axes (upper right) may be useful. The bottom graph combines gap sizes and frequencies to show how much of the total area of gaps occurred for each size of gaps. Gaps covered a grand total of 6% of the forest area. **Source:** Based on Kane et al. 2011.

The Ecology of Gaps is Not Binary

The idea of a location in a forest falling within a gap, or outside a gap, is too binary to capture what actually happens in forests. More sunlight may fall in the center of a gap than beneath the crown of a large tree, but the size of this difference depends on size of the gap, time of day, and whether values are calculated for the whole gap, or the sunny versus shady side of the gap. Gaps also have edges, and the non-gap portion of a forest may be influenced by adjacent gaps.

Another way to consider the spatial patterns of forests would be to simply gauge the proportion of full sunlight that reaches the ground across a landscape. This has been done in a number of studies around the world, and a clear pattern emerges (Figure 8.14). Tropical forests have the highest proportion of area with very low light reaching the ground (which means high light capture by the canopy). Temperate forests also have high proportions of area with very little light reaching the ground, but boreal forests commonly have many patches receiving at least 20% of full sunlight.

Forest structure across landscapes can include gaps within the forest, as well as edges where adjacent forest have a difference in structure. In all cases, a spatial pattern of microclimate and resource supply would be present across the boundaries. A clearcut in an old-growth forest in Washington, USA was monitored for the microclimate pattern along transects more than 200 m into the uncut forest (Figure 8.15). Soil surface temperatures were much higher at the forest edge than in the interior, with midday peaks 15 °C higher. The effect extended only a few meters in from the boundary. The edge effect for air temperature was much lower, reaching maximum values of 5 °C

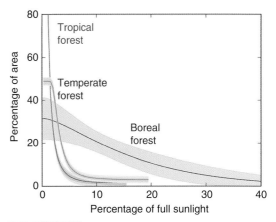

FIGURE 8.14 Most of the ground area beneath tropical forests receives only a few percent of full sunlight, but the ground beneath boreal forests commonly receives 10% or more of full sunlight. The pattern derives from the difference in canopy leaf area and structure, as well as the varying angles of incoming light. **Source:** Based on Messier et al. 2009.

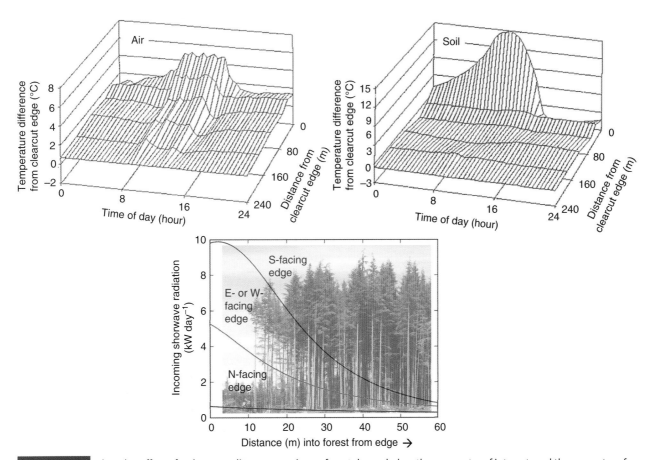

FIGURE 8.15 The edge effect of a clearcut adjacent to an intact forest depended on the parameter of interest, and the geometry of the boundaries. The edge effect for soil temperature was very large, but reached only a short way into the forest. The air-temperature effect was smaller, but reached farther into the forest with movement of air. The penetration of light from the clearcut into the forest depended strongly on aspect, owing to the angle of the sun through the day at this mid-latitude site. **Source:** Based on Chen et al. 1995.

higher than the forest interior. The energy contained in air masses can move into the forests with wind convection, unlike energy in soil that would be restricted to only conduction. Midday air temperatures in the forest were raised up to 100 m from the forest edge. The driver of these changes in temperature was of course the increased solar radiation reaching the ground. The aspect of the boundary between forest and clearcut strongly influence light penetration into the forest. North-facing boundaries showed essentially no extra light reaching into the forest, but south-facing boundaries had increased light up to 50 m into the uncut forest.

The Ecology of Gaps and Edges Affects Animals, and Is Shaped by Animals

Animals respond to forest structure in many ways, with impacts that shape forest structure. A dense forest may provide hiding cover for herbivores, and a gap may provide good forage, so herbivore use of the area may be high near forest edges. A boundary developed in a forest in Cape Breton Highlands National Park in Nova Scotia, Canada after spruce budworm caterpillars killed a patch of balsam fir trees (Franklin and Harper 2016). The microclimatic gradient across the forest edge led to high cover of grasses in the open, and high cover of mosses in the forest (Figure 8.16). The impact of moose browsing on saplings of balsam fir was low in the forest, and high in the openings out to a distance of about 50 m. The intact forest had low numbers of fir seedlings, probably as a result of low light and nutrients. The density of seedlings increased close to the edge of the forest, but was low in the open. The low seedling density in the open may have resulted from both browsing impacts (by moose and other animals), and by competition for resources with other understory plants.

The Location of Each Tree Allows a Wide Range of Assessments of Forest Structure and Processes

Spatial patterns of tree locations have interested forest scientists for more than a century. Aaltonen (1919) mapped the sizes and locations of trees in a number of Finnish forests. The maps illustrated some general tendencies, such as tree locations being more clumped than uniform or random, seedlings also being clumped but away from the vicinity of trees, and zones around trees devoid of seedlings (despite available light; Figure 8.17).

Interest in the details of forest arrangements has continued, and modern instrumentation allows for greater insights and quantitative analysis. Forests are often described with summaries that tally all the trees, such as the number of trees per hectare. The overall tallies can be subdivided, with classes of counts by species. An old-growth forest in western Washington, USA had only 30 Douglas-fir trees ha^{-1} compared with 240 hemlock ha^{-1}, but the total biomass of the two species was similar (Figure 8.18) because the Douglas-fir trees were so much larger (1.1 m diameter at breast height, compared with 0.4 m for hemlock). These forest totals and averages can be

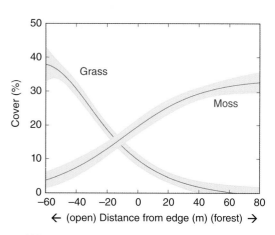

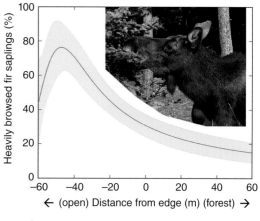

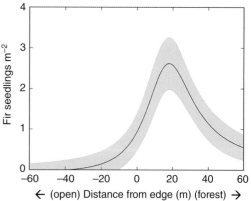

FIGURE 8.16   An outbreak of spruce budworms created dead patches and edges in balsam fir forests. The changes in resource availability (and perhaps microclimate) foster growth of different sorts of plants (upper), and browsing on saplings by moose (middle). The combination of browsing and competition among plants leads to a distinct trend of fir seedlings across the edge. **Source:** Data from Franklin and Harper 2016.

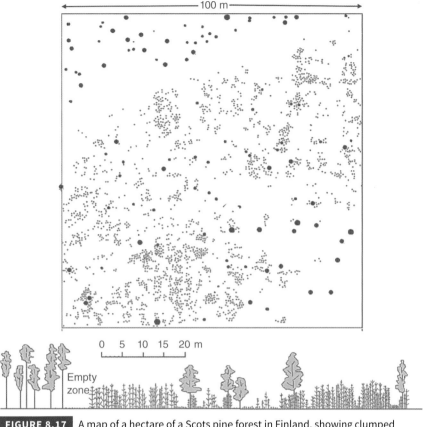

FIGURE 8.17   A map of a hectare of a Scots pine forest in Finland, showing clumped patterns of trees (solid circles, sizes proportional to tree diameters) and seedlings (open circles), and seedling-free zones around clumps of large trees. **Sources:** from Aaltonen 1919 and Kuuluvainen and Ylläsjärvi 2011 / Taylor & Francis; see also Figure 4.11.

broken down in a variety of ways. The biomass was higher in portions of the forest that contained very large trees. At a scale of 25×25 m (0.06 ha), 25% of the forest had more than 815 Mg ha^{-1} of stem mass, and 25% had less than 490 Mg ha^{-1}. Distributions can also be examined based on sizes of individuals (Figure 8.18).

What insights might be gained by breaking down forest-level averages and totals? The distribution of sizes of Douglas-fir trees suggests that most of the trees established at some time in the past, likely when environmental conditions were different (such as after a major fire). The distribution of hemlock sizes would be consistent with some establishment during the period when Douglas-fir established, but also continued recruitment across the centuries under the influence of the current forest. The pattern for western redcedar would indicate a recruitment story similar to hemlock, though with far lower numbers of trees.

Forest-Level Information Can Be Dissected Down to the Level of Individual Trees

The sums and averages of forests can be broken down even further to include the spatial location of each tree. Knowledge of locations can be used to evaluate competition between trees (as in the neighborhood interactions described in Chapter 7). Deeper insights are also possible. The old-growth forest in Washington, USA was stem mapped, with the species, size, and location recorded for each tree (Figure 8.19). Once the location of individual trees is mapped, the distribution of values such as stem mass can be mapped for each species, and for combinations of species (Figure 8.20). Such maps illustrate how the variation is distributed through the forest, which would be important for a variety of management aspects. Areas with high stem mass of one species tended to have low stem mass of the other species.

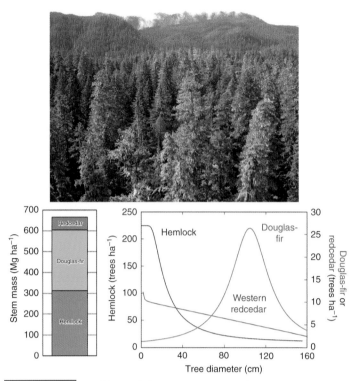

FIGURE 8.18 An old growth forest in western Washington, USA averaged almost 700 Mg ha⁻¹ of stem mass, with roughly equal mass of Douglas-fir and western hemlock trees. The Douglas-fir trees were far larger, but the hemlock trees were more numerous. The size distributions indicate the Douglas-fir trees established long ago (likely after a major fire), but hemlock and redcedar trees continued to establish across the centuries. **Source:** based on Chen et al. 1995, 1999.

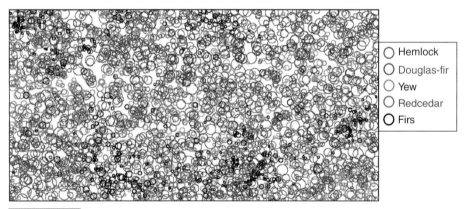

FIGURE 8.19 The species, diameter, and location of all stems were mapped in the old-growth forest (8 ha shown here). Larger circles indicate trees with larger diameters (**Source:** based on Chen et al. 2004). This information could be further sliced for crown structure as a function of heights above the ground (Song et al. 2004).

The spatial patterns of trees can be examined in even more detail (Chen et al. 2004). Stem masses tended to correlate at distances less than about 30 m, but showed random spatial trends at greater distances. If a location had a hemlock tree, other hemlock trees were likely to occur at higher-than-random probabilities up to about 90 m away (Figure 8.21). A location with a Douglas-fir tree was very likely to have other Douglas-fir trees at all distances up to 150 m, but less likely than random to have hemlock trees within 90 m. A Douglas-fir tree taller than 60 m was less likely than random to have another tall Douglas-fir within 20 m. Whether knowledge of these patterns is useful for science or management would of course depend on the questions being addressed.

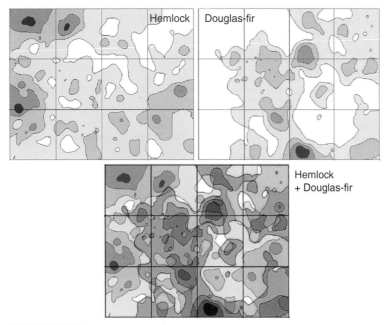

FIGURE 8.20   The map of species, sizes, and locations can lead to other maps such as domains of high or low stem mass for each species (upper two maps) or for combinations of species (lower map). The stem biomass for each species is divided into five equal levels, with darker colors for higher stem mass locations (**Source:** Based on Chen et al. 2004). Areas with high mass of one species tend to be associated with low mass of the other.

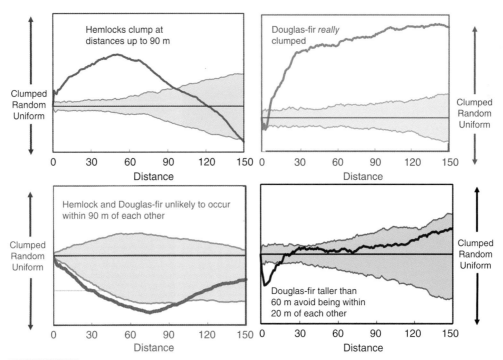

FIGURE 8.21   Statistical analyses can examine how clumped, random, or uniform the spatial arrangements of trees may be, and whether the occurrence of trees is more or less likely to occur near another tree. Both hemlock and Douglas-fir trees were more-than-randomly likely to be near trees of the same species (heavy lines represent probabilities; if the heavy lines fall within the shaded areas, probabilities were not different from random). Hemlocks and Douglas-fir trees were less-than-randomly likely to occur within 90 m of each other; this does not mean they did not occur within 90 m, only that the associations occurred less frequently than would happen in a random mix of trees. **Source:** Based on Chen et al. 2004.

Riparian Forests Are Special and Important, for Different Reasons in Different Forests

"Riparian" is a word used to mean streamside, and riparian forests are important parts of most forested landscapes. The availability of water is higher near streams, of course, and in dry environments this makes the difference between forests versus grasslands or shrublands (Figure 8.22). Water supply is not the only ecological factor behind differences in riparian and upland forests. The long-term flow of water downhill tends to be associated with inputs of sediments, nutrients and alkalinity (countering soil acidity) that drive large differences in biogeochemistry and plant nutrition.

Studies in Scandinavia have documented the important role of soil fertility in the development of riparian forests, even where water supply does not limit tree growth. Aaltonen (1919) diagramed forest changes from upland areas with low densities of trees and understories dominated by lichens and heather, downslope to riparian forests with high densities of faster-growing trees and understories of blueberries and mosses (Figure 8.23). Another topographic sequence was studied in northern Sweden in the late 1990s. The Betsele transect occurred on almost flat topography (about a 2% slope) on sandy glacial outwash material. Over a distance of about 100 m, the vegetation shifted from slow-growing pines with small woody shrub understory plants, to a highly productive forest of spruce with tall, dense herbs. The biogeochemical story behind the gradient included low availability of nitrogen in the upslope pine forest, and very high N (and low phosphorus) in the spruce forest (Giesler et al. 1998; Högberg 2001).

Some sites have soils that are too wet for maximum forest growth. Types of wetland forests include fens with a moderate rate of water flow through the soil, and bogs with very slow water movement. Swamp forests may alternate seasonally between dry, well-aerated soils and water flowing overland. Mangrove forests dominate many riverside and coastal areas in tropical regions. Many ecological stories and details are important for these types of wetland forests, and reading about them in other texts and research papers would certainly prove fascinating.

FIGURE 8.22 Riparian forests differ from upland ecosystems in dry regions, where precipitation alone cannot provide enough water to support trees (upper left, diverse streamside forest in West Clear Creek Canyon, Arizona, USA). In wetter areas, riparian forests still tend to differ from upslope forests because of increased water availability, and typically more fertile soils that develop at lower-slope positions (lower left, China Creek, Vancouver Island, Canada). Riparian forests also develop even in high rainfall environments, as lower-slope soils differ in biogeochemistry and support different communities (and larger trees) than upland sites (right, aspen/spruce forest in northern Sweden).

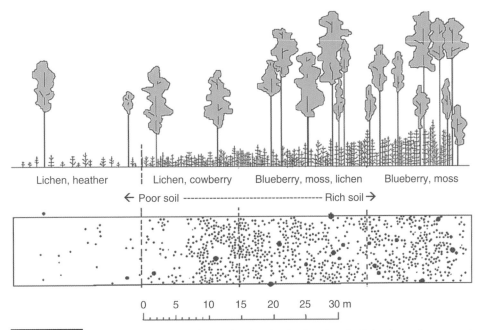

Lichen, heather   Lichen, cowberry   Blueberry, moss, lichen   Blueberry, moss

← Poor soil -- Rich soil →

0   5   10   15   20   25   30 m

FIGURE 8.23 A large gradient in soils from upland to riparian locations in Finland was diagrammed a century ago by Aaltonen (1919; **Source:** Kuuluvainen and Ylläsjärvi 2011 / Taylor & Francis). Note the density and size of understory trees was much higher where density and size of overstory trees was highest, indicating the dominant importance of forest soils (not light) in shaping the forest sequence.

Spatial Patterns Are Important, Even in the Most Uniform Forests

Plantations of fast-growing eucalyptus trees are the most uniform forests in the world. The trees are clones, with all stems being genetically identical. Soils are intensively managed with fertilization (and often plowing), and competing understory vegetation is controlled. The trees are planted in very uniform rows and columns, and they look like identical copies of each other (Figure 8.24). Despite all these reasons for expecting trees to all be the same size, operational plantations do have distributions of tree sizes. The distributions can be illustrated with graphs that describe the population of trees in a site, and in maps that show patterns of tree sizes across the plantations. Larger trees in the plantation in Figure 8.24 grew faster than smaller trees, and trees surrounded by larger than average trees grew more slowly than those surrounded by small trees. After accounting for these patterns in growth, the uniformity of neighborhoods also influenced growth (see also Figure 7.13 for more on the role of uniformity of tree sizes and forest growth). For a given average size of neighboring trees, a focal tree grew faster if those neighbors were all the same size than if neighbors included larger and smaller trees. The subtle differences in uniformity of neighborhood tree sizes altered the growth of large trees by up to 10%, and total plantation growth was about 5% lower than would be expected with perfectly uniform tree sizes.

Forest Classification Is Different in the Twenty-First Century

Many forest classification systems were developed in the twentieth century, some based on empirical patterns that were useful for management, and others based on hopeful (though weak) analogies of hierarchies borrowed from the taxonomy of species. The development of geographic information systems (**GIS**) in the late twentieth century provided opportunities for improving representations of forests, and changing how (and whether) classes of forests might be defined. GIS represents locations with multiple layers of information, often including topography, soils, and vegetation characteristics. Much of the information is derived from remotely sensed data. The spatially explicit systems can be coupled with field measurements in creative, powerful ways. For example, a large area burned in a wildfire of varying intensity across varying terrain would likely have different amounts of exotic species in different locations. Field projects can measure exotic plants at sites selected to represent the large area. The results of the field sampling can be related to site properties within a GIS, and then GIS methods can be used to give the best estimate for every location on the landscape (Figure 8.25). Classifications could be developed to identify high risk and low risk locations, or to estimate risk with 5 or 10 levels. The same approach could be used to estimate the diversity of understory plant species, the species composition of overstory trees, the growth rates of trees, soils, and habitat suitability

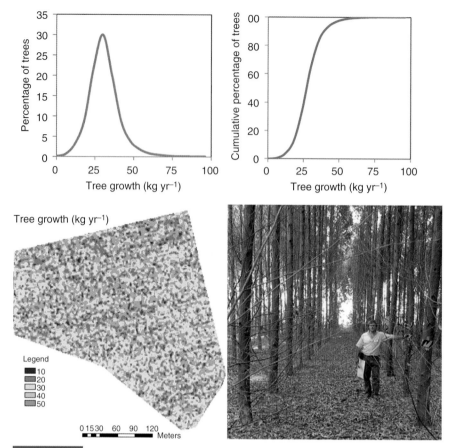

FIGURE 8.24   Clonal plantations of eucalyptus are some of the most uniform forests, but some trees grow faster than others as a result of microsite differences in soils or non-tree competition. The distribution of sizes can be examined in bulk, as a frequency distribution (upper left), or as a cumulative distribution that shows the proportion of all trees with a higher or lower rate of growth (upper right). Even in a three to four year-old plantation with uniform spacing of trees, spatial location information (lower left) revealed the effects of tree sizes on competition, and the loss of growth resulting from imperfect uniformity in sizes. **Source:** based on Trung et al. 2013.

for wildlife species. Each question raised about a forest landscape can be analyzed in a spatially explicit way, without a need to define stands or forest types. Locations with similar conditions for bluebirds and deer may line up differently for tree growth, wildfire risk, or invasive species risk. The old goals of forest classification can now be met better in the twenty-first century, with information and evidence, without resorting to hopeful analogies and single-system classification schemes. Returning to the quotation from Gregory Bateson, modern techniques of spatial information have greatly lessened the necessity of dividing forest landscapes into parts versus wholes.

Ecological Afterthoughts: When It's Not About the Trees

Consider working with these pieces to assemble a puzzle:

- Ponderosa pine trees have thick bark and usually survive low-intensity surface fires.
- Low intensity surface fires occur when grassy fuels dry out and are ignited.
- Grazing cattle eat grass.

Historically the first two pieces of this puzzle would have come together, and forests like the one in Figure 8.26 would be common. What happens when the third piece is added to the puzzle? What might new consequences be? What would be vital if the owner of the forest wanted to take the forest back to the structure that would have occurred two centuries ago?

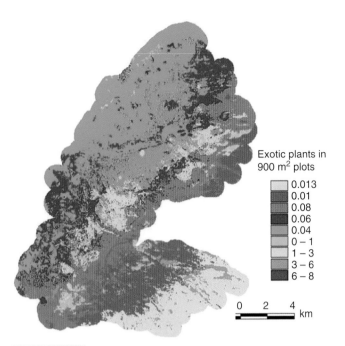

Exotic plants in
900 m^2 plots

0.013
0.01
0.08
0.06
0.04
0 – 1
1 – 3
3 – 6
6 – 8

0   2   4
km

FIGURE 8.25 The probable risk of exotic plant encroachment was estimated across the 20 000 ha Cerro Grande fire in northern New Mexico, USA. Field surveys were matched with layers of site information in a geographic information system, leading to maps of invasive species potential that did not require any assumptions about classification of forests into stands or types (**Source:** Pedelty et al. 2003 / Thomas J Stohlgren). Other questions about the same landscape could be addressed in the same way, leading to different groupings of locations. An approach based on stands or ecosystem types would generally line up predictions for various features, but continuously varying systems may lump some locations similarly for one aspect of forest ecology, and separate them for another. As always, accuracy is a key question for any map.

FIGURE 8.26 The pieces of the puzzle include thick-barked old trees that survive low-intensity surface fires, and grass (and other fine fuels) that dry out and burn once every few years when dry conditions are matched with an ignition source. If cattle were added to the puzzle, grasses (and other low vegetation) would be eaten.

CHAPTER 9

Forests Through Time

The second of the three questions in the core framework asks how a forest came to be the way it is now. Answers to past forest conditions can be investigated with a wide array of methods, and this chapter focuses on the sort of slow, observable changes that happen every year in forests. The next four chapters focus on events that bring rapid changes, and the final chapter tackles "What's next?" in the future of forests.

Sometimes a Classic Story Comes True

In the Rocky Mountains of Canada and the US, aspen and lodgepole pine frequently occur on the same landscapes and in the same forests. Both species can thrive after severe wildfire, with aspen sprouting vigorously from surviving root systems and lodgepole pine regenerating from seeds released from serotinous cones. Aspen grows faster initially than lodgepole pine, benefiting from C stores in the roots. Lodgepole pine can grow taller after some decades. A classic story expects that aspens would dominate the early development of a new forest after fire, and that lodgepole pines (or other conifers) would eventually become dominant and outcompete aspen (Figure 9.1). This story is logically possible, and supporting evidence can be found. A key question for any possible or observed story in forest ecology would be how commonly does it occur? In the Front Range of Colorado, USA, an unbiased survey of forests with aspen found this story might apply to about one-third of aspen forests, based on the presence of an understory cohort of conifers. About half of the aspen forests had understories of aspen stems, with too few conifers for the classic story to apply. The rest of the forests had too few understory trees to provide insight on what species might dominate after overstory aspen trees eventually died.

If the changes in many forests across a landscape were followed over time, there would of course be a statistical average that described the whole landscape. That average would always have a variance, and the size of the variance determines whether knowing the average provides much insight for any given forest. Sometimes factors can be identified that might shift a given forest to deviate in one direction or another from the average. This core method for understanding forest change over time has a higher probability of being useful for real forest landscapes than a hopeful expectation that any single story could be expected to represent most forests.

This chapter considers the changes over time in individual trees, and in forests of many trees of different species. Changes in a single tree result from a single set of genes, and the responses to environmental factors that influence physiology or damage a tree. The possible future states of any single tree are very tightly constrained, limited to features such as how big the tree may become, how many seeds it might produce, and how long it might live. The changes in a forest are far less constrained because genes come and go as species arrive or disappear (as noted in Chapter 3), and the possible future states develop from strong, non-linear interactions between species that change over time.

Forest Ecology: An Evidence-Based Approach, First Edition. Dan Binkley.
© 2021 John Wiley & Sons Ltd. Published 2021 by John Wiley & Sons Ltd.

FIGURE 9.1 A classic story of forest change in the Rocky Mountains begins with a stand-replacing fire in a conifer-dominated forest. Aspen shoots grow rapidly from surviving root systems, along with lodgepole pine trees that develop more slowly. After several decades, the forests have large trees of both species, but the pines overtop the aspen and many aspen stems can be found dead on the ground. After more than a century, dead aspen stems verify that a forest now dominated by pines and other conifers used to have many aspen trees. This classic pattern happens, but it is not a general trend in this region; half the aspen-dominated forests have understories dominated by aspen saplings, not conifers (Kashian et al. 2007).

Long-Term Experimental Forests Provide Knowledge at the Scale of Tree Lifetimes

Trees live for decades to millennia, and a period of a century might begin to be a substantial portion of the lifetime of many trees. Some of the long-term records of forests now exceed a century, giving clear historical records that describe how trees and forests change over time.

The first experimental forest set up by the U.S. Forest Service was in Arizona. The timber industry profited from harvesting old trees, but a lack of regeneration led industry leaders to lobby for the creation of an experimental forest. The Fort Valley Experimental Forest has provided some of the most detailed records of forest change, spanning more than a century. The long-term records from the experimental forest provides answers to two of the core questions in forest ecology: what's up with this forest, and how did it get that way? The current ponderosa pine forests are multistoried, with emergent, large trees that are over two centuries old, and high numbers of smaller, century-old saplings (Figure 9.2) and many smaller trees. The records contain details such as tree sizes, spatial locations, ages, and current growth rates. Long-term plots reveal how the current forest developed over the course of a century. Prior to European settlement, the density of trees was low and the understory vegetation was dominated by C4 grasses (warm-season grasses, with a four-carbon sugar as the first product of photosynthesis) and C3 grasses (cool-season grasses, with a three-carbon sugar). The climate includes summer thunderstorms that provided frequent ignitions of dried grasses and pine needles, with low-intensity fires recurring at intervals of 3–10 years. The recruitment of new pine trees was limited by the frequent fires, commonly restricted to patches that might have escaped for several fire events.

The introduction of heavy livestock grazing removed much of the grass and herbaceous vegetation, interrupting the frequent fires. Without fire, pine seedlings thrived and the grassy matrix of the forest was replaced by a dense forest. Understory vegetation was reduced by severe competition from the high dominance of trees. The annual productivity of the understory dropped by about half, with the C4 grasses and forbs declining by more than 75% (C3 grasses declined less).

What drove these changes in the understory? The changes could have resulted directly from livestock grazing, or from the lack of fire. Fenced plots were established in the early 1900s to see how removal of grazing would affect vegetation (Figure 9.3). The fence line shows a strong difference, though the appearance of late-summer vegetation differences may be larger than the actual differences in plant species composition and growth. By the time the grazing experiments were started, the forest density had already begun to increase substantially after several decades without fire. After a century, the increasing density of trees in the forest had a much larger effect on understory vegetation than grazing (especially as grazing intensity decreased through the century).

The changes recorded in long-term records and experiments give insights on some major interactions in forest ecology. The livestock grazing changed the frequency of fires; the change in fire frequency allowed an explosion of tree seedlings; the denser trees further reduced the understory and fine fuels that might carry low-intensity surface fires (Laughlin et al. 2011).

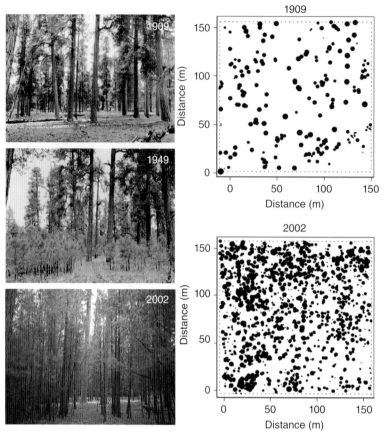

FIGURE 9.2 The ponderosa pine forests in the Fort Valley Experimental Forest in Arizona, USA changed dramatically over time, from the combined effects of livestock grazing (removing fine fuels that carried low-intensity fires) and the absence of frequent fires. **Sources:** 1909 photo by W.R. Mattoon; 1949 photo by G. Pearson; 2002 photo by J. Waskiewicz; stem maps based on Moore et al. 2004.

FIGURE 9.3 After 20 years with no grazing, the impact of heavy grazing by livestock was very clear (upper photo, G.E. Glendening). Differences were sustained over subsequent decades, and a key driver of differences in understory vegetation was the density of overstory trees (**Sources:** photo by J.D. Bakker; from Strahan et al. 2015).

The current condition of the forest, combined with insights on how it got that way, set the stage for asking what is likely to happen in the future. Not all possible future outcomes are equally likely. A low-intensity surface fire is unlikely, owing to a lack of the fine fuels that were historically important (Chapter 11). The old-growth (legacy) trees have persisted through several centuries, but their survival may be impaired by competition from the high density of mid-story trees, which might lead to greater susceptibility to droughts and bark beetles. The probability of a high-intensity, stand-replacing fire is much higher in the current forest than a century ago. Modeling of potential fire behavior indicates that a wind speed of more than 85 km hr^{-1} might be needed to carry a fire through the historical forest canopy, whereas a relatively common wind of only 45 km hr^{-1} might be enough for a severe fire to spread through the current forest (Moore et al. 2004; see also Figures 2.21, 11.14).

When Recorded History Is Not Enough, Tree Rings Can Provide a Record of Both Age and Size

FIGURE 9.4   Henry Thoreau measured the width of the first 50 years of growth on a pine stump (4 in.), and then succeeding 50-year intervals (6 and 7 in.). He calculated that even though the radial increment was greatest in the first 50 years, the basal area growth was highest from age 50–100 (sketch from Thoreau's journal entry on Nov. 1, 1860 superimposed on a pine stump).

Information is needed to understand changes in trees and forests over time, and tree rings are one of the major sources of information in forest ecology. The size of cells formed in xylem varies in response to environmental conditions. Periods of slow growth (such as in late summer or autumn in temperate and boreal forests) are often followed by faster growth (in late spring and early summer). These regular, repeating boundaries between slow and fast-growing cells record information on the age of trees, allowing patterns of growth with age to be estimated. Henry Thoreau measured growth rings on stumps of pines, noticing that the width of the annual rings decreased as trees got older (Thoreau 1860; Figure 9.4). He applied basic geometry to deduce that even though radial growth was highest in a pine tree's first 50 years, the basal area growth was highest from age 50–100 years. Trees in tropical settings typically form visible rings between periods of slow growth and faster growth, but these may reflect patterns in rainfall and not map annual cycles precisely.

Dendrochronology Developed Because There Are No Canals on Mars

At the beginning of the 1900s, scientists had so little knowledge of the universe that speculations and beliefs about the environments on other planets were unbounded. Astronomer Percival Lowell was convinced that Mars had some sort of canal-like features, indicating the likely presence of life and even civilizations (Webb 1979). Andrew Douglass was employed by Lowell, and after some years of peering through large telescopes, he concluded the canals were probably illusions of some sort, either tricks of the eye or of the minds that hoped to see canals. Douglass pushed to have Lowell abandon his writings about civilization on Mars, and so preserve the Lowell Observatory's reputation for astronomy and science. Douglass was summarily fired. Ophthalmologists now suggest that Lowell likely did see real images of faint lines as he stared into his telescope, but they were lens reflections of the blood vessels in his own retina.

Douglass had passed through a sawmill log yard on his daily walks up to the observatory, and noticed the annual variations in ring widths that were not simply related to tree age. He expected the widths reflected the weather in good years and bad years, and came up with a hypothesis that he could extend astronomy's records of cycles in sunspots by measuring cycles of tree-ring widths (expecting that sunspots might influence weather). Like many good hypotheses, this one was refuted because sunspot cycles do not produce strong signals in Earth's weather.

The science of dendrochronology was launched as Douglass refined his quantification of tree-rings. Multiple trees showed similar relative patterns, as they all responded to good years with wide rings, and poor years with narrow rings. Sequences of wide versus narrow rings corresponded to specific sequences of years, allowing tree rings to be dated even in trees where some rings might be missing (owing to very poor weather or other factors) or where some years might have two rings (reflecting weather

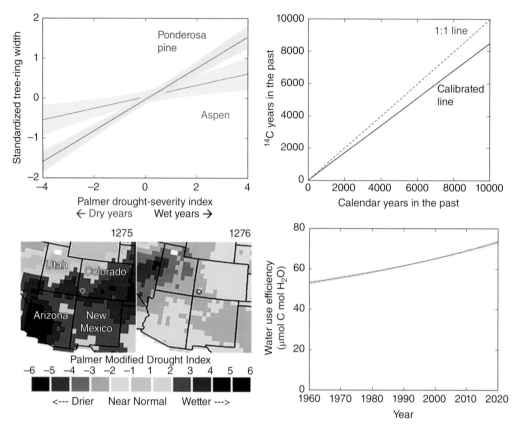

FIGURE 9.5   Tree rings contain information that can be extracted with dendrochronological techniques. The width of rings in ponderosa pine trees on the Kaibab Plateau, Arizona, USA related strongly to drought severity (upper left), while aspen showed less response to droughts or wet periods (**Source:** Based on Binkley et al. 2006). The ^{14}C in tree rings provided necessary calibration to interpret ^{14}C dates (which vary with changes in atmospheric ^{14}C); without this calibration (upper right; **Source:** based on Reimer et al. 2013), archeological dates would be off by about 15 years each century backward in time. Regional drought patterns can be reconstructed, with single-year resolution, far into the past (such as these patterns for two consecutive years over 700 years ago (lower left; **Source:** Based on Bauer 2018). Reconstructions of drought severity showed a prolonged drought coincided with abandonment of ancestral pueblo villages. The ^{13}C stable isotope concentrations in tree rings revealed that the efficiency of water use by trees around the world increased by about 3% per decade in response to rising atmospheric CO_2 (see also Figure 14.11; **Source:** Data from Adams et al. 2020).

fluctuations). The patterns of ring widths could also go farther back in time than the age of living trees, because the early patterns in living trees could be overlapped with the old-age patterns in logs (from either the forest or in buildings).

The history recorded in tree rings can reveal both the tree's response to climate, and once calibrated, can reveal historical climates beyond the period of weather measurements (Figure 9.5). Tree rings of some species are more responsive to climate fluctuations than others; the widths of rings in ponderosa pines track drought very clearly, but rings in aspen do not. Regional droughts can be tracked at annual time scales across hundreds, even thousands of years into the past, providing insights on forest dynamics (such as wildfire years) and archeology. The chemistry in tree rings also contains information on how trees respond to changes in atmospheric CO_2, revealing major increases in photosynthesis per unit of water used (Chapter 14).

Dendrochronology Can Explain Past Forest Structure and Dynamics

Over the years, forest scientists come up with new ideas that could take decades or centuries to challenge with experiments. Sometimes opportunities can be found where information recorded in tree rings can be used for insights without waiting for long-term experiments to mature. For example, an outbreak of a fungal pathogen (*Gremmeniella abietina*) led to widespread damage to

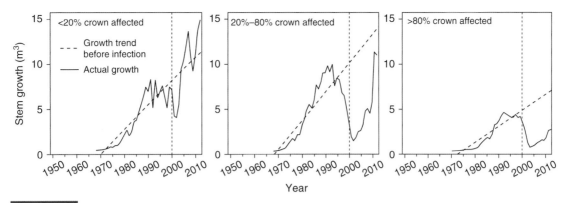

FIGURE 9.6 Tree rings reveal the impact of an outbreak of the fungal pathogen, *Gremmeniella abietina*, on growth of Scots pine trees. Low levels of damage in 2000 were followed by rapid recovery of growth (solid lines) to the level expected based on the pre-outbreak trajectory (dashed lines; left). Trees with damage throughout half their crowns took a decade to recover (middle), and trees with more severe crown damage had not recovered after a decade. **Source:** Based on Wang et al. 2017.

crowns of conifer trees in Sweden. Some trees died, but many survived. How much impact would this sort of pathogen outbreak have on the growth of surviving trees? Is growth reduced for only a short time, rapidly returning to pre-outbreak levels, or is there a long legacy of the damage? Wang et al. (2017) took advantage of records of visual crown damage at the time of the outbreak, and used tree rings to examine tree recovery. Trees with less than 20% of crowns showing damage showed reduced growth for only one or two years (Figure 9.6). Crown damage of 20–80% led to a decade of reduced growth, but the trees did appear to recover to their pre-outbreak level of expected growth. Trees with over 80% of crown damage had not recovered within a decade.

The historical interactions between trees of different species can also be examined with information from tree rings. Painstaking work on cores from trees of different species and sizes examine patterns (and challenge ideas) in relation to species, tree sizes, and also to the spatial arrangements of trees. A 200-year-old forest in New York, USA was dominated by eastern hemlock trees, along with trees of yellow birch and sugar maple (Figure 9.7). Overall, the growth of all three species was affected equally by weather. Large hemlocks grew much faster than smaller hemlocks, but tree size had less influence on growth for the other two species. Yellow birch trees were most sensitive to the numbers and sizes of neighboring trees, and hemlock was the least affected.

The differences in growth between large trees and small trees, young trees and old trees, are fundamental to understanding how trees and forests develop over time. These "background" trends need to be understood before questions about unusual factors and events (such as pollution or outbreaks of pests) can be gauged and answered. One of the most impressive dendrochronological assessments of long-term tree growth comes from very intensive work on cores from giant Sequoia trees in California, USA (Figure 9.8). The diagram examines three types of growth: ring width, cambium surface area, and stem volume. An overall average is computed for each index, and the average is set at 0. Growth that falls below average falls below 0, and above average growth

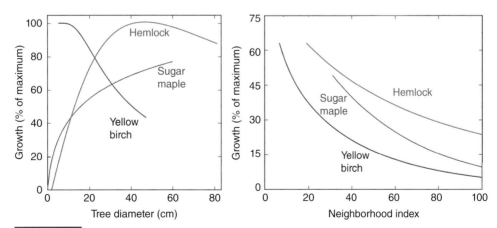

FIGURE 9.7 Sixty to eighty trees of each of three species were cored in a 200-year-old forest in New York, USA. The growth of each tree was related to its size (left) and to the sizes and number of neighboring trees (right). Smaller yellow birch trees grew faster than larger trees, whereas hemlock showed the opposite pattern. Hemlock was also less responsive to neighboring trees than the other two species. This case study illustrates just a small portion of the information that can be derived from tree rings. **Source:** Based on Bigelow et al. 2020.

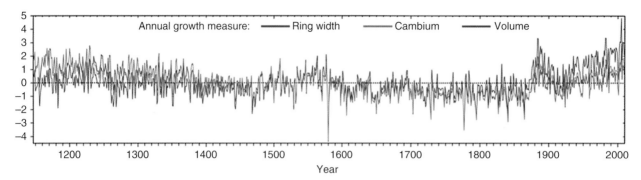

FIGURE 9.8   The annual growth of eight giant sequoias was determined based on cores taken at a range of heights through the trees. Overall average growth was set at 0, and periods of higher or lower growth were scaled in relation to the standard deviation across the centuries. Growth in terms of ring width (radius) and cambium surface area were higher in early centuries, when trees were smaller. Growth in volume (and mass) was highest in the most recent 150 years, when trees were largest. **Source:** from Sillet et al. 2015b / John Wiley & Sons.

rises above 0. The units on the Y axis are standard deviations. About two-thirds of all observations fall within ± 1 standard deviations, 95% fall within ± 2 standard deviations, and more than 99% fall within ± 3 standard deviations. The growth in terms of ring widths was higher in the early centuries of record, which is typical for smaller-diameter trees (as in Figure 9.2). The surface area of a cylinder (such as a tree stem) would follow the same trend as diameter (but would include the effect of increasing height), and so the cambial surface area under the bark grew fastest in early centuries. The increase in stem volume, however, occurs in three dimensions, and careful tracing of ring widths through time throughout the height of the trees showed that volume (and mass) growth was highest in the most recent century. As noted in Chapter 7, it's common for the largest trees in a forest to continue to increase in growth even at very old ages.

Darwin's Ideas Contributed Very Little to Early Ideas of Forest Change (Unfortunately)

Forests are always changing, either so slowly it takes most of a lifetime to realize how much has changed, or so rapidly that it seems unfortunate or even catastrophic. If all forests change, there must be averages to the rates and types of change. The averages could result from deterministic processes that work uniformly across forests, constraining forest changes to a narrow range. Or the averages could just be statistical, with forests showing broad ranges of changes. A century ago, ecology had such high hopes for deterministic explanations that the field of vegetation ecology was swept down a path away from mainstream science.

The path of thinking about how systems change over time might be traced back to the Scottish economist, Adam Smith. Part of his foundational book on *The Wealth of Nations* was an idea that a market has many competing sellers, and sellers with more effective products would thrive and others would go out of business. Charles Darwin made an analogy between Smith's market and the biological world (blending in some thinking from Thomas Malthus), and developed his ideas about species descent with modifications. Variations among individuals would confer higher and lower probabilities of survival and reproduction, gradually leading to the evolution of species. As noted in the preface, analogy is one of the weakest forms of argument, and the domains of economics and biology might be quite different systems. Darwin did not stop with an appealing analogy. He spent more than a decade compiling evidence that might support or refute his idea. How could the pieces of an eyeball evolve, since vision depends on a complete, functional eye? Darwin finally presented his idea in a book, and the book focused primarily on the weight of evidence; he did not rely heavily on the analogy with Smith's market.

Decades later, ecologists tried to figure out what scheme lay behind the changes in forests and other ecosystems (the word "ecosystem" would not be invented until a couple more decades passed). These ecologists had the clear example of Darwin, who started from an idea rooted in a weak analogy and then built up massive evidence, including evidence that might prove the idea to be faulty. Unfortunately for the history of ecology, the prevailing choice of ecologists was to believe in the power of an analogy between the growth of an organism and the changes over time in forests. In North America this analogy was captured in the work of Frederic Clements who asserted: "Succession is the process of reproduction of a formation, and this reproductive process can no more fail to terminate in the adult form than it can in the case of the individual." Entire books were built on this analogy, elaborating ideas, terms and piling up examples of how the core process works. The evidence compiled was used to illustrate and elaborate the conceptualization, not challenge its validity, or to ask if there really was any core process in reality. For the most part, these ecologists did not emulate Darwin's approach. They did not emulate his painstaking method of examining evidence (both for and against), and they missed the implications of evolution by natural selection for how forests actually function and change. The appeal of a compelling, simple story was so strong that the words "succession," "seral" (part of a series), and even "climax" still pop up in most ecology textbooks and even some scientific journal papers.

The actual changes in forests include maturation of individual trees, plants, and animals, but overall the changes of the whole system entail too much flexibility (of genes coming and going with species arrivals and removals) and indeterminate conditions for the idea of maturation to have much value. Sometimes confidence in an analogy can be worse than having no idea at all. A belief that forests have a core trajectory that must be followed leads to expectations that deviations from the core trajectory are undesirable and may lead to collapse of what "should" have been a balanced, healthy forest. This chapter summarizes evidence on how forests actually change, and how these dynamic, indeterminate systems shift across time (and space, Chapter 8) to produce real forests.

Chronosequences Are a Shortcut to the Future, But They May Be Unreliable

Changes in forests across space are easy to sample, and indeed all the forests of the world have been measured or scanned, at varying levels of resolution. Changes across time present different challenges, as many major changes develop only slowly over very long periods. After more than a century of intensive work in ecology and forestry, a large body of evidence has accumulated about types and rates of changes in some forests. Before long-term evidence became available, though, scientists hoped to speed up time by substituting space for time.

A scientist may have a question about how a forest will change over two centuries, and she might hope that examining forests of different ages in different locations would give a good representation of how a particular forest (or set of forests) would likely change over time. This space-for-time substitution is a chronosequence, and it could work well for situations where the future state of a forest is not influenced by historical legacies or future contingent events that may or may not happen. A classic example would be studies of how forests developed in Coastal Alaska, USA, as warming climate in the 1800s led to retreat of glaciers. Melting ice exposed new gravelly outwash slopes and plains that plants colonized, and it's possible that an area laid bare a century ago might depict how a recently opened area will develop over the next century. William Cooper used this approach to document how newly open areas would be colonized by small plants, then by shrubs of willow and nitrogen-fixing alder, and finally spruce which might persist as young spruce trees replaced old trees.

Chronosequence studies hope to identify trends over time by comparing sites of different ages, and each site in a chronosequence can be followed up with resampling at later dates to see if the expected trend was actually followed. Cooper established permanent plots that have now been followed for a century, and the chronosequence trends generally did not happen (Buma et al. 2017). Small plants did establish more quickly than some larger shrubs or trees, in part because of seed dispersal and the fact that small plants attain full size faster than large plants. Willows colonized most of Cooper's plots within a few decades, but further change seemed to be limited to willows getting bigger (Figure 9.9), rather than willows giving way to spruce. Spruce trees established only in a site where willow was largely absent, suggesting (not surprisingly) that competition between individuals of different species likely impedes establishment of new individuals (willows inhibited spruce).

Another sort of chronosequence has been studied intensively near Fairbanks, Alaska, where flooding of the Tanana River creates riverside benches that are colonized by plants (Figure 9.10). When floodwaters recede, colonization by plants depends on the arrival of seeds. Colonization is less challenging than after glacier melting, as the river-deposited sediments already have a notable amount of nutrients and good water holding capacity. The benches receive more sediments from floods in later years, especially as flood waters swirl around vegetation, losing speed and dropping suspended sediments. The higher the benches rise above the water, the less frequently are floods large enough to add new sediment. Moving away from the river from one bench to a higher bench gives the impression of moving through time. These bench sequences are not just about time, however, as higher benches are also farther above the water table and have massively larger amounts of soil. Low benches are dominated by willows, N-fixing alders and poplars, and the highest benches by white spruce. Will the vegetation of a low bench shift over two centuries to become a forest dominated by spruce? Would spruce be able to colonize the benches if soil development was not first driven by other species?

Three decades of measurements, coupled with experimental treatments, found that some simple features of the chronosequence were supported, but other trends were not. Tall trees did appear to reduce the presence of some short-vegetation species (such as some species of willows). After the first several decades, forests on different benches (of the same age) tended to diverge over time, especially when beavers removed most of the balsam poplars from some (but not all) of the study locations. The density of white spruce declined over two decades, and younger benches would not have enough white spruce to reach the densities found on older benches. The expected recruitment of black spruce (along with incipient permafrost formation) did not happen.

View downslope:

View upslope:

FIGURE 9.9 Photos from Cooper's research plots in Glacier Bay, Alaska, USA. Two decades after original measurements, willows were becoming well-established, and after 100 years willows dominated most of the landscape (with a few isolated spruce, and some N-fixing alders). The expected shifts in dominant vegetation predicted from the chronosequence mostly did not happen (**Source:** Buma et al. 2017, used by permission).

FIGURE 9.10 Benches are created on river floodplains, and newly deposited sediments are colonized by plants. The velocity of water in later floods is slowed by the vegetation, leading to deposition of more sediments. Infrequent large floods can add sediment even to the higher benches. The lowest benches along the Tanana River in central Alaska, USA are dominated by herbaceous plants and colonizing willows and balsam poplars, as well as some white spruce (foreground). After a few decades, the woody plants get taller, with faster-growing poplars dominating over slower-growing spruce and shorter willows (background, left). Benches that have accumulated sediment over two centuries are dominated by white spruce trees (background, right). Long-term measurements and experimentation have shown some trends expected from the chronosequence studies were supported, but others were refuted. Each bench seems to be following a separate trajectory, aside from the simplest patterns associated with tree sizes (most got bigger) and age (surviving trees all increased in age) (**Source:** photo provided by Terry Chapin).

Alder trees were expected to decline over time, but the number of alders actually increased on all benches. Overall it seems each bench had some unique changes (different trajectories for the major species), rather than each following the same common trend (Hollingsworth et al. 2010).

Experiments also failed to support the idea that species that dominate on young benches benefit species that dominate on older benches. White spruce trees that were planted on young benches died or grew more poorly when planted near N-fixing alder (Chapin et al. 2016). Soil enrichment with N might eventually have a benefit for surviving spruce on older benches, but substantial amounts of N are deposited with the flood sediments, and alder-added N may have little opportunity to increase spruce growth.

Strong Chronosequences Require Large Numbers of Replicates

Given all the factors that influence forest growth over time, and the factors that lead to differences in growth from one site to another, chronosequences typically need large numbers of observations before strong confidence is warranted in the overall trends. The lodgepole pine forests in Yellowstone National Park, USA are dominated by a single species, with relatively minor differences in soils across large areas. Most of the forests originate after high severity fires kill all the trees, with serotinous cones providing seeds for the next forest generation. A chronosequence with more than 70 locations identified the likely age-related pattern in tree growth (Figure 9.11; this is the same study used in Figure D in the Preface). Growth rates increased to peak levels in about 20–25 years, followed by a very rapid decline that continued to drop over the next two centuries. The shaded area in Figure 9.11 indicates the 95% probability zone for the "true" age-related trend (subject to assumptions about space and time that are present in all chronosequences). What if the scientists had a small budget, and could only sample the growth rates in five stands? Random subsets of five sites across the age range all showed high growth rates for young forests, followed by declines. However, the five random sets of five sites gave very different estimates of the peak rate of growth and the rate of growth of old forests. Other forests may be more complex, with higher variation among locations over time. Would it be worthwhile to spend even a small budget if the final patterns might be wrong by a factor of two? Sometimes there are no easy paths toward reliable knowledge.

Growth Always Declines in Old Forests

One of the most universal patterns in forests is that growth of wood increases soon after major events lead to the establishment of a new forest. The increase is easily explained by the establishment of roots and leaves to obtain resources, supporting increasing

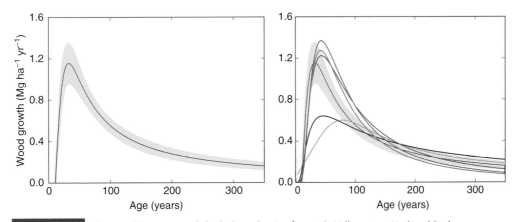

FIGURE 9.11 The trend in tree growth for lodgepole pine forests in Yellowstone National Park was developed from over 70 locations, with varying times since the last stand-replacing fire. Growth increased rapidly as new trees developed large crowns, and then declined by more than two-thirds over the next two centuries (left). Five random subsets, with five sites each across the full age range, showed that the trend of a peak in growth followed by a decline was robust, but the actual estimates for maximum growth, and the growth of old forests, differed by as much as twofold. Low numbers of samples might lead to poor conclusions. **Source:** Based on Kashian et al. 2013.

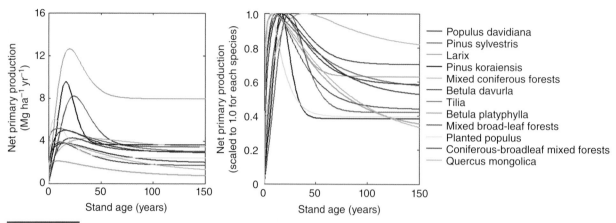

FIGURE 9.12   The growth of young forests increases as roots and leaves increase the acquisition of resources. Growth rates peak reached when trees have full crowns and root systems, but this level is not sustained. Growth declines by one-third or more, almost as rapidly as the increase in growth in early times. This example comes from a dozen forest types on sites with medium productivity in Heilongjiang Province, China. **Source:** Based on Yu et al. 2017.

rates of photosynthesis. The increase is followed by a peak (or sometimes brief plateau), followed by a decline of typically 30–60% (Figure 9.12).

If the decline in wood growth is large and happens in all forests, what factor (or factors) is behind it? The production ecology approach in Chapter 7 can be used to formulate testable hypotheses. A decline in growth might result from:

A. a decrease in resource supply, such as declining soil nutrient supply;

B. a decrease in use of resources, such as a decline in leaf area to capture sunlight;

C. a decline in the efficiency of using resources, such as declining photosynthesis per unit of water transpired; or

D. a shift in partitioning of C, such as away from wood toward roots.

These possibilities were tested in an experimental plantation of eucalyptus trees in Hawaii, USA (Figure 9.13). The full C budget was measured across a six-year rotation, as the trees grew taller than 30 m. One treatment involved repeated fertilization to test whether a decline in soil nutrient supply might be important. Wood growth showed the expected pattern of peaking and then declining by more than half. The fertilization treatment prevented any nutrient limitation, and did lead to substantially higher wood growth, but it did not prevent the decline in wood growth. The high rainfall ensured no water limitation, and there was no trend in rain or sunlight through the rotation, eliminating hypothesis A. Measurements of the use of water and light showed no decline over time, eliminating hypothesis B. The production of roots declined along with stem growth, as did the respiration of stems, eliminating hypothesis D. The only hypothesis consistent with the data was that wood growth declined because of the decline in efficiency of resource use (C, lower photosynthesis and wood growth per unit of resource). The largest trees continued to increase in growth and use resources efficiently, so the decline in wood growth resulted from lowering efficiency of resource use by medium- and small-sized trees.

An experiment such as the Hawaii growth investigation leads to more questions. One of the most important is whether the outcomes from this experiment would apply to other eucalyptus forests, or other types of forests in general. The key role of declining resource use efficiency by non-dominant trees leads to questions of how trees could achieve higher efficiency when younger, and why that efficiency wasn't sustained. Good experiments often show that some ideas are not consistent with evidence, while others remain plausible. Good experiments also tend to raise new questions that dig deeper into how forests work.

People Change How Forests Change Over Time

All forests across the planet are influenced by people. Some extreme examples are intensively managed plantations, with trees of genotypes blended for fast growth, resistance to insects and diseases (Chapter 12). Competition with other plants is limited, and soil fertility is enhanced with fertilizers. The least impacted forests might be in remote locations where people have limited engagement, but even these places are affected by warming climate, rising CO_2 concentrations, and typically by altered wildlife populations (with follow-on effects on tree species). The remotest boreal forests no longer have browsing mastodons and mammoths, likely as a result of human hunting.

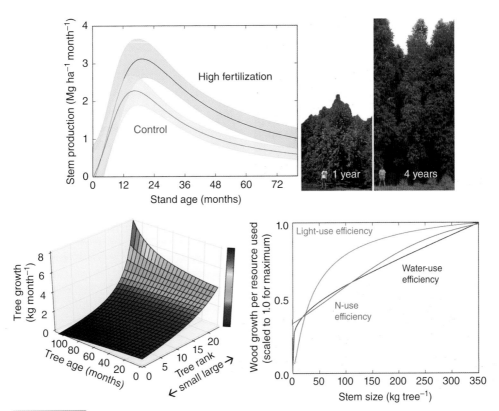

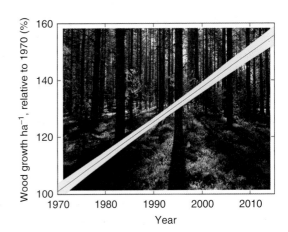

FIGURE 9.13 Stem production of a eucalyptus plantation in Hawaii, USA declined by more than half after reaching a peak; high fertilization increased growth, but did not prevent the decline (upper figures). The decline in growth was very large for small and medium trees (lower left), and the largest trees showed the highest efficiencies of growing wood per unit of light, water, and nitrogen used (lower right; **Sources:** Data from Binkley et al. 2002; Ryan et al. 2004).

Managed forests around the world typically increase in growth rates as a result of improved management techniques. The boreal forests of Scandinavia have increased in growth rates by 1% or more annually over the past 50 years (Figure 9.14). Foresters may not deserve all the credit for increasing forest growth rates, because other factors have also changed. Growing seasons have lengthened, concentrations of CO_2 in the atmosphere have increased, and air pollution has increased inputs of some limiting nutrients (such as N). One evaluation of forest inventory data across Finland concluded that about two-thirds of the increase in productivity came from silviculture, and one-third from changing climate and other factors (Henttonen et al. 2017).

Forests also change as a result of changing land use practices. Many parts of the world experienced deforestation in historical times, as demands for food subsistence led to replacing forests with croplands. Connections between farms and markets often had a complex effect on forests: farmers could sell any crops that exceeded their own needs, but consumers could also gain access to food produced far from towns and cities, reducing reliance on local farms.

A 250-year story from New England, USA illustrates some of the common features when forest and agricultural land uses meet. European settlement of a forested landscape that would become part of Harvard Forest led to removal of the white pine/hemlock/broadleaved forest by cutting, by burning, and by laborious stump removal (Figure 9.15). After a century most of the landscape was meeting human needs through livestock grazing of meadows, some was used for plowed row crops, and a small portion remained forested (with trees occasionally harvested). The land use was mostly for subsistence of farm families, with some income generated by selling agricultural products in local villages. The development of canals and roads connected villages and cities with agricultural lands at greater distances. More productive and efficient agriculture in distant

FIGURE 9.14 Growth of forests across Sweden increased consistently by about 1.2% annually over the past five decades. **Source:** Data from the Swedish National Forest Inventory, Nilsson et al. 2014.

FIGURE 9.15   A total of 250 years of change are illustrated in these diorama models at the Fisher Museum at Harvard Forest. The landscapes originally covered by white pine, hemlock and broadleaved species were largely deforested to make pastures for cattle and sheep, and plowed fields for crops. White pine trees dominated the forests that regrew on abandoned agricultural lands. When the pines were logged, surviving understory broadleaved species dominated the next forests. **Source:** Images from Harvard Forest archives, of dioramas created in the 1930s to illustrate changes in landscapes.

places lowered commodity prices and led to abandonment of poor-quality farms. Trees reestablished in the meadows and especially the plowed fields, with impressive growth of valuable white pine trees. Other species also established in the understories over the decades. Many of the white pine stands were harvested in the early 1900s, and rather than regenerate to pine stands, the surviving understory trees flourished and produced forests dominated by broadleaved species.

The deforestation and reforestation illustrated in the dioramas in Figure 9.15 can be joined with historical records to provide overall quantification of land use, and spatial distribution and dominance of tree species to support two major conclusions. Forests in this region occupy about 80% of the landscapes that were forested before European settlement, documenting the ability of forests to regrow even after decades (or centuries) of agricultural use of the soils (Figure 9.16). But the forests are not repeating the composition, age, or structure of the historical forests. Beech was once a major species, especially in the western portion of the state of Massachusetts, but it's now only a minor species. Oaks also declined, but are still abundant, while maples expanded considerably. These changes of course influence wildlife species, and wildlife species influence both seed distribution and which tree species persist to transition from seedling to tree.

Forests around the world are influenced by the same social, economic and ecological factors that determined the changes in Massachusetts. Just like ecological systems, the more complex social/economic/ecological systems do not tend to repeat strongly over space or over time. Each situation has particular events and characteristics that interact in non-linear ways to determine the overall dynamics of the systems. Deforested landscapes across much of Europe also included later reforestation, as the economic and social aspects of land use altered. The landscapes of the Belgian Ardennes have been influenced heavily by people for thousands of years, and the changes continued to be dynamic in historical times (Figure 9.17). In the mid-1800s, some heathlands were converted into agricultural land uses and others shifted to broadleaved forests and then coniferous forests (Figure 9.16). Improved agricultural management led to higher crop yields, and the region increased in specialization in raising cattle for milk and meat. Wetter portions of the landscape that were in peatlands and wet meadows declined. Changing forests (and changing land use) are not limited to history; the forest landscapes represented in Figures 9.15–9.17 will continue to shift into the future. As with most aspects of forest ecology, details always matter and generalizations need to be done with humility.

One final point about deforestation, agriculture, and reforestation is that human population is not a simple driver of the changes. The reestablishment of forests in Massachusetts occurred while the population doubled, and human populations soared in Europe as many marginal agricultural lands returned to forests. Human populations, demands for products from landscapes, and forest changes are related in strong ways, but the interactions are too complex for simple ideas about cause and effect to be useful.

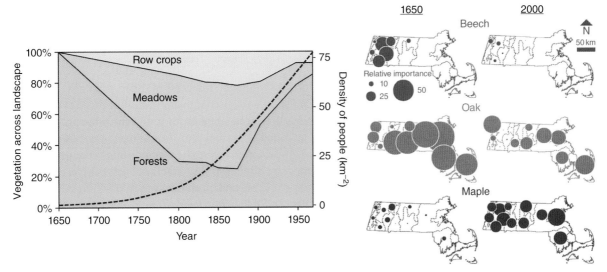

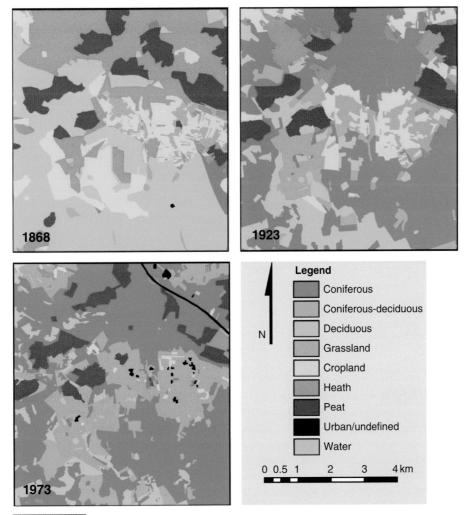

FIGURE 9.16 Deforestation removed two-thirds of Massachusetts' forests within 200 years of European settlement (left). The economics that favored row crops and livestock grazing in meadows shifted as transportation systems brought competition from more efficient farms (on better soils) from outside the region, leading to farm abandonment and reforestation. The reestablished forests differ substantially in species composition (right), age and structure from the historical forests.
Sources: land use and population density based on Foster and Motzkin 1998; species importances based on Hall et al. 2002.

FIGURE 9.17 Land use in the Belgian Ardennes changed the forests in major ways over the course of a century. Changes were driven by economic opportunities for a wide range of products and uses, and these changes will no doubt continue to be large in the future. **Source:** from Ramankutty et al. 2006 / Springer Nature.

Time Scales of Forests and Human Planning Do Not Always Match

Management plans for short-rotation forests typically produce forests that match the goals of the forest plans, because the time frames of management decisions, investments, and operations are short enough that markets and the goals of the forest owners do not shift much. Less-intensively managed forests develop across time scales of a century, and the influence of people comes from a blend of intentional management and changes that are unintended and unforeseen. A large company may enter contracts to manage forests sustainably on publicly owned land, but a century-long rotation is certain to include large changes in markets, ecological conditions, and the goals of the forest's owners. Large companies often own large tracts of land, but few companies persist as long as a century-long rotation of a forest. When the time frames of people and businesses are so different from the time scales of forests, responsible forest management is a complex system that cannot fit within fixed, static management plans (see Franklin et al. 2018, for some insights that evolved over decades of wrestling with these challenges).

Over the Long-Term, Forests Have Not Changed As Predicted

Change in forests is not random, as some changes are clearly more likely than others. On average, forests increase in biomass over time, and longer-lived species usually gain dominance in forest canopies. Even these typical trends get interrupted when events such as fires and storms kill large trees. Other expectations could be speculated, and in some cases data might be available to challenge the speculation. The Duke Forest in North Carolina, USA has many long-term forest plots with more than 80 years of data. Scientists working on forests in the region advocated ideas about forest change, based on concepts and some evidence from space-for-time substitutions. The long-term evidence showed the early speculations were largely unrelated to what really happened in the forests over time. One speculation was that young forests may differ somewhat in species composition, but that strong ecological influences would lead the forests to become more similar (converge) over time. The species composition of the long-term plots in Duke Forest actually became less similar over time rather than more similar (Figure 9.18). Plots located on drier portions

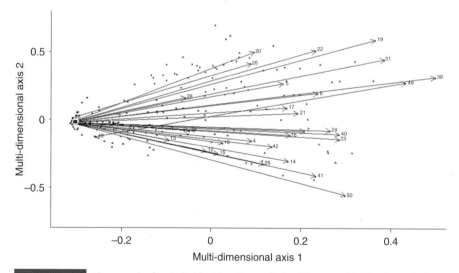

FIGURE 9.18   The growth of an individual tree is constrained by unavoidable, deterministic interactions among the parts of the tree. If forest change was similarly constrained, we might expect that differences in species composition would decline over time as the unavoidable processes shaped the forest toward the somewhat-deterministic future state. Forests are not like organisms, and eight decades of change in plots in the Duke Forest that were initially dominated by pines showed increasing divergence in the composition of overstory trees. The two axes are abstract representations of the species composition of canopy trees, the points represent individual samples in time for each permanent plot (numbers identify plots), and the arrows show how each plot "moved" in the abstract space over the eight decades. The distance between plots increased on both axes, emphasizing the great increase in dissimilarity rather than the speculated convergence. Clearly no process or set of processes pushed the forests to become more similar. **Source:** Based on Payne 2018.

of the landscape shifted in species composition toward the composition of forests on wetter landscape positions, rather than to the expected composition for older forests in drier locations. Plots dominated by loblolly pine trees were expected to shift in composition to match forests that were originally dominated by broadleaved species. The dominance of pine trees did decline, but the suite of broadleaved species did not match that of broadleaved forests decades in the past. Some species that once were abundant across the landscape became less common (across all forest types), while others increased. Oaks and hickories were expected to increase in dominance over time, but these species actually declined overall. Young beech and red maple have increased across the landscape, both locally and regionally, for reasons that are not well understood. The early idea that a suite of strong ecological interactions would push future forests along conceptual pathways turned out not to be useful, though the long-term plots established by the early scientists provided a wealth of information about actual changes in forests.

Why did the landscapes in Duke Forest not follow the conceptual pathways envisioned decades ago? There are two direct reasons, and one other broad, important reason. The first reason deals with how well the ecology of individual species was understood. The sciences of tree physiology, tree growth, and competition between trees of different species were far less advanced than today. The second reason is that unexpected events had large impacts and long legacies on the forests. Two major hurricanes toppled large trees, changing the understory environments in ways that shaped later development of the forests. The population of white-tailed deer was near zero decades ago, but it exploded after the 1970s, leading to near total removal of herbaceous plants and severe reductions in tree seedlings of palatable species. The deposition of nitrogen in precipitation likely increased soil fertility, and rising CO_2 concentrations might benefit some species more than others.

The broad, important reason that the forests did not change as expected was that forest changes are generally not deterministic. Some changes may be more likely than others, but trends are not very robust, and year-to-year interactions among trees of the same or different species routinely are joined by unpredictable (yet not unexpected) events that have large impacts and long legacies. Predictions about the future state of a forest can be made with confidence only if the predictions allow room for small and large surprises. So once again, details always matter and generalization needs to be done with humility.

Ecological Afterthoughts

Changes in forests over space and time were considered in separate chapters in this book, but of course both dimensions operate in the same forests. The two dimensions of space and time come together in an example of lodgepole pine forests in a large watershed in Yellowstone National Park, USA. In the 1730s, the landscape was dominated by forests that were more than two centuries old (Figure 9.19). The forest was a mosaic of patches, ranging in size from about 1 ha to over 100 ha. A few years after these studies were published, a very large fire burned across more than 25% of the Park, and all of this watershed shifted into a single age class of young forests. Some thoughts to consider include: why would the age structure across a watershed be important? What might be the implications of a matrix of many-aged patches of forests compared with a very large, single-aged forest? How might species other than lodgepole pine respond to the changes in space and time?

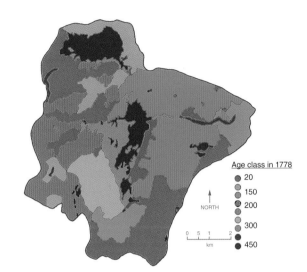

Age class in 1778

- 20
- 150
- 200
- 300
- 450

NORTH

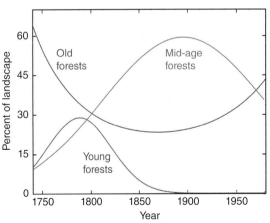

FIGURE 9.19   The vegetation in a 73 km² watershed in Yellowstone National Park, USA in 1738 was dominated by old-growth forests of lodgepole pine, with mid-age and young forests each accounting for about 10% of the landscape. Large forest fires reduced the extent of old-growth forests to less than 30% for 150 years, and the landscape was dominated by mid-age forests. **Sources:** based on Romme and Knight 1982, map based on Romme 1982 / John Wiley & Sons.

CHAPTER 10

Events in Forests: Wind, Insects and Diseases

Winds, insects and diseases affect the background, routine changes in forests. Winds affect trees by cooling leaves, thinning the boundary layer of air next to leaf surfaces (increasing uptake of CO_2 and release of water), and shaping the taper of tree stems. Trees and forests host hundreds to thousands of thousands of species of insects and other arthropods in each hectare (Basset et al. 2012), and many of these eat pieces of trees at one life stage or another (Figure 10.1). All forest trees are infected by mycorrhizal fungi, and these generally helpful fungi are joined by a very diverse array of other fungi that can decay wood and leaves while trees are still alive. These features of winds, insects and diseases influence the slow changes that occur in forests over long decades (Chapter 9), but this chapter considers the rapid events that create major changes in the structure and function of forests.

Recalling Bateson's quotation about parts and wholes (Chapter 3), it may be convenient to separate the factors driving rapid changes in forests into categories of wind, insects and diseases. This does not mean these driving factors work independently. The effectiveness of insects in eating and killing trees often depends on weather, including whether a recent windstorm occurred (see later). Winds topple trees more readily if they are infected with wood-rotting fungi. When trees die from drought, the final agent of death is often insects or diseases (or both). Fires that burn with strong winds are more severe and cover larger areas. The coverage of winds, insects and diseases in this chapter takes a case study approach, illustrating some case-specific details along with some important interactions that apply more broadly.

It's Remarkable That Trees Can Stand Up to Strong Winds

After one has seen pines six feet in diameter bending like grasses before a mountain gale, and ever and anon some giant falling with a crash that shakes the hills, it seems astonishing that any, save the lowest thickset trees, could ever have found a period sufficiently stormless to establish themselves; or, once established, that they should not, sooner or later, have been blown down.

(John Muir 1894)

Annual average wind speeds are commonly between 4 and $10 \, \mathrm{m \, s^{-1}}$ ($14–36 \, \mathrm{km \, hr^{-1}}$). Winds that can break or uproot trees generally need to be higher than $25 \, \mathrm{m \, s^{-1}}$ ($90 \, \mathrm{km \, hr^{-1}}$; Figure 10.2), and severe forest damage occurs with wind speeds above $35 \, \mathrm{m \, s^{-1}}$ ($125 \, \mathrm{km \, hr^{-1}}$). Of course the level of damage from a given wind speed depends on details, including tree species, soil factors (including moisture content), and forest structure.

FIGURE 10.1 Insects (and other arthropods) comprise a large part of the species in forests, and many of them eat parts of trees. The aspen leaf (left) shows the track of frass (insect poop) left behind as it ate layers of cells with the leaf. About 80% of leaves of European beech show damage from insects (right), with holes comprising about 6% of leaves (Gossner et al. 2014). Wood-rotting fungi commonly decay the heartwood in aspen trees, with larger trees showing more "heartrot" than smaller trees (lower), and some regions showing more decay than others (**Source:** based on Binkley et al. 2014). Each of these represent typical, on-going effects of insects and diseases in forests, not the rapid events that are the subject of this chapter.

Tree Stems May Break or Uproot

Why did the trees in Figure 10.3 blow down? They both withstood storms for more than a century, and the final winds that brought them down may have come in storms that were not more severe than earlier ones the trees survived. The tree on the left was in a forest where most of the large surrounding trees had died during a bark beetle outbreak. When the tree was surrounded by crowns of neighbors, the force experienced from storm winds on its crown was lower. The wind velocities and forces dramatically increased when this tree's crown became the only sail in the neighborhood. The tree on the right blew down even though the canopy from surrounding trees was still intact. In this case, the tree's anchorage in the soil declined in recent years as wood-decaying fungi ate through the roots of the tree. Some storms have winds that are stronger than any previously experienced by the current forest, taking down many trees rather than isolated individuals.

Wind stresses on the crowns of trees are resisted by the strength of stems, and by the ability of the root system to remain secure as the tree transmits the wind force to the soil. Trees sway in the wind, deflecting some of the horizontal force of the wind. Palm trees are particularly flexible; videos of palm trees in hurricane-force winds show the crown bends over to nearly horizontal, and the leaves press together, reducing the lateral forces on the stem and avoiding breakage.

Where root systems are deep or well-anchored in rocks, severe winds break tree stems, leaving a stump (short or tall) rooted in the soil. Where the resistance of the stem to breakage exceeds the strength of roots in the soil, trees are uprooted. The legacy of broken-stemmed trees comes from the effect of the fallen stem and crown on surrounding vegetation, and to some extent the soil changes that occur as the stem decomposes. Uprooted trees have a much larger effect on soils, creating pits where the tree once stood, and adjacent mounds of soils as the soil material sloughs off the upraised root ball. In some forests the uprooting of trees is common enough, with effects that last long enough, that the ground surface has a distinctive "pit and mound" structure.

Severe windstorms damage and kill many of the large trees in forests, but the impacts need numbers to be understood clearly. Hurricane Hugo hit the Island of Puerto Rico as a severe Category 4 storm in 1989, bringing both high winds and rain. The

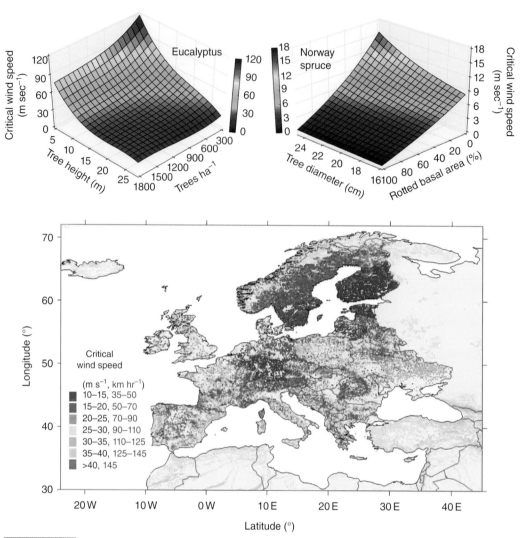

FIGURE 10.2   The risk of wind damage to trees and forests can be examined in relation to a variety of factors. The wind speed needed to snap eucalyptus stems in a simulation model was about 25–30 m s⁻¹ for trees taller than 20 m, regardless of the density of trees in the forest (upper left). Forest density was more important for critical wind speeds for shorter trees; trees growing at low density have greater stem taper and greater resistance to wind stresses (**Source:** Based on Locatelli et al. 2016). The critical wind speed for breaking a 20-m tall Norway spruce tree depended on tree diameter, and the percentage of the stem cross section that was rotted by fungi (upper right; **Source:** from data in Honkaniemi et al. 2017). Regional mapping of wind risk across Europe shows a general pattern of lower critical wind speeds for areas with denser forests to the north, though local details of species composition, forest structure, and soils influence local risks (lower figure; **Source:** Based on Gardiner et al. 2013).

combination is particularly important for the risk of uprooting by trees, as the strength of soil declines as it becomes wet and saturated. The forest impact of the storm was considered to be very large, even though 85% of the trees survived. About half of killed trees were broken off, and half uprooted (Figure 10.4). The average effect across a large forest of course has variation around it, and with windstorms it's common for patches (a few hectares) to have high mortality, and others to very low mortality.

Storms Blow in with a Wide Range of Wind Speeds

The distribution of wind speeds through the days of a year, and across years, shows a typical pattern of mostly light-to-moderate winds, and periods of high winds. Within a given time, the average speed of measured winds is a combination of background wind speeds, and gusts that may be two to four times the background rates. These wind patterns drive background patterns of tree mortality, with perhaps 0.25–0.50% of the large trees in a forest blowing over each year (Franklin and DeBell 1988). When a

FIGURE 10.3 Severe winds break the stems of well-rooted trees (left, lodgepole pine), especially after the loss of neighboring trees increased exposure to wind. The anchorage of a tree depends on many soil factors (soil depth, rock content, and water content) as well as root factors. The tree on the right (Engelmann spruce) blew down because the root system was too weak to withstand the force of wind, as a result of fungal decay of the previously sufficient root system.

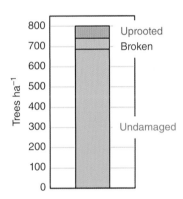

FIGURE 10.4 Hurricane Hugo blew across Puerto Rico as a Category 4 storm (winds up to $45\,m\,s^{-1}$, $160\,km\,hr^{-1}$) uprooting and breaking about 15% of all trees in a 16-ha plot in the Luquillo Experimental Forest. The pattern of tree mortality was very clumped, with extensive damage in small (<0.5 ha) patches, and less damage in the broader matrix. Sources: Based on data from Zimmerman et al. 1994; photo by Allan Drew, from Turner et al. 1997, used by permission.

tree is toppled by wind, the wind may only be the final agent following years of weakening of stems and root systems by wood-rotting fungi.

Severe storms are generated by atmospheric conditions that range from very local to continental. Thunderstorms can have microburst downdrafts that blast winds across several ha to a few thousand ha. Wind speeds in microbursts can range from about 40–$70\,m\,s^{-1}$ (150–$250\,km\,hr^{-1}$), more than enough to topple trees. Tornados can have wind speeds that are even higher, exceeding $80\,m\,s^{-1}$ ($300\,km\,hr^{-1}$). The wind speeds in hurricanes and typhoons are lower than in the most severe tornados, but the vastly larger size of the storms creates in far more damage.

Storm Impacts Can Be Severe in Local Areas

Microbursts and tornados do not affect large areas in any given year, but the impacts on forests are large and long-lasting where they occur. Tornados often leave long linear swaths (from a few hundred meters wide to more than one kilometer wide) across landscapes, where almost all large trees are uprooted or snapped off (Figure 10.5).

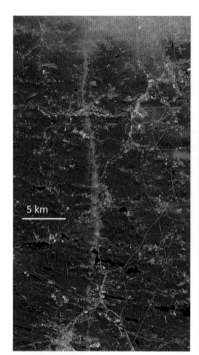

FIGURE 10.5 A tornado in Massachusetts, USA in 2011 toppled most of the large trees in a strip that was about 500 m wide and 60 km long (**Source:** left, from NASA Earth Observatory). A view looking along the path of this tornado, with severe but localized impact on the forest (right; **Source:** photo by Michael Southam, https://commons.wikimedia.org/wiki/File:Tornado_Damage_at_Sturbridge.JPG).

Storms that Are Severe Enough to Be Named Are Strong Enough to Topple Vast Numbers of Trees

A storm named Lothar crossed southwest Germany in 1999, and intensive forest measurements illustrated some typical trends that would apply to other storms (Figure 10.6). Wind speeds were higher near the Atlantic Ocean, and so were levels of forest damage. Taller trees were more likely to be killed than shorter trees (which is not surprising), but the risk associated with being tall declined inland as a result of lower wind speeds. Some species had much greater damage (for a given tree height) than others, and again the patterns varied across the geographic gradient. The impacts of this particular storm can be compared with the impacts of many storms across the same area. Records from more than a century of monitoring (with up to 1000 growth-and-yield plots, Albrecht et al. 2012) showed that the strongest predictor of wind damage was tree height, followed by tree species, with smaller influences of forest structure (recently thinned forests had somewhat higher risks of damage). Factors related to topography and soils did not seem to be very important. The factors that led to higher damage as Lothar blasted across the landscape do indeed seem to be the important factors whenever strong winds occur.

How Large an Area Can Be Covered by a Single Storm?

This simple question may not have a simple answer, and it might not be a very good question anyway. Large storms can cover more than 100 000 ha, but the impacts of the storms are so variable that the distribution of effects (including severity and patch sizes) have more ecological significance than the total area within the perimeter of the storm.

Another large windstorm in 1999 swept across the Boundary Waters Canoe Area of Minnesota, USA, and Ontario, Canada and continued eastward for more than 1500 km in a single day. More than half the trees were blown down in some areas, and light-to-moderate damage was very widespread (Figure 10.7). A severely impacted area in the Boundary Waters Canoe area was 50 km long and 6–10 km wide. From an ecological perspective, what are the impacts and legacies of such a storm? The storm reduced average stem volume across the area from about 80 m³ ha⁻¹ to 50 m³ ha⁻¹. Note that more than half the volume (and mass) of trees survived this high-severity storm. The mass of woody material on the ground increased by several fold, and this could have implications for fire behavior in later years. It might seem that severe damage of overstory trees would increase the diversity of understory plants, but in fact any effect appeared small (and diversity also did not relate strongly to tree basal area even in undamaged plots). What's

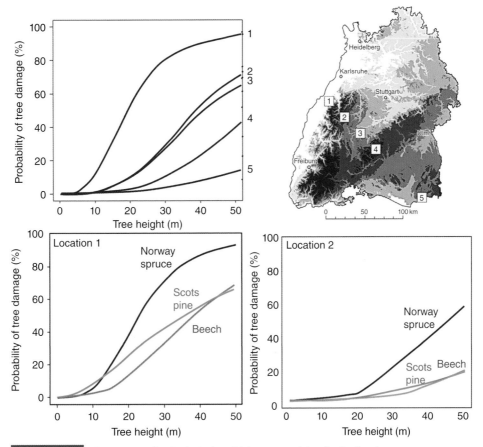

FIGURE 10.6   The storm Lothar had winds as high as $50\,m\,s^{-1}$ ($200\,km\,hr^{-1}$) as it swept across southwestern Germany in 1999. The risk of tree mortality was of course higher near the ocean where wind speeds were greatest, and higher for taller trees. The risk of mortality also varied among tree species, as well as location. **Source:** Based on Schmidt et al. 2010.

FIGURE 10.7   An analysis of over 200 forest inventory plots documented the ecological impacts and legacies of a severe windstorm in the Boundary Waters Canoe area in Minnesota, USA. The lake would not have been visible from the photo point before the storm. **Source:** from Moser et al. 2007.

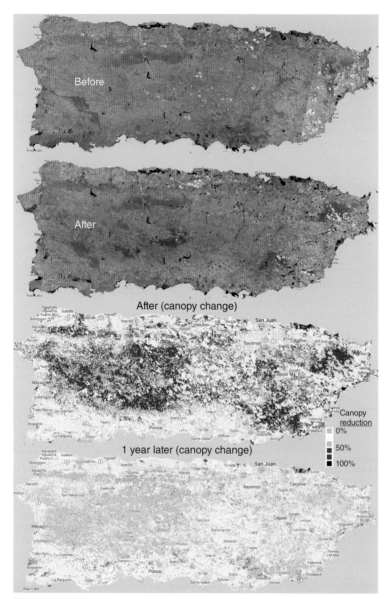

FIGURE 10.8 Hurricane Maria stormed across Puerto Rico in 2017, dropping 100–1000 mm of rain with winds up to 70 m s^{-1} (250 km hr^{-1}). The extent of the storm's impact is apparent in the before-and-after satellite photos (upper pair). Most forests across the Island had 50% or greater loss of canopies but canopy recovery was rapid (lower pair; **Source:** based on Feng et al. 2020, and https://ylfeng.users.earthengine.app/view/forestdisturbancemapinpr and https://ylfeng.users.earthengine.app/view/forestdisturbanceafterhurricanemaria).

next for these forests? Of course that depends on whether droughts and fires occur within coming decades, and whether other factors such as wind, insects and diseases have strong influences. The canopy of the forest likely returned to pre-storm levels in 5–10 years, and it might take about 25 years for the average biomass to reach the pre-storm average.

Another example of the massive area affected by a single storm is Hurricane Maria which swept over the Island of Puerto Rico in 2017 as a Category 4 storm. Satellite imagery showed that forests across the island were severely damaged (Figure 10.8), with most forests losing more than half their leaves. The recovery of forest leaf area is rapid in tropical forests, and within a year the canopies had recovered and forest growth rates were likely close to pre-hurricane levels. The long-lasting effects of the hurricane included massive amounts of woody material on the ground, and the legacies of species composition that were launched by the rapid change in the forest (Figure 10.9). Another long-lasting impact of the storm resulted from huge numbers of landslides. Densities of landslides ranged between 10 and 30 landslides km^{-2} over most of the forest area of the Island.

FIGURE 10.9   Airborne lidar images of Puerto Rico forests before (upper) and 7-months after (lower) Hurricane Maria showed that 40–60% of large trees lost large branches or were toppled in the storm, creating openings in the canopies (and soils?) that provide opportunities for shifts in species composition and structure in subsequent decades. **Source:** from NASA's Earth Observatory, https://earthobservatory.nasa.gov/images/144441/a-haircut-for-puerto-ricos-forests; for animation, see https://dl.acm.org/doi/abs/10.1145/3302502.3319428.

How Massive Can a Storm's Impact Be?

Again, this simple question should not be expected to have a single or simple answer, but another case study can provide some context. A mid-winter storm dubbed Gudrun blew across Sweden, Denmark, Latvia and Estonia in 2005, with sustained wind speeds of over $35\,\mathrm{m\,s^{-1}}$ ($125\,\mathrm{km\,hr^{-1}}$). More than 80 million $\mathrm{m^3}$ of stems were toppled (Gardiner et al. 2013), enough wood to build more than 2 million wood-frame houses. The impacts can be examined more closely for more insights. The volume of wood blown down in Sweden equaled about 1–2% of the average annual harvest across the whole country. A focus on the hardest-hit region showed the damage equaled about 3% of the annual harvest rate. At the smaller size of a county, the damage equaled about 6% of the annual harvest, and the worst-hit forest district (smaller than a county) had damage equal to about 20% of the district's annual harvest. The impacts of the storm cascaded beyond the simple volumes of wood to include lowering of wood prices (as the high supply of salvaged wood leads to lower prices), and even more injuries to workers employed in salvage logging (which is more hazardous than normal tree harvesting). Other effects could include reduced opportunities for harvesting non-timber products such as mushrooms and berries, and perhaps reduced access for wildlife hunting. A decade after the storm, the forest biomass across southern Sweden had fully recovered; the annual area planted with pine seedlings more than doubled, but spruce remained the species of choice for 75% of the planting despite higher risks of wind damage (Valinger et al. 2019).

The impact of wind on forests and forest industry can be large, and indeed wind is responsible for the death of more trees in Europe than any agent other than harvester machines. Across the 200 million ha of forests in Europe, the annual mortality from storms averages less than 50 million $\mathrm{m^3}$ (Figure 10.10) compared with over 400 million $\mathrm{m^3}$ harvested by machines.

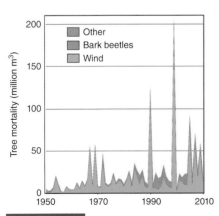

FIGURE 10.10   Most of the non-harvest mortality of forests across Europe results from windstorms, with much lower amounts resulting from bark beetles, and even less from fire and other events. **Source:** Based on Gardiner et al. 2013.

When Will the Next Storm Come?

Chapters 8 and 9 presented case studies of Harvard Forest and Duke Forest, where long-term expectations of forest change were sidetracked by hurricanes. Long-term predictions of forest change that focus only on chronic, annual processes are likely to miss the big events that rapidly change forests and launch very long-term legacies. There is no way to predict when big events will come to specific areas, but even random processes can inform some expectations of the future.

A frequently used idea about random processes is the return interval for an event. A once-in-a-century storm might be expected to happen about once a century. A more useful way to characterize such a situation would be to say that the probability of the event in a single year is 1:100, or a 1% probability. This sounds like it is a statement about probability of occurrence in time, but such expectations also need to have a clearly defined spatial component. If an event has a typical size of 1000 km^2, then a forest of 100 000 km^2 would be likely to have more than one such event each century (perhaps even one each year).

A random event that has a probability of 1% of occurring this year might seem to imply a certainty that one event will occur sometime over 100 years, but this is not how random probabilities work. An estimate of the probability of an event sometime within the century would be calculated as the series of probabilities of it *not* occurring in each year. If the probability of occurrence is 1%, then the probability of not occurring is 99%. The probability of a sequence of years all having no occurrence would be calculated as $0.99 \times 0.99 \ldots$ for each year of the period of interest. So the probability of a once-in-a-century event failing to occur within a century would be $(= 0.99^{100})$, or 37%. So once-in-a-century events happen one or more times in about 2/3 of all centuries, and fail to happen even once in about one-third of the centuries.

What if we would like to estimate the return interval for an event when the length of records is much shorter than the return interval? This is of course tricky, but one approach is to use a version of the space-for-time substitution that was described for chronosequences in Chapter 9. An example comes from the Government Land Office surveys of the United States in the 1800s. Surveyors crisscrossed large territories, recording tree species and forest conditions at fixed intervals. Schulte and Mladenoff (2005) used those historical records to summarize forest conditions in northern Wisconsin, prior to any substantial influences of European settlement. The surveyors occasionally entered areas showing somewhat recent blowdowns, and the blowdowns occupied about 0.01–0.2% of the survey locations (varying among regions and forest types). This would scale up to a return interval for blowdowns of 500–10 000 years. A key assumption of such calculations is that events are random and independent from each other. If the survey had been repeated in 2000 for the Boundary Waters Canoe Area (Figure 10.7), the temporal return interval based on a spatial pattern would have been far shorter than that calculated from a survey of the same area in 1998.

The Next Storm Will Be Different Than the Last One

Big events like major windstorms cannot be predicted in space and time, and it turns out that even if it were possible to know when one would happen, any forecast of the impacts would remain very unclear. Wind speeds vary among storms, and much of the damage to trees depends on the gusts rather than storm averages. The direction of winds also matters, especially if topography is not flat. The Island of Jamaica experienced 12 major hurricanes in one-and-a-half centuries, with each one differing in important details such as general direction of the hurricane, the direction of topographically modified local winds, and the speed and duration of the storms (Figure 10.11). The impact of storms also depends on the structure of forests; some are more wind-firm than others, and legacies of past storms include differences in forest structure that partially determine the impacts of the next storm. As with many aspects of forest ecology, the best that can be done with understanding the effects of severe winds is to characterize the current structure of a forest, understand the legacies that led to the current structure, and then be very cautious indeed when trying to predict the future.

Trees Provide the Dominant Structure of Forests, But Small Insects Can Play a Very Major Role

All forests contain many more species of insects than trees. The life cycles of most insects contribute to the background processes in forests, including flower pollination and feeding on various plant tissues. Many species of insects (and arthropods in general) interact with species of bacteria and fungi, either mutualistically or antagonistically. Sometimes the life cycles of insects amplify to the point of having major effects on forests, such as caterpillars consuming most of the leaf area of a tree, or a beetle boring into the cambium and introducing a fungus that kills the tree.

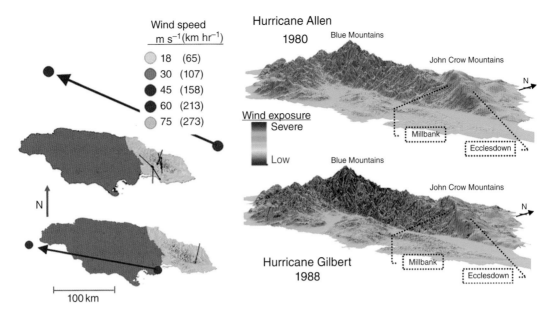

FIGURE 10.11   Hurricanes occur frequently on and near Jamaica, but the storms do not repeat each other. Variability in the hurricane winds combine the effects of topography and forest composition and structure to ensure forest ecology does not follow simple repeating cycles. The maps of the Island for two hurricanes (left) show similar directions of hurricane travel (brown vectors), with Hurricane Allen passing to the north, and Hurricane Gilbert passing across the Island. The spinning storms combined with topography to shift local wind directions. The wind exposure of forests was much more severe for Hurricane Gilbert.
Source: modified from McLaren et al. 2019, used by permission.

The forest pictured in the left photo of Figure 10.12 was dominated mostly by lodgepole pine trees that established after a very large, intense fire in the mid-1800s. The forest could be characterized by measuring many traits, such as the average wood volume per ha, and the spatial patterning of tree patches across the entire landscape in the photo. The history of the forest could be examined based on historical records during the period after European settlement began, perhaps historical photographs, and a wide range of information hiding within rings of living and dead trees. Such an investigation would show the forest reached a peak rate of growth near age 50 years, and that the current growth rate would be about 60% of that maximum. Other tree species would be present, including scattered subalpine fir and Engelmann spruce, along with stems and patches of aspen. Moving on to the third question of the core framework, we might ask, "What's next?" for this forest. One expectation might be that the trends over the past 150 years would continue for decades, but tempered by the realization that forest-replacing fires are routine for this type of forest. Probably no one would have expected that five years later most of the large lodgepole pine trees would have been killed by bark beetles, and that many lodgepole pine forests throughout the Rocky Mountains would have the same rapid change. Mountain pine beetles are just one species out of the almost uncountable suite of insects in forests, but this single species can illustrate some of the factors that are important in the ecology of insects and forests.

The autumn of 2020 was unusually dry in the Rocky Mountains of northern Colorado, and a severe windstorm (winds in excess of $25\,m\,s^{-1}$, $100\,km\,hr^{-1}$) propelled a human-ignited wildfire across more than $70\,000\,ha$. Would the fire have behaved differently if the ignition and windstorm had occurred when the forest structure matched that of 2002 (before beetles) or 2007 (during the beetle outbreak)? This is a difficult question for developing evidence. The fires in 2020 did in fact burn primarily through forests with large numbers of trees that died from beetle attacks in the previous 20 years. However, most of the forests in the area experienced similar beetle impacts, so there would not be an opportunity to compare fires in areas with and without beetle-killed conditions.

How Do Tiny Insects Manage to Kill Large Trees?

Mountain pine beetles bore through the bark of trees, and trees produce resin that might force the beetle back outside. A tree that can produce less resin might not overcome the beetle, leading to a successful invasion. Two key aspects of this balance would be how much resin a tree can produce, and how many beetles simultaneously bore in and trigger resin release. Trees that are stressed by drought or heat may produce less resin, and invading beetles release chemical pheromones to attract other beetles and help overwhelm the tree's defenses. Trees can produce more resin to make up for the resin lost in pushing out the beetles, but beetles can counter this by introducing fungal spores (carried in special pouches on their legs) that grow hyphae (fibers) that penetrate the

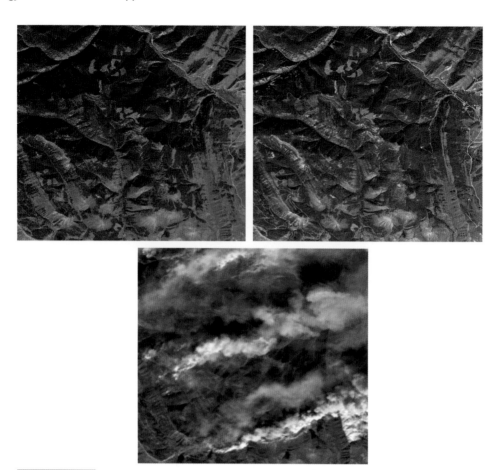

FIGURE 10.12   The photo on the left (10 x 10 km) shows a 150-year-old forest of lodgepole pine in Colorado in 2002. Five years later (2007), most of the large trees were dead as a result of fungal infections spread by mountain pine beetles. Thirteen years after that (2020), a severe wildfire burned across 70,000 ha, including the area in this image (**Source:** images from QuickBird images from Digital Globe).

tree's sapwood and interfere with water supply to the crown (Figure 10.13). Low populations of beetles are present in most forests, with occasional population outbreaks leading to the death of patches of susceptible trees. Sometimes the population grows so large that the defenses of even healthy trees are overwhelmed, and the population declines only after most of the available host trees have died.

Which Trees Are Most Vulnerable to Mountain Pine Beetles?

This question is answered in Figure 10.14 for ponderosa pine, showing that the more vulnerable trees were either the larger ones or the smaller ones. The seeming contradiction arises when mixing up apples and orchards. Within a single forest, the larger trees (the level of apples) tend be more susceptible to beetles. When comparing across forests (the level of orchards), those that have larger trees (because of thinning) may have healthier trees, and show lower beetle mortality.

Which Forests Are Most Susceptible to Mountain Pine Beetles?

Mountain pine beetles interact with single pine trees, but some patterns emerge at the forest level as a cumulative result of many individual tree responses. Figure 10.15 illustrates some of the results from a forest thinning experiment where four decades of repeated thinning led to forests with contrasting average tree sizes and densities. The experiment was within a landscape with very high populations of beetles. Beetles killed about one-third of the pines in the lowest density plots (with the largest trees), and almost all the trees in the highest density plots (with the smallest trees).

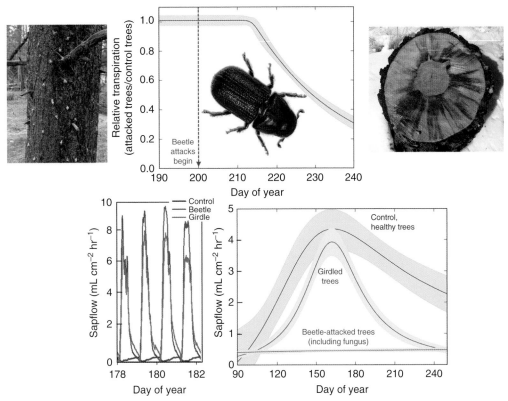

FIGURE 10.13 Bark beetles harm trees primarily by interrupting water flow from roots to crowns as the blue stain fungus clogs the sapwood. Attacked trees often have lumps of resin where the trees attempted to repel the beetles (top left), and the fungus brought in by beetles permeates sapwood and kills trees by stopping water flow to the crown (top right; **Source:** USDA Forest Service). Transpiration of beetle-attacked trees began to decline about two weeks after the attack began (upper graph; image of 5-mm long beetle by Steve Clarkson). An experiment the following year tested the effect of tree girdling (along with continued monitoring of beetle-attacked trees from the year before). Simple girdling of trees had much less effect on water transport than the combination of beetles and fungi (**Source:** based on Hubbard et al. 2013).

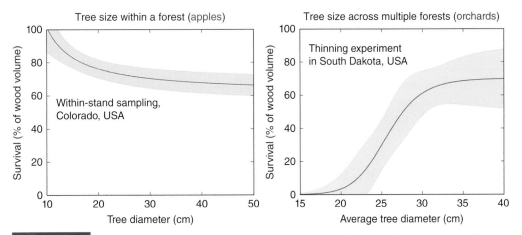

FIGURE 10.14 Larger trees may be more susceptible, or less susceptible, to bark beetles than smaller trees. This apparent riddle comes from confusing apples with orchards. Within an individual forest, larger trees are often more susceptible than smaller trees (left, a comparison among apples; **Source:** from data in Negrón 2020). A tree that is large because it is healthy with few competing neighbors is less susceptible, so thinning experiments that give plots with few, but large stems have higher survival than plots with high densities of smaller trees (right, a comparison across orchards; **Source:** from data in Graham et al. 2019).

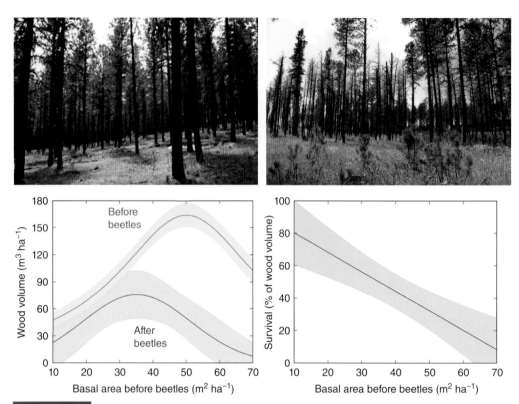

FIGURE 10.15 A four-decade experiment examined the effects of repeated thinning to maintain various densities and tree sizes of ponderosa pine in South Dakota, USA. Mountain pine beetles decimated the study site, as shown in the photos of a moderate-density plot before and after the beetle outbreak. Survival of trees (and forest volume) depended very strongly on basal area; low-basal-area forests had about 25% mortality, whereas the highest basal-area forests had over 90% loss of volume (**Source:** based on data of Graham et al. 2019).

Mountain Pine Beetle Impacts Are Consistent When Scaled Up to Regional Areas

Patterns and processes in forest ecology may be quite different in a local case compared to a regional scale. In the case of mountain pine beetle impacts on forests dominated by lodgepole pine, the patterns are actually quite consistent. A sampling of forests across the Rocky Mountain region of the USA revealed a very strong pattern of increasing mortality in denser forests, though this pattern depended on how mortality was assessed (Figure 10.16). Forests with more stems had higher numbers of stems killed by beetles, but the percentage of killed trees actually declined. The risk of death increased strongly with increasing tree size, showing a very tight overall relationship across the region.

Tree Death Alters Environmental Conditions at Local Scales, But Less at Watershed Scales

The death of a large proportion of trees in a forest might lead to substantial changes in environmental conditions, from the scale of individual trees up to watersheds. The loss of canopy leaf area would be expected to lead to more light reaching the soil surface, less transpiration from trees and more evaporation from soil surfaces. Forest harvesting typically increases stream flow, so death of trees in beetle outbreaks might be expected to increase stream flow too. These sorts of expectations warrant confidence if they are backed up by actual measured responses. The ground does get more light after beetles have killed trees, and soils warm up (by perhaps 1.5 °C) and sustain higher moisture levels (perhaps 50% higher; Reed et al. 2018). Along with more shortwave radiation reaching the ground, more longwave radiation is emitted to the sky. Less snow accumulates in crowns in winter, and more on the ground, but the evaporation losses to the atmosphere may not be very different. At the scale of watersheds, streams draining valleys with

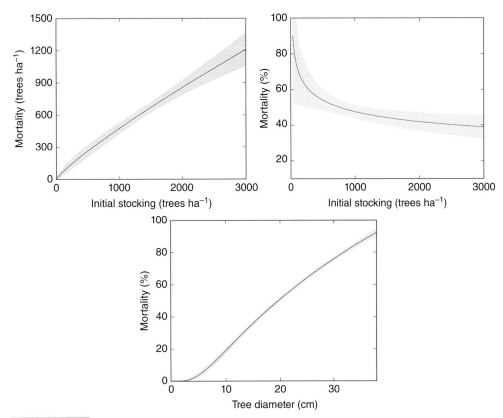

FIGURE 10.16   Across the Rocky Mountains in the USA, tree deaths from mountain pine beetles increased with increasing forest density (upper left). The percentage of trees that died was actually higher in lower density forests (upper right). Across the region, larger trees had far greater probabilities of death than smaller trees (lower; **Source:** from data in Audley et al. 2020).

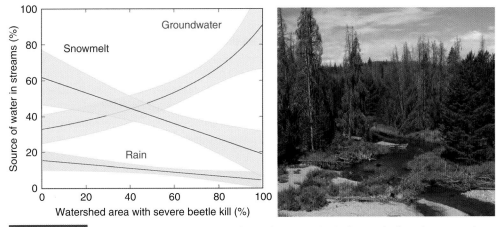

FIGURE 10.17   Twenty-five watersheds across the Rocky Mountains in the USA had varying proportions of their area affected by beetle-killed pines. With the death of trees, the water entering streams shifted from rapid runoff of recently melted snow to water that had lingered longer within the soil and subsoil (groundwater). These patterns can be traced based on isotope ratios of hydrogen and oxygen in water (Sources: from data of Wehner and Stednick 2017, photo by Karl Malcolm, US Forest Service).

higher mortality from beetles were fed primarily by ground water, and low-mortality streams contained mostly recent snowmelt water (Figure 10.17). Do all these changes add up to changes in streamflow, as is common for forest harvesting? Probably not. The available evidence does not point to any widespread, consistent changes in streamflow (Biederman et al. 2015; Slinski et al. 2016). Why don't streams draining beetle-mortality forests show the same responses as harvested forests? The possibly important differences include differences in the amount of disruption of understory vegetation and soils, and the absence of impacts of heavy machinery and roads. A confident answer to this question would require good evidence.

Why Don't Beetles Kill More Trees?

This question does not have a simple answer, but some of the important aspects could be framed for insight and investigation. The evidence presented above showed that larger trees within a forest are more susceptible to beetles, though a forest with larger trees might be less susceptible if the larger size resulted from better growing conditions (such as fewer competing neighbors). Historically the higher risk for larger trees was guessed to relate to the thickness of the phloem, either as suitability for beetle (and fungus) habitat or as a covariate with the ability of trees to produce resin to impede beetles. Direct assessments have not quite supported this cause-and-effect explanation for the commonly seen pattern. Trees differ in many ways that affect beetles, from the smoothness or roughness of bark (which might affect the ability of beetles to begin boring), to resin production rates, to the chemistry of resins, and to the ability of the blue-stain fungus to plug xylem. These features might be influenced by environmental factors, and by genetic factors that vary among individuals.

If outbreaks of beetles are common, why don't populations of lodgepole pines evolve to be less susceptible? Questions about evolution need to be considered carefully, as it's easy to have fuzzy logic and unrealistic expectations. Evolution does not select among competing organisms as a way of aiming toward any outcome; natural selection just means that organisms with poor traits are more likely to be taken out of the gene pool (along with those with bad luck). Lodgepole pine trees can produce viable seeds before the trees are large enough to be very susceptible to beetles, which constrains the chance for selection based on resistance traits. A bigger point is that beetles (and their fungi) go through many generations for each generation of tree, so natural selection can be complex in relation to timescales of generations among interacting species. A final point is that some pines may already be genetically more resistant to bark beetles than other pines, but the complete genotype of an organism covers a wide range of traits and the genes that confer beetle resistance may be associated with lower growth, lower reproductive output or other traits that pose some disadvantages. Evolution does not lead to optimal outcomes across all organism traits, it only ensures that disadvantageous traits are less likely to span generations.

Why don't bark beetle outbreaks happen all the time? Single questions can have answers with many pieces. If beetles require older trees, then landscapes dominated by young forests would not support large beetle populations. Many landscapes have (or have had) vast areas of old forests, so it might seem that beetles missed many decades of opportunities. Some broad ideas have been speculated, but available evidence cannot provide a very sharp set of answers. When beetle populations are low, routine factors such as beetle-eating birds and beetle-killing diseases might "trap" the population at low levels. Maybe these agents lose the capacity to limit the population in some years, and outbreaks ensue. The potential host trees might also be important. Perhaps trees experiencing average climate years have less risk from beetle attacks than when stressed. Indeed, records of mountain pine beetle outbreaks in Colorado over a 350-year period showed that drought was the best predictor outbreaks occurrences (Hart et al. 2017). And of course if beetle success depends on having large trees to attack, then outbreaks might occur only at intervals required for trees to get large enough. Lodgepole pine trees are a prime host, but they grow slowly on relatively low-productivity sites, and it might take about 200 years to reach 35 cm in diameter and provide ideal habitat for beetles (Negrón and Huckaby 2020).

Favorable climates could be an issue for beetles too. Beetles are remarkable for withstanding below-freezing temperatures. Their bodies accumulate compounds that have lower freezing points than water, such as glycerol which can be one-quarter of the (dry) mass of beetle larvae in midwinter. Sustained temperatures below freezing kill many beetles, and temperatures below about −20 °C kill almost all mountain pine beetle larvae (Bleiker and Smith 2019). Beetle larvae overwinter under the bark of live trees, and an air temperature of −20 °C would take some time to chill a larva. Beetle populations might plummet after long chilly winters, or winters with short frigid periods.

Is This a Healthy Forest?

The idea of "health" makes sense for individual organisms. Many traits might indicate poor health of a tree, such as loss of branches or roots, the development of fungal decay inside stems, or the consumption of leaves by herbivores. One outcome of poor health might be decreased growth, or even loss of mass. Poor health in trees is often followed by death, in a very straightforward way.

It's common to hear about healthy and unhealthy forests, but the idea of health may have little meaning (and great confusion) when applied to forests rather than to individuals. Does Figure 10.18 picture a healthy forest or an unhealthy forest? Areas dominated by old trees experienced high mortality from mountain pine beetles, but does this mean the forest was somehow unhealthy before the beetles (and therefore vulnerable to beetles)? Or is the forest now unhealthy because there are so many dead trees? At a landscape scale, does the presence of younger patches (where trees regenerated after small clearcuts) increase the health of the forest?

What traits might characterize an unhealthy forest? High mortality might seem like an obvious sign, but when one tree dies, other trees gain more resources and prosper. If most of the overstory trees in a forest die, the trees in the understory rapidly increase growth and they might almost be heard celebrating the death of overstory trees. If most trees in the overstory die, and there are few

FIGURE 10.18   This picture of the "red hand of death" comes from a watershed experiment at the Fraser Experimental Forest in central Colorado. The green areas were cut 60 years earlier to test responses of streamflow, and the forests were too young to provide good habitat for mountain pine beetles. The surrounding older forests have many trees that died recently, with red needles still attached, and others that died a few years earlier (standing gray stems). Is this a healthy or unhealthy forest? Many opinions are possible, but clear evidence is most helpful when applied to very clear and suitable questions (**Source:** photo by Chuck Rhoades, US Forest Service).

trees in the understory, then the vegetation may shift with increasing shrubs, grasses, and forbs. A landscape where a dense forest is replaced by a mosaic of patches without trees might provide improved habitat for some species of animals. An apple may be tasty, but the idea of "taste" would not apply to an orchard. A tree may be healthy or unhealthy, but the idea of health does not apply well to forests. Conversations among people who care about forests may often include the word "health," but great care is needed to minimize risks of misunderstanding, of confusing the forests with the trees.

Forests Often Thrive When Insects Kill Trees

Mountain pine beetle outbreaks may appear to be devastating, when landscapes turn red (Figure 10.18). The impacts can indeed be large, but a closer look at the surviving trees usually shows the forest remains well stocked with trees that launch the next generation. Even when the loss of wood volume is more than 50%, the residual trees meet the definition of a forest that is well stocked with trees (Figure 10.19). The species composition and structure of the next generation forest will likely be more diverse than the previous generations that developed after forest-replacing fires.

Should Forests with Lots of Beetle-Killed Trees Be Logged?

The amount of beetle-killed wood in a forest could be as high as 100 Mg ha^{-1}, and logging could serve several goals: providing wood to local economies, reducing hazards of falling trees, and reducing intensity of fires that might come later. Other concerns could be important, however. If a forest already had most of its dominant trees killed, would impacts of logging add additional challenges for forest recovery? Reasonable speculations could be offered in support or opposition to salvage logging, but evidence from strong experiments would be valuable for informing choices.

A forest with extensive beetle kill could be divided into four plots, with each plot assigned to a treatment of no-cutting, harvesting only tree boles, harvesting whole trees (boles plus branches and foliage), or whole-tree harvesting with scarification (exposing

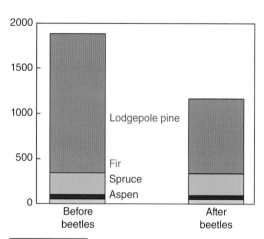

FIGURE 10.19 A set of 20 forests in southeastern Wyoming, USA were dominated by lodgepole pine before an outbreak of mountain pine beetles killed about half the overstory pines. Other overstory species survived, and the post-beetle forests averaged more than 1000 overstory trees ha^{-1}. The understories contained another 3500 trees ha^{-1}, so the future forests were developing strongly despite the massive overstory mortality. **Source:** from data of Kayes and Tinker 2012.

mineral soil to foster conifer regeneration). This design could show that one or more treatments led to satisfactory outcomes, but it's likely that some of the differences in apparent responses actually resulted from differences among the designated plots that were present before the treatments (since no two forest plots are identical). A stronger design would have replicate plots for each treatment in the forest, with randomization of treatments reducing the risk of confusing site differences with treatment effects. This design is not as strong as it could be, however, because the statistical population of inference would only apply to one particular forest. There would be no information on whether the treatment effects (good or bad) would apply to other locations or other times.

A stronger design might choose a dozen different sites, and install one plot with each of the treatments at each site. This design would have a population of inference that covers the whole area represented by the 12 sites, giving a solid basis for inferring how treatments would generally respond. This is the design that was used for the data summarized in Figure 10.20. The harvesting treatments did bring some changes to the development of the next forests, particularly shifting the balance between aspen and pine. This is a clear illustration that there is no single way that a forest needs to be treated, and no single future state for a forest. All the treatments (including the uncut treatment) led to very high densities of trees for the next forest.

A more complete consideration of the effects of post-beetle logging could include assessments of likely influences on fire (Figures 10.12 and 10.18). Removing logs would reduce risks of trees falling on firefighters, and lower the intensity of longer-lasting, smoldering fire behavior. On the other hand, the whole-tree harvesting fostered pine and reduced aspen, and the differences in species composition could have substantial effects in future fires. A landowner who was particularly concerned about future fire risks might choose to go with whole-tree harvesting and scarification to promote less-fire-prone aspen. And of course the outcome from that decision might depend on elk populations (with the potential to browse aspen and prevent development of aspen trees), and the broader availability of browse across the landscape. Decisions about forests can be aimed at moving forests in particular directions, but unless management plans include conversion to an intensively managed tree farm (Chapter 12), the future is bound to present some unexpected outcomes.

Other Dynamics of Forests and Beetles Occurred Across the Region Too

This overview of mountain pine beetles and how they affect forests with lodgepole and ponderosa pines can be joined with stories of the ecology of other forest types and other species of bark beetles. The post-2000 era has been one marked by widespread outbreaks of a variety of beetles in forests (Figure 10.21). The coincidence of the outbreaks of various beetles is interesting, perhaps reflecting regional trends in drought. The state of the forests might also be important; most forests in the region are not harvested, and the average age of forests likely increased over recent decades. However, the outbreaks did not peak in the same years, which might prompt caution in ascribing cause and effect.

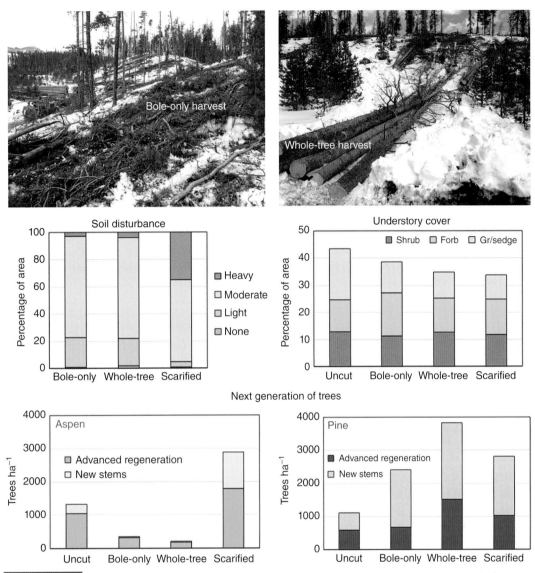

FIGURE 10.20   Treatments were applied across 12 sites where dominant lodgepole pine trees were killed by bark beetles. The experiment tested whether removing trees (boles only, or whole trees) would affect the development of understory vegetation (shrubs, herbaceous forbs, grasses and sedges), the survival of small trees (which survived the beetle outbreak), and the establishment of new stems. Harvesting operations disturb the soil, especially mixing O horizons with mineral horizons, and simply shoving soil around. The intentional moving of the O horizon (scarification) did provide more soil disturbance, which might aid conifer seedling establishment (upper left). Six years after treatments, the understory vegetation remained somewhat lower in harvested areas (upper right). The next generation of aspen stems were highest in the scarification treatment, which stimulated sucker sprouting from roots (lower left). The next generation of pine trees was dense across all treatments, including both surviving small trees (which survived the harvesting) and newly established seedlings (lower right). A choice to harvest wood after a beetle outbreak should be expected to lead to some changes in the next forests, but there was no evidence that harvest impaired the next forests. **Source:** from data of Rhoades et al. 2020; photos by Chuck Rhoades.

Other Forests and Other Insects Have Other Stories

The stories of mountain pine beetles and forests illustrate some of the ecological features of insect/forest interactions. Other cases could be added to the chapter, but other stories would need to be developed beyond this forest ecology text. If this chapter had focused on spruce budworms and forests of eastern North America, the stories might have included discussion of what factors influence population cycles (including evidence for and against a variety of hypotheses from the last century), perhaps focusing on the role of natural enemies in driving down budworm populations after outbreaks, or the influence of fungal endophytes on spruce

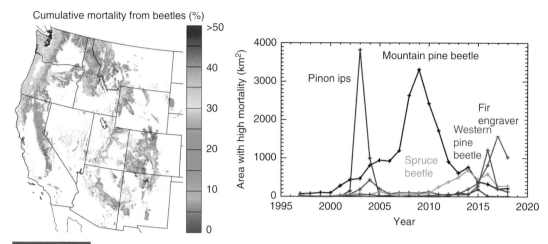

FIGURE 10.21   Across the western USA, several species of bark beetles showed high levels of forest impacts after the year 2000. The most severe impacts were from mountain pine beetles, primarily in forests of lodgepole pine (**Source:** from Hicke et al. 2020 / Elsevier).

that impair caterpillars. Other stories could focus on very long-lived trees, and how they manage to survive so long in a world full of insects that are adapted to life in trees. Most of the stories would be fascinating, but frustratingly incomplete and indeterminate for any attempt to simplify stories about the fantastic complexity of ecological interactions in forests.

Tree Diseases Are Reshaping Forests in a Globalized World

A wide variety of diseases affect trees, including fungi that rot wood in roots and stems, fungi that digest leaves (and parts of leaves), and even some bacteria that wilt leaves and develop cankers on branches and stems. Most of the ecology of diseases and forests fits within the "background" of interactions and slow changes that happen gradually over time. Diseases rarely lead to dramatic, widespread changes in forests, with one major exception: non-native diseases (Chapter 14).

Chestnut blight (a parasitic fungus) was mentioned in Chapter 1, where the arrival of this exotic disease decimated one of the major species of forests in eastern North America in a matter of decades. A similar story followed the arrival of Dutch elm disease, which removed American elm trees from most forests in the same region as the loss of chestnut. North America has several species of pines with 5-needles in their fascicles; these closely related species have been heavily impacted by cankers that develop from the infections with exotic white pine blister rust. European ash is being lost from most European forests as a result of an exotic rust disease from eastern Russia. Sudden oak death in North America results from the invasion of an exotic fungus-like disease. The ecology of each of these diseases has unique features (such as a life cycle for white pine blister rust that includes a required life stage in currant shrubs), and few generalizations would spread across the stories. Non-host species might generally be expected to benefit from the loss of competing species to a new disease, at least until those species are targeted by a newly invading disease. A key point for forest ecology is that major disease impacts in forests should not be surprising when diseases that developed on closely related species in other parts of the world experience opportunities to exploit new species with low resistance. A second key point is that with increasing movement of plants and soils around the world, the opportunities for major new impacts on native tree species are almost unbounded.

Major Events May, or May Not, Influence the Probability of Other Major Events

It seems obvious that beetle-killed trees, such as those in Figures 10.12 and 10.18, have a higher risk of burning than live trees would. This is would be true while the dead needles remain on the trees, because dead needles have lower moisture content and ignite more easily. Once the needles have fallen off, though, would beetle-killed trees pose a higher risk of fire? This would seem to be possible, but it would also be possible that the lack of needles (living or dead) would have to lower the risk of crown fires. When logically possible ideas lead to opposing expectations, evidence is particularly important for providing insights about forest ecology.

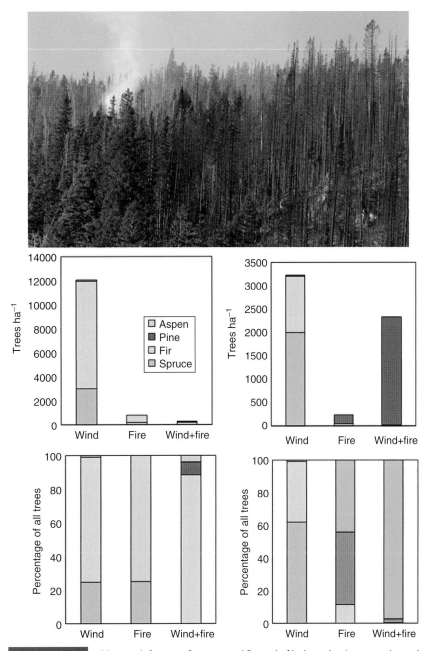

FIGURE 10.22 Old-growth forests of spruce and fir, and of lodgepole pine, experienced a severe windstorm and blow down. Five years later, some of the area burned in a severe fire. The future forests will be very different on these portions of the landscape, depending on the prior forest type and which events occurred (and did not occur). Forest development has some features that are broadly predictable, but events have such large legacies that variety is always broad indeed. **Source:** from data of Gill et al. 2017.

Vast areas of lodgepole pine forests experienced bark beetle outbreaks in the past few decades, and climate across the region also had a variety of normal years and droughty years. If beetles increased fire risk, we would expect that fires would have occurred most often in beetle-affected forests. If droughts were the major factor, then fires should relate more to drought periods, burning across landscapes with and without beetle-killed trees. With three decades of fires available, and hundreds of thousands of km² affected by beetles, the evidence should provide some clear insights. The evidence clearly points toward climate as the major predictor of fires in lodgepole pine forests, with very little influence of beetles. The only influence of beetles that was apparent in the data was that fire occurrence was lower in forests that had been affected by beetle outbreaks (Mietkiewicz and Kulakowski 2016).

Forests with extensive beetle-kill do not have a higher risk of fire, but forests that have burned may have a lower future risk for beetles. Post-fire forests have trees that are too small to be susceptible to beetles for a century or more (Kulakowski et al. 2016).

FIGURE 10.23   Avalanches are major events in some mountainous areas. An avalanche swept down this mountainside 16 months before this photo.

Events in Combinations Can Have Drastically Different Legacies

The composition of a forest after a major event is very different than before the event (or else it wouldn't be an event), and so the impact of a subsequent event would not be the same as if it were the initial event. A combination of windstorm and fire illustrates the size of these legacies. Portions of old growth forests dominated by spruce and fir, or by lodgepole pine, were blown down by a severe windstorm. Five years later, a severe wildfire burned in the same area, covering some areas that had been blown down, and some that had not (Figure 10.22). The number of regenerating trees was assessed 10 years later. Blown-down patches of old spruce-fir forest regenerated with seedlings of spruce and fir, with more than one tree for each m^2. The blown-down pine forest also regenerated to spruce and fir, though at only a quarter of the density of the old spruce/fir blowdown. When fire was the only event, the density of regenerating trees was far lower, with a higher proportion of aspen (and of pine in the former lodgepole pine forest). When the primary event of wind was followed by the secondary event of fire, only fir seedlings dominated in the former spruce/fir forest, and aspen in the former lodgepole pine forest. The species compositions of these forests are among the simplest in the world, with only a few potentially dominant species, and yet the range of future forest composition, structure, and dynamics was very broad indeed. The future of forests is not tightly constrained, even for simple forests like these.

Ecological Afterthought: The Ecology of Avalanches

An avalanche swept down from above timberline across this slope a year before the photo in was taken (Figure 10.23). This chapter did not consider the ecology of avalanches, or what might happen after avalanches occur. What insights might be developed by applying the three questions in the core framework (what's up with this forest, how did it get that way [which is a bit obvious, but some nuances might be considered], and what's next?). What would be the key aspects to consider? Do avalanches resemble blow-downs from windstorms, or are there important differences? How would the hydrology change after the avalanche?

CHAPTER 11

Events in Forests: Fire

Forest Growth Sets the Stage for Rapid Return to Chemical Equilibrium

Carbon dioxide is the most energy-poor state for carbon atoms. Photosynthesis boosts those atoms up to a very high energy state as organic molecules in plant material. Forests store phenomenal amounts of this energy: each kilogram of biomass has about 16–20 MJ of energy that is released rapidly when environmental conditions allow fires to ignite and burn. A better way to picture this amount of energy might be to express it in terms of gasoline. A low-intensity surface fire in a forest might burn about 1.5 kg m^{-2} of organic matter, releasing the amount of energy contained in 1 liter of gasoline. More intense fires that burn rapidly through canopies, or slowly in piles of logging slash, release the energy of 15–30 l of gasoline for every m^2.

It's fortunate that biomass can be stable, even though it's so far from chemical equilibrium within an oxygen-rich atmosphere. A fire needs an ignition source to get started, because the initial burning of organic matter requires an "energy of activation." Pyrolysis is a term for this initial ignition step, where added energy breaks apart large molecules, which then burn and release even more energy, unleashing the positive feedback that promotes further burning. The ability of an ignition source to provide enough energy to bring about burning of biomass depends on the water content of the biomass. Water absorbs large amounts of energy in evaporation, holding back the rise of temperature until all the water is gone. Dry wood ignites at temperatures above about 250 °C. Combustion of solid material develops complex physical features, including accumulation of blackened carbon on the surface of woody materials that might inhibit further burning by restricting oxygen access to the unburned wood beneath the char. Wood converts to charcoal from sustained heating with limited oxygen, but temperatures that exceed about 1100 °C will burn charcoal and sustain very high temperatures indeed.

Fire of course combusts plants (and animals and microbes), but unburned organisms can also die from direct heat damage to cells (most die with exposures of 60 °C or higher), and from loss of water.

Thick Bark Protects Cambium from Heat

Fires that burn on the forest floor kill trees if the transfer of heat to the cambium (the living, growing layers of cells that form bark to the outside, and wood to the inside) is too great. Cambial cells die at temperatures of about 60 °C, and this temperature is reached when fires are very hot, burn or smolder for a long period, or when bark is thin. Not surprisingly, thicker bark means more resistance to heat transfer, and bark thickness is a good predictor of which trees survive fires.

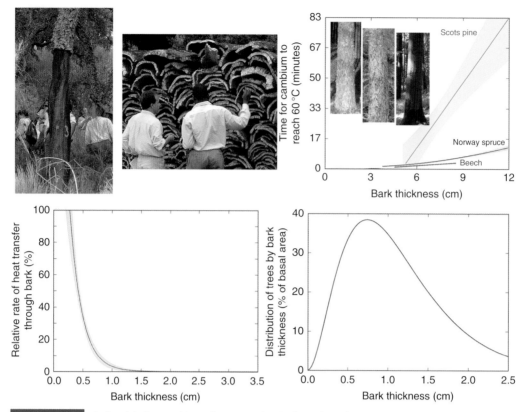

FIGURE 11.1 Cork oak in Portugal is a prime example of a fire-adapted tree with thick bark that insulates the stem's cambium from fire (upper left photos). The bark (cork) can be harvested about every 10 years for a dozen cycles over the productive life of the tree. The insulation provided by bark depends on bark thickness, but also differs substantially among tree species (upper right; **Source:** based on data in Bär and May 2020). The rate of heat transfer through bark of many species in a Brazilian rainforest depended strongly on bark thickness, and about half of the basal area of trees in the forest had bark thin enough that heat transfer rates during a fire would indeed be high (**Source:** lower graphs based on data from Brando et al. 2012).

The evolution of tree species has led to a broad range of resistance to fires. A classic example of a very fire-adapted species is cork oak from southern Europe (Figure 11.1). Cork bark is especially good at insulating cambium from the heat of fires, and the bark grows rapidly after a stem has been damaged, or the bark (cork) has been stripped off as a non-timber forest product. The bark on cork oaks grows outward in thickness by about 1–3 mm annually for over 100 years, and stripping off the bark at 10 year intervals is a sustainable forest practice that can be repeated about a dozen times. Most cork oaks survive wildfires, even in landscapes where plantations are routinely stripped, with survival exceeding 80% if the cork layer is 2 cm or more thick (Moreira et al. 2007).

The insulation properties of bark differ among species, and the species that are adapted to more frequent fires have better insulation properties (per unit of thickness). Scots pine forests in Europe historically experienced more fires than those dominated by Norway spruce or beech, and the bark on pine provides much better protection for the cambium (Figure 11.1).

Tropical rainforests may seem like they should be so wet that forest fires would not be important, but in fact fires burn extensive rainforest areas during dry periods. The fires tend to move across the ground (rather than crown fires moving through canopies), so tree survival usually depends on the ability of the bark to insulate the cambium. One study with experimental use of fire documented the importance of bark thickness in limiting the rate of heat transfer through the bark, and also documented the pattern of bark thickness among all the diverse species of trees in the rainforest (Figure 11.1). Almost half of the trees (based on basal area of the trees, rather than the number of trees ha^{-1}) had bark thin enough that heat transfer would be relatively rapid (and bad for the trees).

The coverage of bark around a tree stem may not be uniform, especially if the stem has been injured by a falling neighbor, removed by a herbivore, or damaged in a harvesting operation. Areas of thin or missing bark can be very susceptible to fire, especially if resin has accumulated around the injury. A fire may kill the cambium at injury points, yet the tree survives. Indeed, some trees can endure having the cambium killed across more than half their circumference and continue to survive and grow well. The

FIGURE 11.2   The thick bark of large ponderosa pine trees protects the growing cambium of the stem from the heat of fires, but a fire in 1994 damaged the base of the tree on the cover of this book and a fire in 2020 then burned through the base, toppling the tree (upper). Sequoia trees have even thicker bark, but successive fires can still lead to torching of the entire tree (lower photos). (**Source:** Sequoia photos by Ellis Margolis, USGS).

areas of dead cambium can pose a larger threat to the tree's survival if a second fire comes along. The second fire might burn and char the exposed wood of the stem, sometimes smoldering long enough to burn a large cavity into the stem (and perhaps toppling the tree; Figure 11.2). This is indeed what happened to the four-century old ponderosa pine tree on the cover of this book. The cohort of small lodgepole pine trees around the old tree developed after a fire in 1994 that killed the prior lodgepole pine forest but not the old ponderosa pine. The ponderosa pine tree also survived fires in 1671, 1690, 1842 and 1896. The stems of the pre-1994 lodgepole pines lay on the ground, with an understory of nitrogen-fixing snowbrush shrubs (see Figure 11.6). Another fire burned the hillside two months after the picture on the cover was taken, and the old ponderosa pine had enough exposed dead wood in the fire scar it burned through at the base and toppled over.

The adaptation of tree species to fires goes beyond just bark characteristics. Fires can move from the forest floor into tree crowns, if trees retain branches (fine fuels) low to the ground. Some species have leaves that are ignited more readily than others, and taller trees have crowns that are less likely to ignite than crowns of shorter trees. The fire risk for a species can also depend on the chemical and physical traits of the litter it produces, which can accumulate at the bases of trees. Multiple traits such as these can be condensed into an overall index of resistance to fire (Figure 11.3). The value of such ranking might be high or low, depending on the question or management situation under consideration.

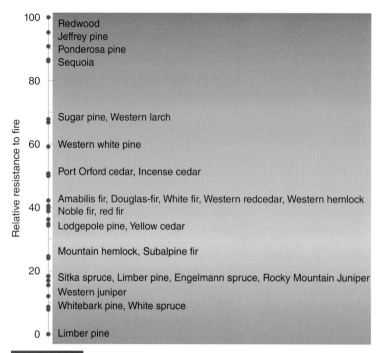

FIGURE 11.3   Indexes of risks from fire can be created by combining multiple traits of species, such as bark thickness, tree height, and flammability (in this case, scaled to 100 for the lowest risk species among common conifers of the western USA. **Source:** Based on Stevens et al. 2020.

The Post-Fire Forest May Be Dominated by Resprouting Vegetation

Fires may scorch or burn leaves and branches off trees. Some fire-adapted species have epicormic buds that sprout after fire, rebuilding new crowns (Figure 11.4). This recovery allows individuals to retain the advantages that come with being tall. Many other species simply send up new shoots from surviving root systems. Only a few conifers can resprout from either epicormic buds or surviving roots, but the ability is very common in broadleaved tree species and understory plants (Figure 11.5). Grasses in particular thrive even when grass shoots are eaten or burned; their growing meristems occur at the base of the plants.

Post-Fire Environments Can Be Good for Seedling Establishment

After fires, soils are typically warm because of high light input and dark soil surfaces. Absence of plants leads to accumulation of nutrients and water in the soil, and full sunlight can support high rates of photosynthesis for the colonizing plants. With abundant resource supplies, a key challenge for seedling establishment is simply the availability of seeds on the site. Seed availability after fire occurs in seed banks in the soil, seed banks in the canopies of trees, and transport of seeds into the burned area.

The seeds of most tree species remain viable in soil seed banks for only a few months or years. Some species, such as pin cherry across North America, have seeds that can remain viable in the soil for decades, or even a century. Some non-tree species also develop soil seed banks that last a very long time, including ceanothus species (shrubs, including snowbrush) that germinate after fires and increase soil nitrogen pools through biological nitrogen fixation.

Seed banks in forest canopies are important of course for trees that survive the fire, and for some trees that perish. Some pine species store ripe seeds in unopened cones that remain on branches for decades. In North America the classic "serotinous" pines include lodgepole pine and jack pine (Figure 11.6). Sometimes the release of seeds from serotinous cones is so prolific that more than 100 000 trees establish per hectare. Within a few decades, the trees can begin to accumulate a new seed bank in serotinous cones. This ecological story is not universal, as many trees of species that are characterized as serotinous also produce cones that open and release seeds without waiting for fire.

FIGURE 11.4   Some trees sprout new branches and leaves from buds in the cambium (epicormic sprouting). This response might convey greater competitiveness relative to trees that resprout from root systems or establish from seeds. Karri (a eucalyptus species) after a fire in western Australia (A); an English oak hillside in southern Portugal 1.5 years after fire (B); Canary pine forest a few years after fire (C); and close-up of epicormic sprouting 3 months after fire (D; **Sources:** photos by G. Wardell-Johnson, F.X. Catry, and J.G. Pausas, from Pausas and Keeley 2017, used by permission).

Seeds arrive on burned sites as a result of gravity (falling from surviving trees), wind (especially for light, aerodynamic seeds), and animals. Surviving trees can be very important as sources of seeds for the next generation of a forest, especially for heavy-seeded species. The regeneration of conifer seedlings may be higher near the edge of a burned area, declining with distance into the burned area (Figure 11.7). The reestablishment of a forest may be much slower when burned areas are so large that surviving trees around the edge cannot provide seeds into the center of the burned area.

The Spatial Scale of Forest Fires is Important, But Not Simple

Forest fires are often described by a number of hectares or square-kilometers, but putting a total, cumulative number on a fire condenses many details to the point where a great deal of information about the real landscape has been lost. A single number does not capture the range of fire intensities that occurred across an area, how much of the forest experienced an intense crown fire, and

FIGURE 11.5 Many species resprout from surviving root systems or bases of stems. Aspen sprouts (suckers) come up prolifically after a fire in a mixed-conifer/aspen forest (upper left). Redwood trees can sprout from the base after fire, even if the parent tree survived the fire (upper right, photo by Steve Norman, US Forest Service). A high-intensity fire killed all the ponderosa pine trees and completely consumed the O horizon (and downed logs, which died in a fire 15 years previously). Three years later, resprouting Gambel oak and New Mexican locust revegetated the severely burned landscape, and reestablishment of ponderosa pine would be unlikely because of the great distances to surviving seed trees, and the competition that would be offered by the resprouting woody plants (**Source:** photos by Craig Allen).

how much a low-intensity surface fire. Sometimes a burned area may be divided into categories such as lightly burned, moderately burned, and severely burned. This level of description still condenses a great deal of variation to the point where substantial information is overlooked. For example, if 30% of a burned area was designated as having burned severely, some of the area may have experienced a much higher severity (as a result of more fuels, drier fuels, different topography or more wind during the fire) than other portions (Figure 11.8). The ecological impacts of a fire might be very different for a landscape which was half burned severely in one large patch compared to the same area burned in a number of small patches.

The words used to describe forest fires have a hard time capturing the reality of fires, as well as the wide range of ecological outcomes. The ability of terms to capture important aspects of forest fires depends in part on how clearly the terms are defined, and whether the aspects of the fire being described can actually be captured by simple terms. Some conventional vocabulary in forest fires includes using the word intensity for describing the physics of burning (such as energy released per unit area or time), and severity for aspects of the ecological impacts of a fire. A fire that burns mostly across the forest floor would be a surface fire, and one that burned within organic layers in soils would be a ground fire. The upper part of a single tree is called a crown, and the tops of a group of trees is a forest canopy, but a fire burning through a forest canopy is often called a crown fire. Coming back to the question raised in this section's heading, the size of a fire is often estimated as the minimum perimeter that would encompass all the burned portions of a landscape, including interior pockets that did not burn.

The idea of putting a single number on the size of a fire has value if it conveys the intended information; a fire described with an area of 1 ha is clearly different from one that is described as 1000 ha. As long as the limited ability of a word or a number to represent a real situation is kept in mind, the size of areas burned in a fire can be described in useful ways.

7 years after fire   25 years after fire

FIGURE 11.6   A forest of lodgepole pine on a glacial moraine in Colorado, USA burned with an intense crown fire. Seeds released from serotinous cones (upper row, two months after fire) reforested the site. A decade after the fire (left photos), some portions of the burned area had pines with understories of grasses and forbs, and other areas had N-fixing snowbrush shrubs that germinated after lying dormant in the soil for almost a century. Twenty-five years after the fire, the trees have begun to redevelop a seed bank in serotinous cones (right photos), and snowbrush had replenished its soil seed bank. The forest burned a year after the latest photos, and it will be interesting to follow the fate of the next generation of shrubs and trees that arise from the seedbanks.

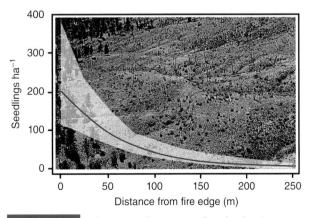

FIGURE 11.7   A few years after a crown fire, the density of seedlings declined exponentially with distance from the surviving trees at the edge of the burned area. **Source:** Based on Chambers et al. 2016, photo by Oscar Rhoades.

FIGURE 11.8 A 2000-ha fire in a spruce/fir forest in New Mexico, USA had patches where fire burned along the ground (upper left), and patches where the flames raced through crowns (upper right); sites differed in topography, but all had complete mortality of the overstory trees. Two years later, vegetation had reestablished across the range of fire intensities, but with very different species composition as a result of pre-fire difference in vegetation, in seed banks, and in post-fire colonization. **Source:** photos by T.N. Gass.

Most Forest Fires Are Small, Though the Uncommon Large Fires Have Great Impacts

Many things in ecology display power function patterns, with some groups being very frequent, and others rare. The number of individuals of each species in a forest usually follows a declining exponential trend, with a few species being extremely common, and a many species represented by only a few individuals. Forest fires also follow declining exponential patterns, with very many small fires and few very large fires (Figure 11.9). As noted above, the idea of the area of a fire conveying clear information can be

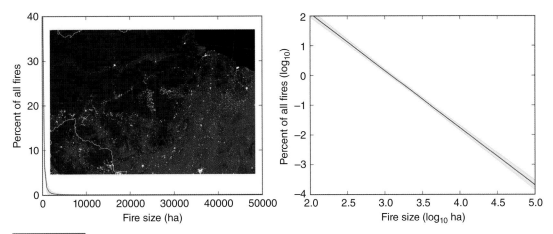

FIGURE 11.9   Most fires in Brazil are small, and few are very large. The two graphs present the same information, but the exponential pattern is hard to see (left) unless presented on logarithmic axes (right; **Source:** from data of Silva et al., 2018; satellite image of fires in Brazil from NASA's Earth Observatory).

problematic, and large fires may differ by more than simple area. Large fires get large by burning under very dry conditions, and often with high intensity. Long-burning fires may span multiple days, and have a high variety of burning intensities that depend on weather conditions when a particular patch burned, as well as the biomass, topography, and other features of the patch.

Fires Burn Differently at Different Places

Broad patterns of forest fires relate to forest structure and composition, forest location, environmental conditions, and the sources of ignition (Figure 11.10). Most forests have enough biomass to fuel a fire, and forests with low biomass typically have high biomass of understory plants that also contribute to fires. Fortunately, most forests are usually too wet to carry a fire, or the occurrences of ignitions is rare enough that even dry forests are not always burning. The frequency of fires often relates to elevation, as lower elevation forests are more frequently dry enough to carry fires than cooler high elevation forests. These patterns also include different types of fires, from frequent fires that burn primarily surface fuels to rare fires that burn through crowns and release so much energy that massive convective storms above fires can collapse and flow like volcanic pyroclastic flows.

Topography influences fire occurrence and intensity. Fires are more likely to spread upslope than across flat topography, and south-facing slopes in the Northern Hemisphere dry out more than north-facing slopes. Soils also influence the structure and amount of biomass that develops in a forest, and patterns of fire movement and impacts across landscapes change when soils change.

Dendrochronologists often find the oldest trees on a landscape survive in areas with shallow soils (and rocky outcrops) where low biomass accumulation has moderated fire behavior over centuries. Severe weather can drive fires across these differences in topography and soils, and when this happens the amount of biomass available to burn can influence fire behavior and ecological impacts more than topography would.

Periods of Gradual Change Are Punctuated by the Large Changes from Fire Events

Forests are always changing gradually, often developing with legacies of past events such as fires. Given that fires do not have simple or consistent effects on forests, and that legacies result from the pre-fire composition and structure of a forest, the overall outcome of gradual changes and the impacts of events generates great diversity in forests over time.

Some major differences that result from differing fire-free periods are illustrated in Figure 11.11 from long-term experiments in longleaf pine forests in northern Florida, USA. It might seem that burning a forest every year would lead to the same vegetation as a burning every two years, but in fact cutting the fire-free period in half leads to very large differences in understory vegetation. Fire-free periods of four years are long enough for some species to establish and persist that could not cope with more frequent fires. If fires do not occur for decades, a vigorous set of broadleaved trees provides a very different structure (and wildlife habitat). The legacies of the periods between fires also depends on the season of burning. Longleaf pine forests develop different understory compositions when fires occur early in the growing season compared to late in the growing season.

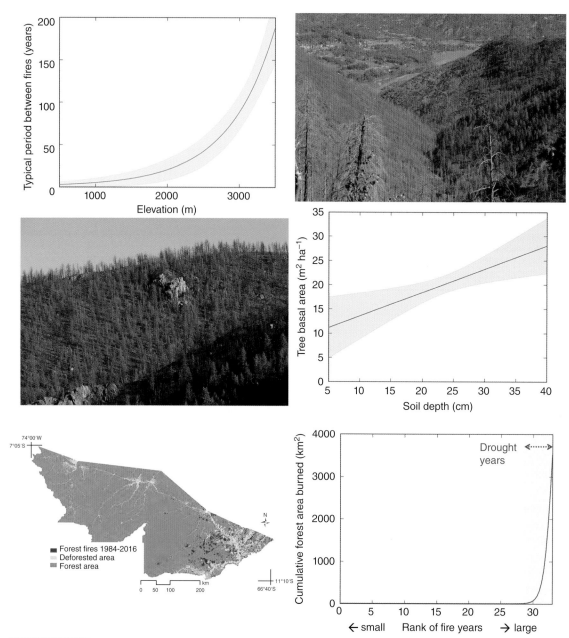

FIGURE 11.10 Fires do not occur everywhere, all the time, or with the same intensity. Fire in the Sierra Nevada mountains of California, USA are very common at low elevations, with only a few years between fires. Higher elevation forests typically go a century or more between fires (top left; **Source:** based on data from Caprio and Lineback 2002). Local topography also influences fires, with drier S-facing slopes (in the Northern Hemisphere) drying out and burning more frequently than N-facing slopes. However, both slopes may burn under severe weather conditions, and greater fuel accumulations on N-facing slopes can lead to more severe impacts (top right photo, San Juan Mountains, Colorado, USA; the N-facing slope on the left side of the photo burned more intensely because of higher fuel loads than on the S-facing slope). Local soil conditions affect the accumulation of biomass, and rocky, shallow soils with less biomass may burn less (middle left photo; middle right **Source:** Based on data in Hasstedt 2013). The rainforests of the Brazilian state of Acre in the Amazon (bottom) burn primarily in dry years (98% of area burned in 33 years occurred during droughts), and primarily in areas with high human population (the easternmost corner of the state; **Source:** Based on Silva et al. 2018 / Elsevier).

FIGURE 11.11 Management of longleaf pine plots at the Tall Timbers Research Station and Land Conservancy in Florida, USA uses different periods between fires to examine a broad range of ecological effects on the forests. These fire intervals have been applied for decades, leading to very different forest structures, wildlife habitat, and of course fuels that would lead to differences in fire intensity in the event of a wildfire (the unburned plot had not experienced fire in over 50 years).

Typical Fire-Free Periods Within Forest Types Vary Across Sites and Over Centuries

Records of fires can often be developed based on information lurking within forests. Some forests have a clear age-cap, perhaps indicating an intense, forest-replacing fire occurred some time just before the year of the age of the oldest trees in the forest. Other forests have trees that survive fires and retain records in their rings. A fire that kills most trees in a forest might lead to a clear increase in ring widths of surviving trees. This approach also can trace other events such as insect outbreaks that affect one species of trees in a forest more than others. Yet one more type of record deals with injuries to a tree that survives and recovers from fire damage. Fire scars develop when a tree survives a fire that killed the cambium around only a portion of the stem. Surviving cambium develops a callus that grows over the dead portion of the stem (Figure 11.12.) A small fire scar can become completely enveloped by new growth, and the fire scar is revealed only by taking a sample out of the tree. If another fire occurs before the wound heals over, the cambium in the overgrowing callus can also be killed (especially if the fire burns intensely with the exposed wood and any resin at the wound site). A second scar might result, and this sequence can be repeated with later fires. Fire-scarred trees record the presence of a fire, but the absence of a scar does not mean a fire did not occur (as it may not have left a new scar).

Records of fires recorded in scars on surviving trees are information at the spatial scale of an individual tree. The size of the fire might have been limited to just the vicinity of one tree, or might have covered a very large landscape. If many trees across a landscape record a fire in the same year, a fire likely occurred across the landscape. A commonly used criterion for a landscape fire

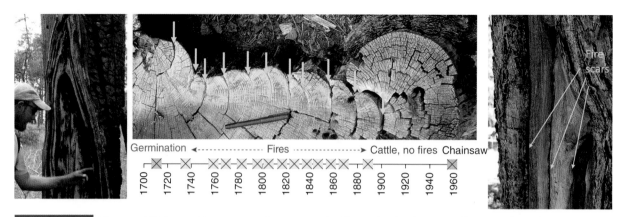

FIGURE 11.12   Fires may kill the cambium on part of a tree stem, leading to the development of a scar with a callus growing partially or fully across the scar. The tree stump shows that a ponderosa pine tree scarred in a fire when it was 25 years old (leftmost arrow in the middle photo). Fire-free intervals lasted for 9–25 years (fire scars marked with yellow arrows), with an average of 15 years. No fire was recorded in the final 70 years of the trees growth, as a result of livestock grazing removing the understory vegetation that carried the frequent fires.

is that 25% of the trees that already have fire scars must record the same fire. These studies have documented a fairly broad range in fire-free intervals for a single type of forest (ponderosa pine) at scale of 1000 km² to 100 000 km² (Figure 11.13). Across northern Colorado, the fire-free periods lasted about 15 years on average, but the variation around that mean spanned from about 5 years to more than 50 years. The fire-free intervals generally lasted longer at higher elevations, though the variation around that elevation trend was broad. Across the continent, the fire-free periods were shorter at southern latitudes and lower elevations, and longer at northern latitudes and elevations. The fire-free interval increased dramatically across the continent when livestock grazing reduced the fine fuels available to support frequent fires, along with increasing fire breaks (such as roads) and active fire suppression.

When Fire-Free Intervals Get Longer, Forests Get Denser

Figure 11.13 showed the fire-free intervals increased across the range of ponderosa pine after livestock grazing removed the understory vegetation that would have carried surface fires. Small trees that would have been killed by fires had a chance to live longer and increase the density of trees, further reducing the understory vegetation. The expanding tree crowns reduce the distances between adjacent crowns, increasing the density of fuels in each cubic meter of the canopy (called crown bulk density). Reconstruction of historical forest structures showed the crowns increased in bulk density by several fold (Figure 11.14). The risk of a running crown fire increased substantially, as the wind speed required to spread fire through the crowns drops from over 90 km hr⁻¹ to less than 30 km hr⁻¹. The continuity of fuels increases too, increasing the probability of intense fires spreading far across the landscape.

The Spatial Aspects of Fires Also Include Patterns Within Burned Patches

Fires can be described as high or low intensity, with high or low severity. Many fires fit one or the other end of this spectrum, but many (most?) fires would fall somewhere in between. Mixed-severity fires might have patches that burned at high intensity and others with low intensity (or unburned). These terms are all relative, not quantitatively clear terms. This ambiguity can make it hard for clear communication about fires and their variation across space.

The vegetation that develops after a fire depends strongly on details of the fire, and the immediate post-fire period. A fire burned the area in Figure 11.15, at a lower intensity on the downhill side, and higher intensity up the hill. Many of the ponderosa pine trees survived the low intensity fire, and none survived the high intensity fire. Nearby surviving trees provided seeds that regenerated a dense pine forest where the fire intensity was high, leading to little opportunity for understory vegetation. Pine regeneration was spare in the low-intensity area, perhaps owing to competition from the resprouting grasses. Why did not understory vegetation resprout vigorously enough after the high-intensity fire? One possibility is that the fire intensity was high because the prior forest was too dense to allow much understory vegetation (and also dense enough to carry a crown fire).

What would happen when the next fire occurs? If a fire burned the slope of the ponderosa pine forest in Figure 11.15, the more open area might have a surface fire that many of the large trees would survive, and the dense area might carry a crown fire that

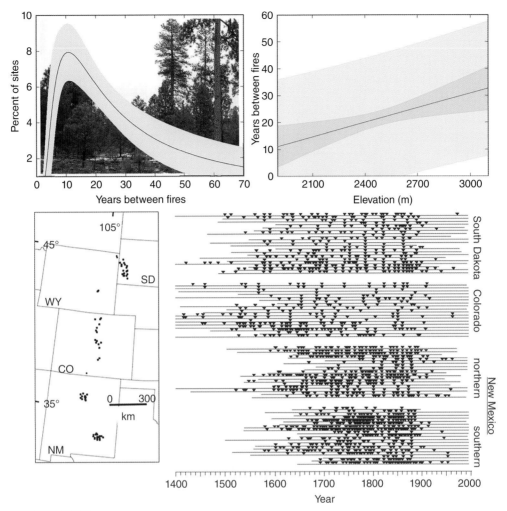

FIGURE 11.13   Fires return to ponderosa pine forests in northern Colorado with an average of 15–20 years, though the most commonly observed interval is about 12 years, and some go more than 50 years between fires (left). Some of this variation relates to elevation, with shorter periods between fires at lower elevations (right). Even though high confidence is warranted that fire intervals relate to elevation, the light green zone shows that 95% of all sites would span so broad a range that some low elevation sites would go three or four decades between fires, and some high elevation sites might have fires returning after 10 years. High confidence in a trend does not mean that the population does not have a high variance (**Source:** Based on data compiled by McKinney 2019). At a continental scale the fire-free intervals ranged from 5 to 18 years at the southern part of the species range, to 15–40 years to the north. A combination of elevation and latitude accounted for about half the variation in the fire-free intervals. Each line represents a single research site, with about 10 fire-scarred pines. The triangles indicate years when 25% or more of the trees within a site had a fire scar. Some years have strong climate patterns that lead to fires across many sites within a region (**Source:** lower map and graph provided by Peter Brown, Rocky Mountain Tree Ring Research Lab).

killed all the trees. What would the vegetation be like a century after the fire? The rapid regrowth of grasses in the surface-fire area would compete with seeds from the surviving ponderosa pines, perhaps limiting the density of new pine seedlings. The crown fire area would have very little understory vegetation to resprout after the fire, and pine seedlings might establish well because of low competition. A century later, the spatial patterns might be different if the post-fire weather was too dry for successful tree regeneration, allowing the site to have a stronger component of understory vegetation. The patterns over space and time depend on contingencies (weather, pre-existing forest structure) in ways that ensure no single pattern can be expected to apply broadly.

Some quantitative approaches might be developed to describe the severity of a fire across a landscape. The patterns of seedling establishment with distance from likely seed sources in Figure 11.7 can be extended by mapping an area relative to distance from trees that survived a fire (Figure 11.16). A low severity fire would have the majority of the area with surviving trees nearby, and high severity fires would have surviving trees (if any) so widely dispersed that the availability of tree seeds would constrain the regrowth of the next forest.

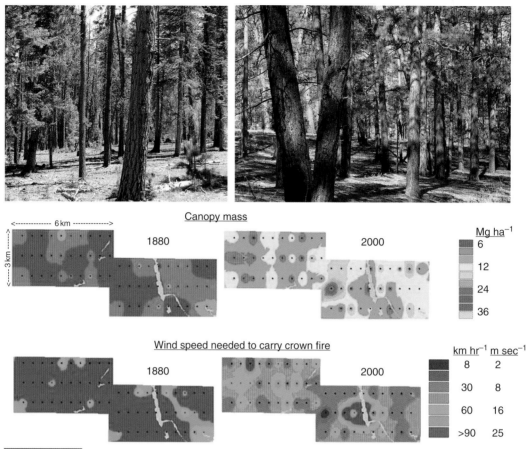

FIGURE 11.14 With a fire-free period of over a century, the canopy of ponderosa pine/mixed conifer forests on the Kaibab Plateau, Arizona, USA, increased in density and the mass of needles in the canopy. Left photo shows a forest structure sustained by frequent fires, and the structure in the right photo develops in the absence of frequent fire. The canopy mass increased by two- to fourfold without fire, and the denser canopy could carry a crown fire at much lower wind speeds. **Source:** Based on Fulé et al. 2004.

FIGURE 11.15 This hillside burned about a century ago, with many trees surviving a low-intensity fire to the right, and no trees surviving the higher intensity to the left. The legacy of the fire resulted in current differences in tree densities and understory vegetation, and these differences in structure (and fine fuels that might carry a surface fire) might influence the intensity of the next fire. **Source:** photo by Tania Schoennagel, from Schoennagel et al. 2011, used by permission.

Fire Ecology Might, or Might Not, Be Described with Fire Regimes

Fires might be characterized by several traits, such as the length of fire-free intervals, how routinely fires occur, the magnitude of fires (including aspects of intensity and severity), the extent of fires, and maybe the seasonality of fire occurrence. These traits might be combined to define fire regimes that capture differences across vegetation types, landscapes and regions. The idea of a

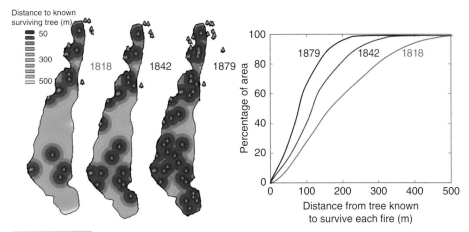

FIGURE 11.16   Locations of trees and their ages allow partial reconstruction of historical forest structure, and by inference historical fire severities. Currently living trees that were present and survived a fire in 1818 served as seed sources for the post-fire regeneration across a 100-ha forest. About half the area was within 200 m of a known surviving tree (other trees may also have been present, but died before 2010). The last fire in 1879 left so many known survivors that more than 90% of the area was within 200 m of a potential seed source (**Source:** based on Hasstedt 2013; see Coop et al. 2019 for a larger analysis).

fire regime may be useful if these traits have clear central tendencies and low variation, but less useful if central tendencies are weak between fire regime types and variability is high.

Forests and landscapes that have predictably short intervals between fires might be usefully classified as having a frequent-fire, low-severity fire regime. Longleaf pine and ponderosa pine forests would typically fit into this fire regime, with many large trees surviving each fire. At the other end of the spectrum, another fire regime might apply to forests of jack pine and lodgepole pine, which typically experience high-intensity crown fires after fire-free intervals of a century or more, and regeneration from seed is a key part of forest reestablishment. Most forests do not fit on either end of the spectrum, so a mixed-severity fire regime might seem to apply, except the variation in the traits of fires is so large that this class may be too broad to be useful.

Fire regimes might be developed as a framework based on expectations about traits, but the value of the framework would depend on supporting evidence. An evaluation of supporting evidence might be particularly useful if evidence from real forests is compared with expectations based on simple random likelihoods. If a set of forests over time differed from random expectations, and if those differences related strongly enough to fire regime categories, then confidence would be warranted that fire regimes were a useful idea and tool.

What would fires across a landscape, over a very long time, be like if only random variation was behind the patterns? A great example of random analysis, or null expectation, is developed in Figure 11.17 (from Lertzman et al. 1998). A model landscape could be defined as having 10 000 cells (100 × 100). A fire ignites and burns 500 cells every 5 years (a 100-year fire-free period for each cell, on average). The fires start at random locations, and burn square patches. The resulting pattern of burned areas, and time since last fire, would bear no resemblance to real forests, so it would be easy for an idea of fire regimes to represent real forests more powerfully than this null model. The null model could be made a bit more realistic by having the same number of cells burning, but let the timing between fires vary by a reasonable, random number of years. This null model also does not look like a real forest landscape. The null model is more interesting if the extent of each fire is allowed to have a complex shape (rather than square), and a combination of variable fire shapes and variable fire-free periods (but same overall average probabilities). This more realistic null model gives a remarkably interesting map of what random expectations might map out to be. A large portion near the center of the modeled landscape had no fire in 200 years or more, and large areas around the edges burned a few decades ago.

The null expectation becomes even more interesting when the random simulations are repeated many times, showing how different the patterns are with each rerun. The grand average across all years and all simulations is 500 cells burned every 5 years (for an average of 100 years between fires for each cell). This grand average has random variation that leads to very different numbers of cells that actually burn in any single century. Some sets of centuries average much larger than average numbers of cells burned, and others have much less.

These simulations provide insights on what sorts of patterns might appear in fire-free intervals across landscapes and over time. Some important questions can be raised based on this null expectation. How different would the histories of fires across two landscapes need to be to support a confident conclusion that the fire regimes were *not* the same? How different would the fire traits need to be for a landscape to indicate that the fire regime changed? These questions might be pondered in relation to patterns such as those in Figures 9.19, 11.13 and 11.14.

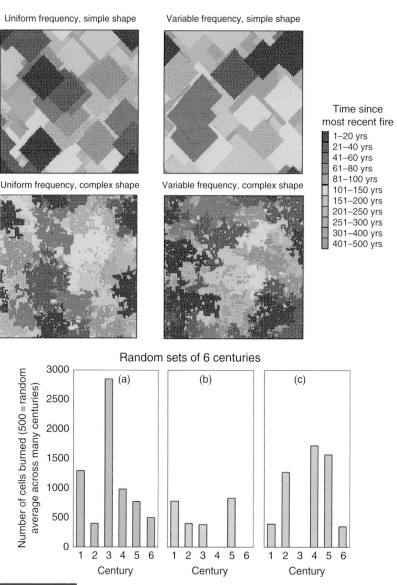

FIGURE 11.17 Confidence in maps of fire regimes would be warranted if a simple, random pattern did not account well for the observed patterns. Some simple random expectations can be combined to show how minor random variations around a given average can lead to maps that might appear to reveal fire regime patterns even where none exists (see text; **Sources:** based on Lertzman et al. 1998, scenario maps provided by Ken Lertzman and Joseph Fall).

Fires Change Soils

Almost every aspect of forest soils changes after fires, with larger effects from more intense fires (Binkley and Fisher 2020). The organic matter on the ground is the O horizon of the soil, and fires usually combust some, or even all, of this horizon. Some of the organic matter in the mineral soil can also burn. These changes in the physical structure of the soil are joined by large changes in chemistry and biology.

The productivity of many forests is limited by low supplies of nitrogen in the soil, and fires burn nitrogen just as they burn carbon. The organic N contained in organic matter is burned (oxidized) to various gaseous forms that are lost to the atmosphere. The burning of N is essentially the same as burning of C, and N losses scale directly with losses of C (Figure 11.18). Over the long term, N accumulates in soils and forests from atmospheric deposition and nitrogen fixation (Chapter 7), but rates of these processes are generally low and many years are needed to restore the levels of N.

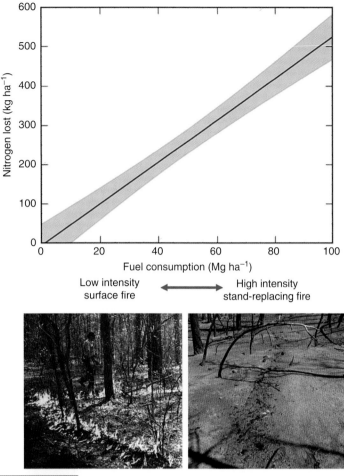

FIGURE 11.18   Nitrogen contained in organic matter burns just like the carbon does, and N losses in burning are directly related to the amount of fuel burned. **Source:** Based on Binkley and Fisher 2020.

The tremendous energy released when organic matter burns can penetrate into the soil, affecting soil chemistry and biology. Dry soils heat more than wet soils during fires, because water absorbs large amounts of energy as it evaporates (Figure 11.19). Heat transfer also depends on how hot the fire is, and how long the high temperatures persist. A rapidly moving fire may release a large amount of energy, but the heating of the soil is only near the surface and it lasts a short time. Burning of large amounts of woody material sends more energy into the soil over longer periods, and the high temperatures reach deeper into the soil. Temperatures above about 60 °C may kill roots and a large portion of the soil microbial community.

Fires Generate Erosion in Areas That Burn, with Sediment Deposition Downslope

The energy of rain falling on a forest is absorbed by tree crowns, and by the spongy organic layer that comprises the upper soil. Fires remove these energy buffers, letting high intensity raindrops hit the ash and other mineral particles. The O horizons of unburned forests can absorb very large amounts of water, which is slowly released into the mineral soil beneath. Without the water-absorbing O horizon, rain can fall faster than the mineral soil can absorb it, leading to surface flow and erosion. A second factor that may increase post-fire erosion is hydrophobicity. This is a water repellency that develops when small organic molecules volatilized (evaporated) during fires move into the soil and precipitate, forming a plastic-like layer. Fortunately hydrophobicity is a very patchy phenomenon, and water that runs off one patch may soak into another patch rather than flow far down a slope (Figure 11.20).

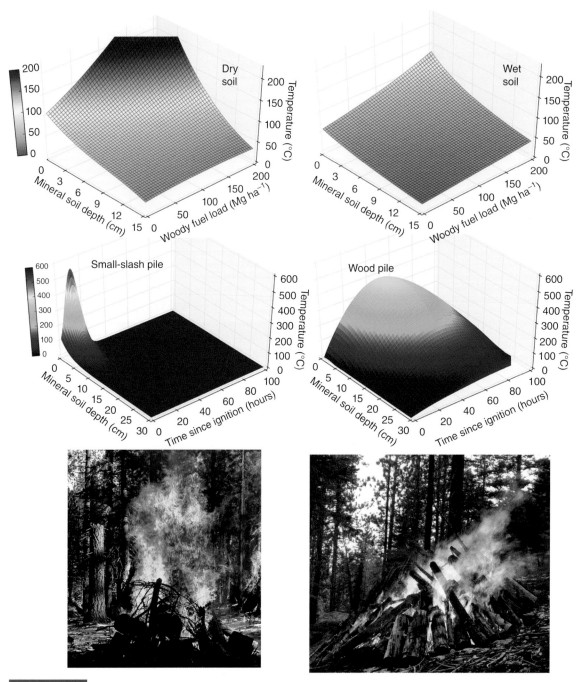

FIGURE 11.19 Fire impacts on soil include raising soil temperatures. The transfer of heat into the soil is greater on dry soils than wet soils (upper graphs). The patterns can be complex, based on how fast organic matter burns, and how long it smolders (middle graphs; **Source:** based on Busse et al. 2010, 2013, photos by Matt Busse).

Erosion After Fire is Usually Not a Problem, But Sometimes It's Very Severe

Eighty percent of wildfires lead to too little erosion to be a problem either from the point of view of soil loss or downstream areas (Figure 11.21). The other fires can lead to such large amounts of soil erosion that gulleys are created, and downstream areas are buried by sediments. Erosion and downslope sedimentation are major changes in forests, but value judgments need to be informed by larger perspectives. Riparian forests are usually the most productive in a landscape, and many riparian systems develop on

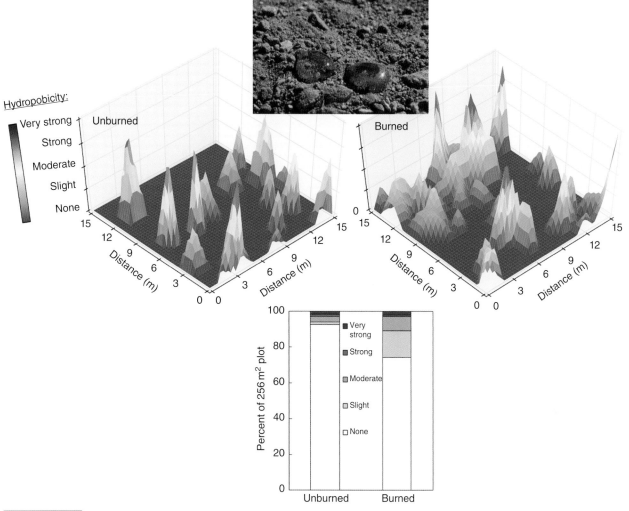

FIGURE 11.20 Soils can have patches that repel water rather than absorb it, and this can increase after fires. Adjacent patches might absorb the water droplets, or they might coalesce, flow down slope, and erode sediments (**Sources:** Data from Woods et al. 2007, photo from DeBano et al. 2008).

FIGURE 11.21 Post-fire erosion is usually minor, but in the less common situations the losses are massive, leading to deposition downstream. The immediate effects seem severe, but the long-term structure of riparian ecosystems may depend on these events. (**Sources:** graph based on data in Robichaud et al. 2000, photo inside graph by John A. Moody, USGS; lower photo by R.H. Meade, USGS).

FIGURE 11.22 A severe forest fire in Warrumbungle National Park in New South Wales, Australia led to very high rates of erosion, and deposition in downstream areas. The photos show how floods led to massive sediment deposition. A few years later, a subsequent flood removed the sediment, and even eroded the channel deeper than before the fire. **Source:** from Tulau et al. 2019, used by permission.

benches, terraces, and alluvial fans that were deposited after fires at some time in the past. The immediate impacts of massive deposition of sediments may seem harmful, but in the long run they are fundamental to future forest development.

The sediments that are added to downslope areas after fire may remain a short time, or may wash farther down the watershed in a later flood. Riparian ecosystems may be influenced by post-fire floods and erosion, and periods of rapid accumulation can be matched by rapid losses (Figure 11.22).

Each Species of Animal Has a Different Response to Forest Fires

Animal species respond to forest fires, and the best generalization is that generalizations do not apply generally. Dead trees benefit some species, such as woodpeckers, which find abundant food when beetles feast on the cambium of dead trees. Decreases in overstory trees can lead to increases in understory vegetation that may be eaten by herbivores. Figure 11.23 gives just a glimpse of how animal species might respond to forest fires, in the short run after fire and across long-term periods of forest change.

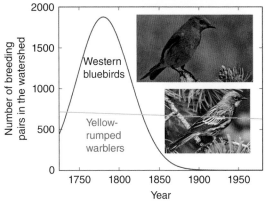

FIGURE 11.23   Many species thrive in the aftermath of fires, including beetles that feed under the bark of dead trees, and others like the blue fungus beetles that feed on fungi on decomposing logs. The longer-term effects of fire include changing habitat suitability for bird species, as the mixture of age classes of forests change across a watershed in Yellowstone National Park. **Sources:** based on data from Romme and Knight 1982; mountain bluebird photo by Elaine R. Wilson, yellow-rumped warbler photo by Jacob W. Frank, National Park Service.

Fires Interact with Other Major Events in Forests

The gradual, long-term changes in forests are often punctuated by not just one major event, but sometimes by multiple interacting events. A glance at Figure 10.12 gives the impression that a landscape covered with beetle-killed pine trees would represent a major rise in the risk of severe crown fires. The dead needles have lower moisture contents than live needles, and they ignite more easily. The needles fall off the trees within a couple years, and the risk of a severe crown fire might be much lower than before the beetles. As dead trees fall, the large fuels on the ground increase, along with the risk of a slow-moving fire that might burn with high intensity. The fallen trees would also pose a very high safety risk to any fire fighters. A fire did indeed sweep across the area in Figure 10.12; the fire burned under severe weather conditions, and it might be difficult to discern whether the presence of post-beetle woody fuels influenced the burning or not. A similar set of complex interactions would apply to cases of major blowdowns.

Ecological Afterthoughts: How Do Slow Changes in Forests Shape the Effects of Fires?

The upper photo in Figure 11.24 of a ponderosa pine forest can be examined for insights about the condition of the forest at the time it burned, and how it came to be that way. What are the key points to describe the current condition of the forest? What clues can be found in the picture about the history of the forest? The lower picture shows the same location after a moderately intense wildfire. The fire was hot enough to kill the trees, though it was not a crown fire (the needles on the large trees were scorched, but not burned in flames). What might the future development of this forest be? How would the future forests develop on other aspects and slopes?

FIGURE 11.24 A ponderosa pine forest in the Spring Mountains, Nevada, USA, before and one-year after an intense fire. **Source:** photos by Scott Abella.

CHAPTER 12

Events in Forests: Management

Forests have been influenced by people for millennia. The Gilgamesh story at the beginning of the book showed that issues about forest harvesting have been around as long as civilization. Human influences on forests probably go back 100 times farther, to the point where human ancestors first started using fire. Forests have been cleared for agricultural use for more than 6000 years, with frequent shifts back into forests. Agricultural use often included soil augmentation with animal wastes and ash fertilizer, and soil enrichment apparently can last for more than 1000 years. Old forests in France that were once thought to be primeval (never cleared for cropping) turn out to have been cleared during Roman times, and soil chemistry and understory plant diversity remain higher at the center of sites of Roman settlements (Figure 12.1).

Harvesting Is the Third Largest Forest Event Across the Planet

Forest harvesting occurs on about 18 million ha each year (Furukawa et al. 2015). This is of course a vast area, but forest harvesting ranks only #3 in major events that change forests (FAO 2020). Fire is #1, with about 67 million ha of forest burned annually. Insects come in second, affecting about 29 million ha yr^{-1}. Storms and diseases each affect about 5 million ha annually. Adding up all the major events, harvesting accounts for about 15% of the area of major events. Just like the other major events, harvesting is influenced in part by the current characteristics of a forest (large trees are targeted for harvesting more than small trees), and harvesting has a very major legacy of influence on the subsequent composition and structure of forests.

Few Forests Are Plantations, But Plantations Provide Most of Our Wood

Aldo Leopold (1949) thought of forest management as fitting into two classes:

> In my own field, forestry, group A is quite content to grow trees like cabbages, with cellulose as the basic forest commodity. It feels no inhibition against violence; its ideology is agronomic. Group B, on the other hand, sees forestry as fundamentally different from agronomy because it employs natural species, and manages a natural environment rather than creating an artificial one. Group B prefers natural reproduction on principle. It worries on biotic as well as economic grounds about the loss of species like chestnut, and the threatened loss of the white pines. It worries about a whole series of secondary forest functions: wildlife, recreation, watersheds, wilderness areas. To my mind, Group B feels the stirrings of an ecological conscience.

The issue of ecological conscience will come back around in the next chapter. A key point for this chapter is that success in managing forests in both group A and group B depends on getting the ecology right.

Worldwide harvests of wood have been roughly constant since 1990, or perhaps there has been a slight increase. The current rate of harvest is about 3.5–4.0 billion m^3 yr^{-1} (~1.8 billion Mg yr^{-1}, 0.9 billion Mg of C yr^{-1}). About a quarter of the wood is simply

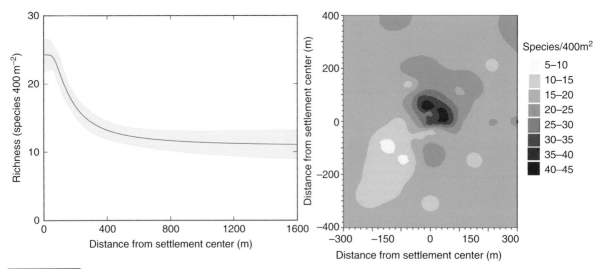

FIGURE 12.1 Archeological relocation of ancient Roman settlements showed legacies of soil changes that have lasted almost 2000 years. About double the species of understory plants occurred at the focal point of settlements, declining about 50 m away, reaching background levels after 400 m from the settlement. **Source:** Based on Dambrine et al. 2007.

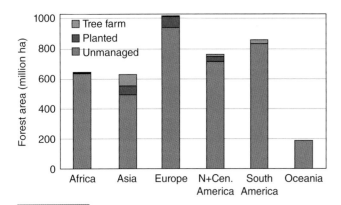

FIGURE 12.2 Almost all forests fall into the category of unplanted, unmanaged forests, but the 7% of all forests that are planted (and managed at low intensity, or as intensively managed tree farms) produce about half of the world's wood supply (Oceania is mostly Australia; **Source:** Data from FAO 2020).

burned to meet cooking needs in areas without access to other energy sources. The rest of the wood goes for solid wood products and fiber-based products, along with minor amounts for bioenergy (FAO 2020).

Prior to 1990, most of the wood came from harvesting trees no one planted. The role of planted trees in managed forests and tree farms increased to the point where perhaps half the current wood supply in the world comes from planted trees (Warman 2014; Payn et al. 2015), even though planted forests account for about 7% of the forests of the world (Figure 12.2). Tree farms are the most intensively managed forests; they account for only about 1.5% of the world's forests, and their high growth rates (and good locations relative to infrastructure and markets) allow tree farms to produce about one-third of the world's non-fuel wood supply. If it was not for the wood supply from intensively managed plantations, the price of wood products would be notably higher, increasing profitability of harvesting wood from unmanaged forests. Unmanaged forests generally have lower volumes and growth rates than tree farms, so larger areas of unmanaged forests would be needed to compensate for reduced plantation areas.

Deforestation Can Be Tallied from Government Reports, or from Satellites

Everyone "knows" the forested land area of the Earth is declining, and the latest summary from the United Nations says deforestation is continuing (FAO 2020). The worldwide assessment is the tally from values reported by governments around the world. Some countries have very reliable databases, and others do not. Definitions of terms (such as forest, forest area, and loss of forest) also differ among countries. Satellite-based tallies of forested land area show the Earth now has about 7% *more* forest land area than in 1980 (Song et al. 2018). The global trends for forest area are actually increasing, but a net increase results from the combinations of increases and decreases in different parts of the world. Many areas of the Tropics are experiencing losses of forests faster than expansion of forests, and most of the Northern Hemisphere is experiencing expansion of forests along with reductions in agricultural land area.

Human Influences on Forests Have a Spectrum from Low to Very High

As noted in Chapter 3, all forests on Earth have been influenced by humans for thousands of years. The first major influence was likely an increase in fires, followed at the end of the last Ice Age by the extinction of megafauna. The third impact was probably forest grazing by livestock, which would cover more area than direct forest harvesting in historical times. European colonization led to conversion of forests to agricultural uses. Widespread forest harvesting in the nineteenth and twentieth centuries changed vast landscapes in the New World. These changes were followed in the twentieth century by transport of insects and diseases from regions (where local species were somewhat resistant) into new territories where novel species were decimated. The world's forests in the twenty-first century are already quite different from historical forests in many areas, and the differences will widen (Chapter 14).

This chapter focuses on the ecology of forests that are intentionally managed, extending some of the material on microclimate and water from Chapters 2 and 7. The variety of management approaches is very large around the world, and even in local areas, so some general distinctions may be useful in examining ecological aspects (Figure 12.3). Decisions about forest management often relate to economic opportunities for wood products, though in some cases large investments are made to reduce risk from fires, avalanches, or other hazards in areas other than plantations.

A small amount of insight into economics provides part of the background for understanding the ecology of forest management. The most profitable forest management might be harvesting large, old trees that did not require any investment other than the costs of logging (Figure 12.4). High profits from logging unmanaged forests can be limited by two things: policies that limit or forbid

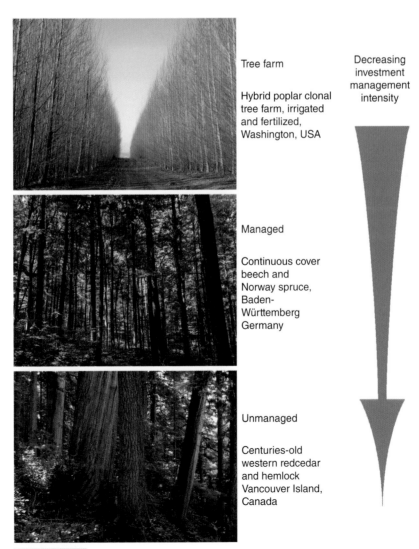

Tree farm

Hybrid poplar clonal tree farm, irrigated and fertilized, Washington, USA

Decreasing investment management intensity

Managed

Continuous cover beech and Norway spruce, Baden-Württemberg Germany

Unmanaged

Centuries-old western redcedar and hemlock Vancouver Island, Canada

FIGURE 12.3   Forests around the world have a wide range of costs and values, leading to different levels of profitability for management. The typical rates of return are very, very low for long rotations, and high for tree farms.

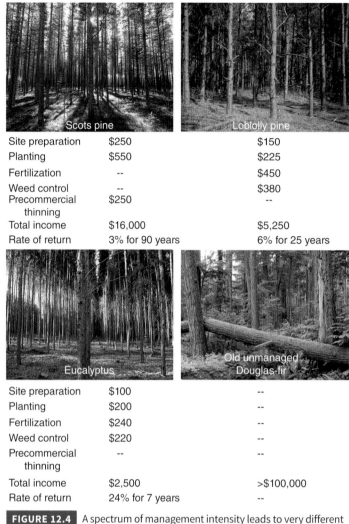

	Scots pine	Loblolly pine
Site preparation	$250	$150
Planting	$550	$225
Fertilization	--	$450
Weed control	--	$380
Precommercial thinning	$250	--
Total income	$16,000	$5,250
Rate of return	3% for 90 years	6% for 25 years

	Eucalyptus	Old unmanaged Douglas-fir
Site preparation	$100	--
Planting	$200	--
Fertilization	$240	--
Weed control	$220	--
Precommercial thinning	--	--
Total income	$2,500	>$100,000
Rate of return	24% for 7 years	--

FIGURE 12.4   A spectrum of management intensity leads to very different forests around the world. Unmanaged forests are comprised of trees no one planted, and social values for the forests typically have many dimensions. Managed forests may rely on natural regeneration (from seeds or sprouts) after harvesting, the intensity of management (and investment) may not be large, and a number of shorter-term goals might be important. Tree farms require heavy investments to support high growth rates, and management typically aims to optimize only the production of wood. Profitability fluctuates with short-term and long-term changes in markets, and the overview numbers here do not include the costs of land or impact of tax policies.

logging in such forests, and factors that limit value (such as large distances to markets, or difficult terrain for logging). Intensively managed tree farms may be the next most profitable, especially when short rotations provide early payoffs for investments. Investments for establishing slow-growing forests are sometimes justified when forest policies require establishing a new forest as a precondition on harvesting large trees. A constraint like this essentially ties together the option of harvesting a valuable old forest with paying to establish a new forest (even if that second step is not a profitable investment).

Tree Farms Are All About Production, Not Broader Ecological Features

Intensively managed tree farms may have as much or more in common with agricultural fields as with forests (as noted by Leopold in his group A characterization). A typical sequence for a eucalyptus rotation illustrates some features that are common for tree farms (Figure 12.5).

1. The previous forest is harvested by large machines that grab the base of the stems, saw off the stem, lay the stem down and then pull it through a cutting head that strips off branches and bark, cuts stems into precise lengths and piles them. Another machine comes along to collect the wood and transport it to a central place for loading onto trucks. The forwarding machine drives across the branches and bark that were stripped from the trees, reducing the compaction of the soil. The branches and bark are taken off the trees and left on the site to reduce the cost of hauling wood to the mill, and to reduce the loss of nutrients from the site.

2. Plantlets are developed in nurseries from buds of selected "parent" trees, and the parent trees are often hybrids of at least two species. Plantations with clones typically grow faster than seed-origin plantations not because of superior genes, but because high uniformity of tree sizes leads to higher forest-level growth.

3. Site preparation may entail treating the logging slash by chopping it, or shoving it aside from the rows where trees will be planted. Back in the twentieth century, the slash was sometimes burned, but the loss of nutrients, depletion of soil fertility, and smoke pollution led to abandonment of burning. Some soils have compacted layers at some depth below the surface (such as 50 cm), and large machines can pull a long, vertical blade to rip through this layer and facilitate access by roots to deeper soil layers.

4. Tree farms always receive fertilizer applications, though the elements, rates, and timing of application differ among soil types. Fertilization before planting is common, followed by one or two additional applications.

5. Understory vegetation competes with trees for soil nutrients and water, so herbicides are sprayed to minimize non-tree vegetation. The spraying may come before planting, or after tree crowns rise a few meters above the ground.

6. Trees are planted, typically at spacings of 3×3 m, and six or seven years later the cycle is repeated. In some cases, the cut stumps are allowed to resprout to provide a "coppice" rotation.

Other types of tree farms around the world are treated similarly, with important differences where site conditions warrant. For example, poorly drained sites might be drained with ditches to improve soil aeration, or soil can be piled 10–20 cm high in bedding rows where trees will be planted to facilitate early root development.

The key ecological features of tree farms include the funneling of site resources only into trees (not understory plants); any ecological opportunities for other plants (and animals) are limited. These forest-level features may be diversified at a landscape scale, where areas of natural forest are blended with plantations.

Two other points are important to keep in mind about the ecology of landscapes with planted forests. Sites used for tree farms are often suited well enough for agricultural use, so that the forests are sometimes replaced by agricultural land use. In some areas, tree farms use non-native species, and some of these (especially acacias and pines) pose high risks of "escaping" and becoming established far outside the tree farms (see Chapter 14).

FIGURE 12.5 Intensively managed tree farms, such as clonal eucalyptus plantations in Brazil, include mechanized harvesting (upper part of the top photo), soil preparation and fertilization (machines in the background of the middle photo), weed control, and careful planting in rows (foreground of middle photo). The ecological features of tree farms depend both on within-forest characteristics, and those of the larger landscape matrix (lower photo; **Source:** Votorantim Institute Collection 2006).

How Sustainable Are Tree Farms?

High growth rates in tree farms result from favorable climate, soils, and investments in management, the same as in agriculture. Empirical evidence clearly shows that growth in tree farms is sustainable, typically increasing across rotations as understanding and techniques improve. Small land holders in Vietnam can supplement their incomes with tree farms, and learning about ecology and silviculture have allowed substantial increases in growth of acacia plantations across multiple rotations (Figure 12.6; Harwood and Nambiar 2014). More intense management gave higher yields than lower intensity management. Even though acacia trees can enrich soils via biological nitrogen fixation, they still depend on soils (and fertilizer-amended soils) for all other nutrients. The optimal level of management intensity of course depends on the economic costs, and the value of the final harvest. The yields of the two most recent rotations in this example were similar, so future increases might be possible only with substantial increases in management intensity.

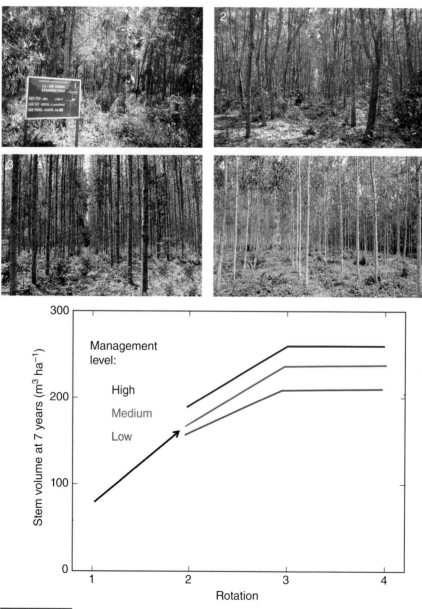

FIGURE 12.6 The productivity of acacia more than doubled across four rotations in southern Vietnam, from less than 90 m³ ha⁻¹ in the first rotation to over 150 m³ ha⁻¹ in the second rotation, and over 200 m³ ha⁻¹ in the third and fourth rotations. Rising growth resulted from improved understanding and management intensity. Levels of management inputs (slash retention, fertilization, weed control), were tested after the first rotation, demonstrating the increased yields with increased inputs. **Source:** from Huong et al. 2020.

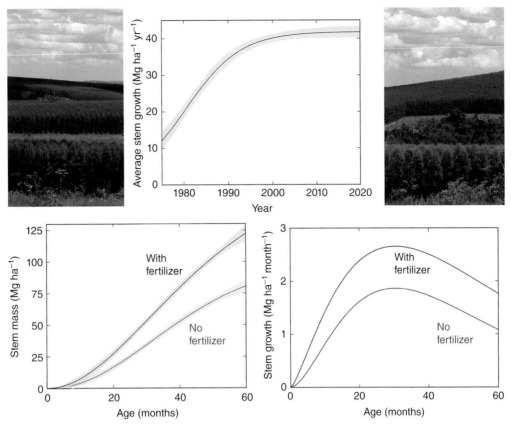

FIGURE 12.7   Growth rates of eucalyptus across a large area of southern Brazil increased by more than twofold across several rotations (upper graph). Rates of increase slowed as technology and management pushed the limits of biological potential (**Source:** data provided by Rodrigo Hakamada). Current rates of growth depend strongly on management inputs; if plantations are not fertilized, growth is not sustained (lower graphs of accumulated wood mass, and current month growth rates; **Source:** Data from Ryan et al. 2020).

Eucalyptus tree farms in Brazil have shown similar patterns. As ecological insights and technology developed over several decades, growth rates increased by more than twofold (Figure 12.7). The insights included understanding the capacity of the trees to use higher rates of fertilization, genetic breeding (including blending species into high-performing hybrids), and shifting from seed-origin to clonal plantations. Government policies in Brazil require landowners to retain (or restore) 20–80% of forest lands as native forests, leading to diverse forest landscapes in areas with vast eucalyptus plantations.

Tree farms are designed to foster the growth of a single species, and sometimes a single genotype. The choice of species depends on both the market values and the suitability of a species at a particular site. Low genetic variation in a plantation fosters high growth rates, but might also pose a risk of an insect pest or disease outbreak. Forests with many species might lose a tree species to a pest that preys on that species, with little change in the residual forest. However, the downside might be lower growth rates and lower market values for wood. Maximizing profits depends on the balance between risks of low product values and the empirical risks of pest and disease problems. The risk of pests and diseases in single-species and single-genotype forests is significant, and tree farms use varieties that have been developed with resistant genes. Substantial research investments are needed to stay ahead of the challenge of novel spread of pests and disease. The bottom line is that billions (trillions?) of dollars' worth of trees and agricultural crops are produced every year from successful monoculture farms around the world, so empirically the risks of monoculture forestry are too small to offset high profitability.

Managed Forests Come in a Variety of Systems

Most of the managed forests of the world are not cropped as intensively as tree farms, and the goals of management are diverse and not always sharply defined. These forests are often considered to be important for watershed values, recreation, wildlife habitat and sometimes grazing by livestock, in addition to producing wood. Many managed forests have seedlings of preferred species planted, and others rely on natural regeneration from seeds or resprouting stumps and root systems.

The variety of managed forests is so broad that no clearly distinguished categories arise naturally. This chapter explores the variety using somewhat arbitrary categories of rotational forests and continuous cover forests, with continuous cover forests also including some varieties that are called "close to nature" and "variable retention" systems.

Rotational Forests Have Birthdays

Along with tree farms, rotational forests begin after a previous forest is harvested. In some cases, rotational forests are established on areas that had other land uses, such as pasture or row-crop agriculture. The legacies of the prior forest or land use are usually important in the early development of the new forest. For example, plantations of loblolly pines in the southeastern USA established on former agricultural fields would experience little competition from understory vegetation, but by the end of the rotation an understory community may be well developed. The next rotation of pines would have substantial competition challenges, unless the understory was reduced by prescribed fire before harvest. Herbicides are commonly used as to boost early development of desired tree species. The intensity of biomass removal in harvest can affect the growth rates of the following rotation, though the effects are generally not large (Figure 12.8).

Harvesting operations usually impact soils in rotationally managed forests, and site preparation activities change soils even more. Site preparation can include the treatments used in tree farms, though longer rotations make it difficult to apply expensive treatments profitably at the beginning of a rotation. Any treatments that substantially affect soils can enhance or degrade the ability of the soils to support tree growth (Binkley and Fisher 2020).

Some site preparation treatments can be extreme in rotational forestry, such as extensive drainage systems in wetland forests. About 15 million ha of forests in Scandinavia and the Baltic countries were drained to facilitate tree growth in the twentieth century (Nieminen et al. 2018). The increase in growth was about 3–10 m^3 ha^{-1} yr^{-1}, which may be too little to cover the cost of the expensive modification of the wetland sites. This widespread practice declined in the twenty-first century, as a result of economic losses combined with concerns about environmental degradation. Indeed, investments are being made to try and reverse the loss in biodiversity values that resulted from the earlier investments in drainage (Similä et al. 2014). It's not unusual for one generation of foresters to wish the previous generation had made different choices.

Less intensive site preparation treatments are profitable for some situations in rotational forests. Treatments such as raised beds for planting, disking the topsoil, and pulling blades through the subsoil might raise forest growth on average by about 0.5 m^3 ha^{-1} yr^{-1}. Other treatments such as fertilization and understory control (raising growth by about 2 m^3 ha^{-1} yr^{-1}) may be better choices for increasing growth in rotational forests (Carlson et al. 2006). Most of the rotational pine plantations across the southeastern USA use herbicides for site preparation (almost two-thirds), though some invest in mechanical soil treatments (less than 20%, Baker 2020).

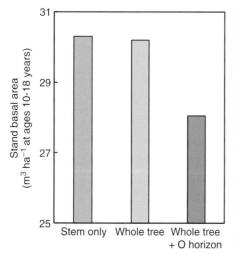

FIGURE 12.8   The harvest that precedes a rotation can have a legacy on growth rates into the next rotation. Harvesting of radiata pine plantations in New Zealand may remove only stems, or may also remove branches and crowns (increasing nutrient removals). The legacy of the intensity of harvest typically either is unimportant, or intensive harvest might lower growth of the next rotation by 5–10%. However, harvesting that is followed by removal of O horizons usually drops growth dramatically (**Source:** Data from Davis et al. 2015). Removing O horizon material happens in some forests, and this can also lower nutrient supplies and tree growth (photo of Graham Will's litter raking experimental site in New Zealand).

After harvesting and site preparation, rotational forests develop either with planted seedlings or natural regeneration. Planting is expensive ($200–$300 ha^{-1}), but the investment may be warranted where seedlings lead to forest composition and structure that meet management goals, including rapid tree establishment, high growth rates, and resistance to pests and diseases. Some of the initial investment in site preparation and planting may be recovered with intermediate cutting (thinning) of some of the trees. Most pine plantations across the southeastern USA produce income from two thinnings (at 14 and 21 years) before final harvest at 29 years (Baker 2020). Rotations also have investments after planting. More than half of these pine plantations are fertilized at both the time of planting and midway through a rotation. Understory control is practiced across the region, with most receiving applications of herbicides or understory burning (or both). Investments in mid-rotation silviculture are less profitable for forests managed at longer rotations.

A major ecological aspect of rotational forestry is the change over time, with little opportunity for long periods without change. The dynamics that follow harvesting and forest reestablishment are rapid, and later changes are slower but still quite fast. The interception of light by tree canopies reaches more than 80% of incoming light in the first decade of most rotational forests, though some may need a second decade to reach that level. Transpiration ramps up from near 0 to account for a very large proportion of incoming precipitation. The understory environment experienced by plants changes from warm (or hot) with moist soil to shaded and cooler conditions. The changes during a rotation also have a spatial component, as the influence of neighboring forests shift around the boundaries of a plantation.

Understories and Overstories Interact Through a Rotation

It might seem that trees control how much understory vegetation develops in a forest, as trees have the opportunity to intercept light and minimize light supplies to short plants. Competition for light is of course important in forests, but trees cannot monopolize the supplies of soil nutrients and water. Without control of understory vegetation, a large proportion of a site's nutrient and water

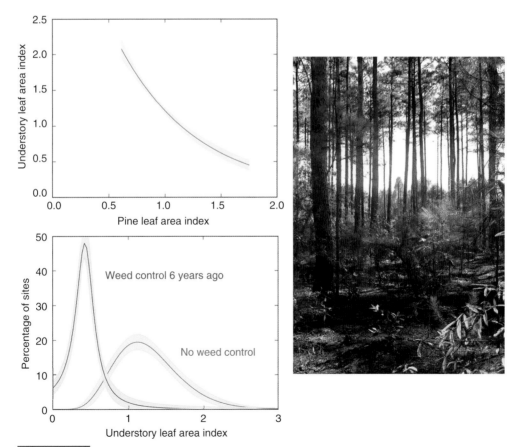

FIGURE 12.9 Understory vegetation is typically very important in managed rotational forests. The leaf area index of overstory pine trees (measured in winter) is low in forests with high understory leaf area (upper). This relationship goes both ways: high understory leaf areas reduce overstory resource use, and high overstory leaf area holds back understory vegetation. Application of herbicides in established rotational forests can substantially reduce understory leaf area (six years after treatment), and increase tree growth by about 25%. **Source:** Data from Blinn et al. 2012.

supply is used by understory plants (Figure 12.9). More productive rotational forests, such as pine forests in the southeastern USA typically receive herbicide applications or understory burning to reduce understory competition, as described above. Other forests retain more understory vegetation throughout the rotations, with benefits for wildlife and perhaps other forest values.

Continuous Cover Forests Have no Birthdays, and Less Change

Rotational forests are always changing through time, whereas continuous cover forests are managed for smaller changes in both space and time. Harvesting all the trees in a forest at one time gives immediate, dramatic changes in environments, vegetation structure, and animal habitat. Water yields increase after all trees are harvested, and opportunities for colonization by plants and animals shift. Continuous cover forestry (sometimes called close-to-nature forestry) mutes these changes by harvesting only a small proportion of all trees at a time, always leaving a landscape with trees of varying ages. The management variations are very broad, with differences across forests in the harvest rates for large and medium size trees, selection of which tree species are favored for removal or retention, and the frequency of harvests.

The frequent harvesting of trees dispersed through a forest gives a different scale of forest change in both space and time. Each harvest entails use of heavy machinery which can have repeated impacts on the soil (and designed systems of roads and trails may be needed to manage soil compaction). Cutting a single tree provides a gap in the canopy (Figure 12.10), and this might seem like it would create an opportunity for understory vegetation (and maybe newly establishing tree seedlings). The creation of a canopy gap may or may not represent much opportunity. Sunlight reaches a forest canopy at an angle, and small gaps in forests with tall trees lead to the "extra" sunlight simply falling on the lower crowns of neighboring trees. Understory trees and other vegetation would only receive increased supplies of light when harvesting removes groups of trees rather than individuals.

Gaps in canopies do not map onto the same ground area as gaps in soils for two reasons. Tree crowns generally don't intertwine (so one tree monopolizes a position in the canopy), and root networks have very high areas of overlap in soils. The roots of one tree usually explore soil far beyond the area covered by the tree's crown. Very large gaps are needed before areas of soil are freed up from the influence of surrounding trees (see Figure 4.11). The second reason is that the zone of maximum uptake of nutrients and water by a tree would be in the area near the stem, but the angle of the sun means that the zone of increased light falls somewhere other than directly below the canopy opening above the tree stump. Indeed, the sun angle changes throughout the day and across the seasons, so the light that was formerly captured by the harvested tree is dispersed rather broadly, not in a concentrated spotlight of opportunity on one area.

Some continuous cover forests are managed with harvests at longer intervals, with tree cutting dispersed through the forest or clumped in small patches. These forests would have periods of higher and lower tree cover, and the inverse pattern (perhaps) of understory vegetation decreasing and increasing. The spacing

FIGURE 12.10 Continuous cover forestry entails harvesting of single trees, or small groups of trees, at somewhat frequent intervals. Removal of a large Norway spruce tree in this forest of spruce and beech created a gap in the forest canopy (upper photo). However, the gap is small enough that surrounding trees capture most of the light that would have been used by the cut tree, leaving the understory heavily shaded (lower photo).

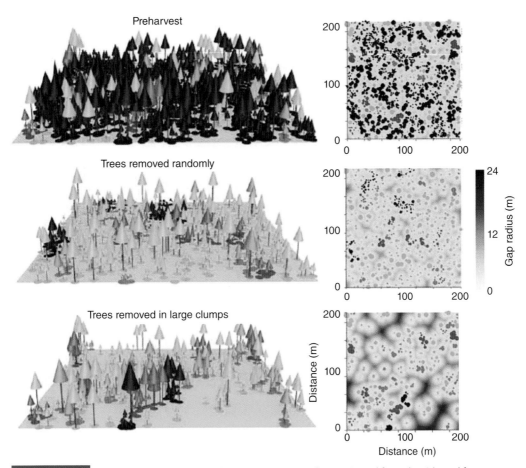

FIGURE 12.11   A computer visualization of a continuous-cover forest, viewed from the side and from above. The preharvest forest (basal area 30 m² ha⁻¹) was very dense, and a 2/3 reduction in basal area was simulated with either random tree removal, or clumped tree removal. The structure of the post-harvest forests would be very different, with clumped removal giving opportunities for understory vegetation with no influence from nearby trees, and for surface fire propagation. **Source:** Based on Tinkham et al. 2017.

between trees probably has a large effect on understories, as narrow (uniform) spacing allow roots and mycorrhiza to occupy the full area of the forest, and clumping of the residual trees provides larger areas free of tree competition (Figure 12.11).

Tree Growth Is Faster in Rotational Forestry than in Continuous Cover Forestry

When all trees are harvested in a forest at one time, some time will pass without full occupancy of the site by the next generation of trees. This period is relatively short in tree farms, ranging from less than a year to a few years. Forests managed on longer rotations may take a decade or more for new trees to reach maximum use of light, nutrients and water. The period of canopy redevelopment has lower growth than the average across a rotation, so it might be that a continuous cover system would achieve higher total growth rates across decades or a century. The loss of some tree crowns might not reduce the total light use by the canopy very much, and intertwining root systems would continue to exploit the soil fully.

This possible pattern of higher growth in continuous cover forestry might be offset by other factors. Rates of growth of individual trees change with tree ages or sizes, and the patterns differ between dominant and subordinate trees. The heterogeneity of tree sizes influences total growth of a forest, and greater heterogeneity lowers forest growth. This heterogeneity penalty might not apply to forests with multiple species.

Given a range of reasons why rotational forests may have higher or lower rates of growth than continuous forests, evidence is clearly needed for any reliable inferences. Not many experiments have tested these two approaches, but there seems to be a clear trend toward rotational forests having 10% higher growth (or more) relative to continuous cover systems (Lundqvist 2017).

FIGURE 12.12 Forests that developed without direct management may have very large, valuable trees that can be harvested without prior investment, followed either by no management or conversion into a managed forest or tree farm (western redcedar/western hemlock on Vancouver Island, Canada).

Management of Unmanaged Forests May Seem Like an Oxymoron

Trees can be harvested from a forest without regard for the future of the forest, and this might or might not qualify for the term "management." In other cases, valuable old trees are logged from previously unmanaged lands and the area transitions into some form of rotational or continuous cover forestry. Clearly the categories for types of forest management are not distinct or simple.

The transition from an unmanaged (typically old-growth) forest into a managed forest entails many decisions (Figure 12.12). Clearcutting may be the cheapest approach, but post-harvest conditions may not provide wildlife habitat or other goals of land-owners. The size of harvest areas can span from single trees (plus the collateral loss of trees damaged as a large tree falls) to very large clearcut areas. Some trees could be retained, ranging from a few per hectare to many. The residual trees might be dispersed across the site, retained in clumps, or in largely uncut blocks surrounded by harvested areas.

With so many options, the choices just don't fit into tidy categories. A catch-all phrase might be useful, such as variable retention forestry, or green tree retention. The single phrase would not communicate much information until it was joined with descriptors such as dispersed, clumped, block retention, or other terms.

The conversion of an unmanaged forest to a managed forest also changes understory vegetation. It might seem that growth of the managed forest would lead to progressive declines in the understory, as the light supply declines beneath the tree canopy. Patterns are not always so simple, though, especially as understory vegetation is typically comprised of a wide range of species with varying resource demands and efficiencies of resource use. A plantation forest that followed logging of an old-growth forest in Oregon, USA reached an impressive biomass of more than $150 \, \mathrm{Mg \, ha^{-1}}$ in four decades, yet understory biomass declined by only one-quarter from the post-logging high (Figure 12.13). The diversity of understory species also declined along with biomass, but only from about 12 species down to 9 species in $3 \times 3 \, \mathrm{m}$ plots.

How Does Retaining Trees Influence the Next Forest After Logging in Unmanaged Forests?

Clearcutting changes almost everything in a forest, from the cutoff of the supply of new carbon to the soil biota, to changing the microclimate near the ground, and altering all aspects of animal habitat suitabilities. Retaining some trees would change the magnitudes of these effects, but experimentation is needed to provide evidence of the size and direction of effects (Figure 12.14).

One experiment on Vancouver Island, Canada has been followed for three decades, documenting post-logging changes in response to systems of retention. The unharvested forest had about 30% cover of understory trees, but most of the understory

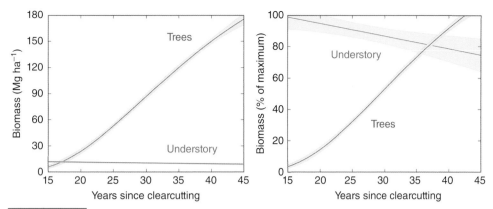

FIGURE 12.13   After clearcutting of an old-growth forest in Oregon, USA, the reestablished trees reached a maximum growth rate after about three decades. Despite the rapid redevelopment of overstory dominance, understory biomass in resampled plots changed by only about 25%.
Source: Based on data in Halpern and Lutz 2013.

FIGURE 12.14   Experiments in variable retention silviculture on Vancouver Island, Canada included testing a range of tree removals, as well as spatial arrangement (clumped or dispersed; **Source:** upper photo by Bill Beese).

trees died in logged plots (Figure 12.15). Regeneration of trees was rapid, however, with cover of understory trees surpassing the unharvested forest after 10–15 years. The cover of mosses dropped by about one-third over the decades in the unharvested forest, while the variable retention treatments saw a near elimination of mosses, with rapid recovery to levels that were about 1/3 to 1/2 of the unharvested forest. The cover of understory herbs increased by several-fold very rapidly after logging, but declined as the trees expanded.

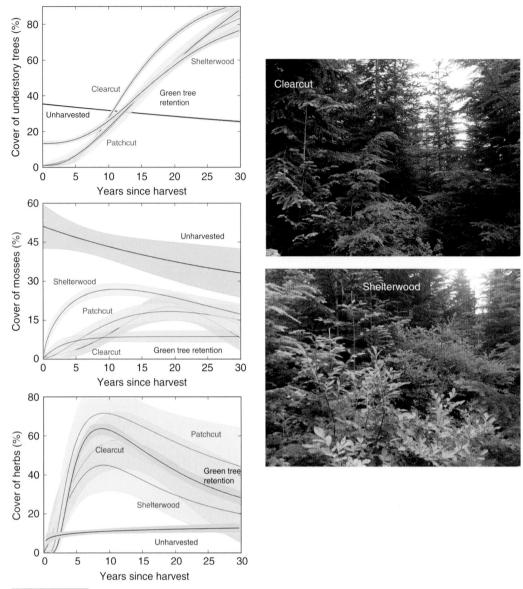

FIGURE 12.15 Understory responses to variable retention treatments in the MASS study on Vancouver Island, Canada. The clearcut treatment occurred across 69 ha, with all large trees removed. Patch cuts were 1.5–2 ha in size, alternating with unharvested patches. Green tree retention left 25 trees ha^{-1} at uniform spacing. The shelterwood treatment retained 30% of the basal area, including some large trees, dispersed across the plots. The responses of cover for understory tree species, mosses, and herbs varied among the variable retention treatments, and they varied even more over time. **Sources:** Based on data from Beese and Sanford 2020, photos by Bill Beese.

The responses of trees to variable retention treatments were also examined in experiments that reduced understory competition and increased nutrient supply. The heights of regenerating trees were similar in the clearcut and the treatments with fewer old trees, but were somewhat shorter in the 30% retention (shelterwood) where competition with older trees remained stronger (Sandford et al. 2020). Regenerating trees of western hemlock and amabilis fir trees had double the biomass at 25 years in fertilized plots as in unfertilized plots, with consistent effects across the harvest treatments. Reductions in understory vegetation gave an even larger response of regenerating trees, and the combination of fertilization and understory control produced the largest trees. The control of understory vegetation gave a smaller response in the 30% retention (shelterwood treatment) than in the others that retained fewer old trees. Some hopeful ideas may be suggested where small trees benefit from big trees, but evidence from around the world clearly points toward competition dominating interactions between trees.

The details of variable retention forestry influence suitability for wildlife. For example, the number of bird species that forage for insects in trees depends on the number of trees retained within a clump. Clumps with fewer than 10 trees had lower numbers of foragers (specialists on the bark, and in the upper crowns) than clumps with more trees (Figure 12.16). Some guilds of birds decline after harvesting, such as cavity-nesting species, but on average the number of bird species in a forest does not show a clear change

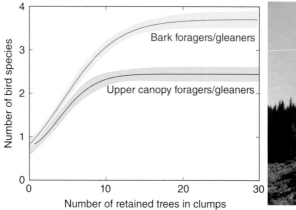

FIGURE 12.16 Variable retention treatments across four sites (from California to Washington, US) showed strong relationships between the number of trees retained in clumps, and the number of bird species foraging for insects (**Source:** Based on data from Linden et al. 2012).

when variable retention studies are compiled from around the world (Basile et al. 2019). This does not mean there are not effects, but only that effects depend on local details, and likely differ among species, and change over time as the new forest develops.

Harvesting Is the End of the Line for Some Trees and Forests, and the Beginning of the Next Forest

Cutting and removing trees creates legacies in soils. Nutrient removals in harvested tree biomass are substantial; a typical harvest of stems removes the amount of nutrients that take several decades or more to be replenished from natural sources of inputs (Binkley and Fisher 2020). Removal of branches, twigs and foliage may increase harvest yields by 10–20%, while doubling nutrient losses (concentrations of nutrients are higher in smaller tissues). The extra nutrient loss from whole-tree harvesting may be most crucial on sites with low nutrient contents for two reasons: the removals are a larger portion of the site's total nutrient stocks, and trees on poor sites have a higher proportion of branches and foliage relative to stemwood (Figure 12.17).

The legacies of harvesting also include changes to soils, at multiple scales. Trees are usually harvested by large machines (Figure 12.18), and the harvesters might compact soil. Compaction risks are lower for machines that move on tracks than on rubber-tired wheels. On soils that are very prone to compaction (such as clay-rich soils under high moisture conditions), wider-tired machines might be used to reduce compaction risk. Soil compaction risks differ across soil types, through seasons (as soil moisture changes), and in response to the traffic patterns of the machines through a forest. Repeated traffic on skid-trail leads to high compaction of the trails, but dispersed movement of machines across the forest may lead to more widespread, but lower compaction. On steep terrain, logs may be moved to a central landing area using cables that either drag them across the soil (with high disturbance and erosion), or with one or both ends lifted above the soil with a cable from a tower.

Removing trees from forests also has impacts at the scale of km^2, as road systems are used for large trucks to reach the landing sites where logs are piled by the machines operating across the forest. The extent of road systems may be minimized, or might account for as much as a quarter of the land area. The area covered by compacted skid trails, landings for log piles, and roads tends to range from about 10% of managed forest's area to more than 30%, with long-term implications for trees. Roads also serve as corridors for people, animals, and sometimes exotic plant species, with a host of effects on forest ecology.

Harvesting Is Not the Only Big Event that Happens in Managed Forests

Management plans lay out goals and treatments that would lead to the desired outcomes for a forest. The desired outcomes would typically include some sort of products from the forest, and structure for the forest that would provide values of recreation and conservation. The plans usually expect that harvesting will be the only major event, but of courses storms, insects, diseases, and fires happen in managed forests as well as unmanaged forests.

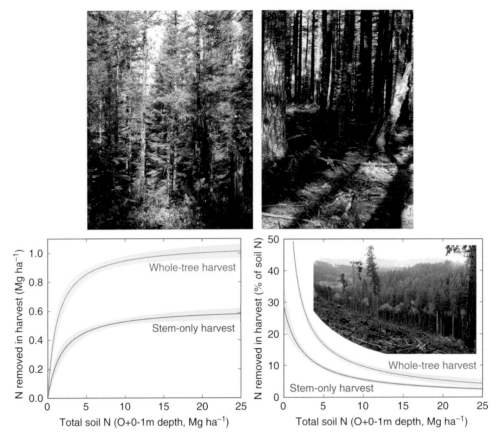

FIGURE 12.17 Harvesting 50-year-old managed forests of Douglas-fir in the northwestern USA removes substantial amount of nutrients, particularly when entire trees are removed rather than only stems. Low-N sites (top left photo) have a high ratio of crowns:stems compared to N-rich sites (top right photo), so whole-tree harvest on poorer sites has the largest increase in N removal compared with stem-only harvest. The amount removed is higher on sites with higher soil N (O horizon plus 0–1 m depth mineral soil, lower left) than on low-N sites, but the proportion of the ecosystem's N lost is lower than on low-N sites (right; **Source:** from data of 68 rotational plantations, Himes et al. 2014).

FIGURE 12.18 Most logging is done by machines that cut trees, and sometimes remove branches (leaving nutrient-rich logging slash on the site; top left). Logs are moved to landing sites by skidding machines (top middle), where they are piled for later loading onto trucks. Wider tires (top right) may be needed for soils with potential compaction risks. Steep terrain may require a high density of roads for access (lower left), or might be harvested with systems of towers and cables (lower right). The impacts and legacies of harvesting occur at a local scale within a forest (a few hundred m²), up to the landscape scale at the size of harvested units, and even beyond as areas near harvested units may experience changes (from altered animal use, increased access by people, etc.)

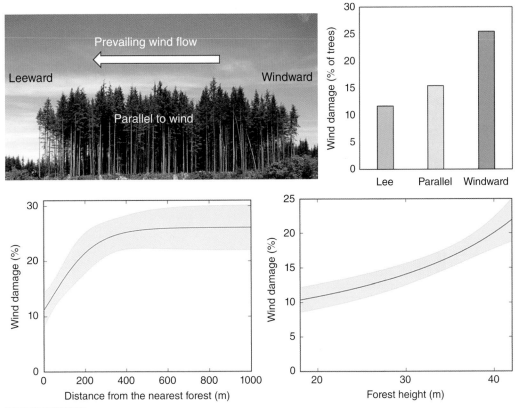

FIGURE 12.19   The risks from wind generally increase after forest harvesting of unmanaged and managed forests. A survey of more than 170 harvested sites across coastal British Columbia, Canada showed that edges of harvest units that are exposed to the highest wind (on the windward side of a forest block) had about 25% of trees blown down, about double the percentage on the leeward (downwind) side (upper). Forests that were within about 200 m of another forest experienced less wind damage than more open blocks (lower left), and tall forests had twice the risk of wind damage than short forests (lower right; **Source:** Data from Beese et al. 2019a, b.

Would the probability of one of these major events be different for managed versus unmanaged forests? Given the differences in forest structure between these two types, and how events depend in part on forest structure, we should expect the probabilities and impacts of events to have major interactions.

Winds shape the structure of trees and forests (Chapter 10), and harvesting changes forest structures in ways that change wind flow. Newly created forest edges experience higher wind speeds, and the changes are more important where edges face the oncoming wind. The interactions of harvesting events and wind events may lead to increases in risks of tree death (by uprooting or stem breakage) by about twofold for a period of several years, as the trees respond to the new wind regimes (Figure 12.19). Some forests show a larger interaction of harvesting with subsequent wind damage; risks are generally (but not always) higher for taller forests, forests near ridgelines, and forests with species that are generally more susceptible to wind damage.

Many of the eucalyptus forests across New South Wales, Australia have been harvested and managed over the past century. Fires are a dominant force across these landscapes. Would managed forests have different extents of fire than unmanaged forests? Major fires burned across the state in 1939. Areas that would later be designated for harvest and management as state forests tended to be at lower elevations and on drier sites, and about 60% of the 1939 fires burned in those areas. Areas that would later be designated for unmanaged parks were higher and moister, and only about 20% of the 1939 fires burned in those regions (Figure 12.20). The year 2009 was another severe time for fires. If the fire risk for managed and unmanaged lands were the same, the fires would be expected to have burned about 60% in the managed forests and 20% in the parks. The actual distribution was about equal for the 2009 fires, so managed forests burned somewhat less than would have been expected (for their lower elevation/drier site conditions), and the unmanaged forests burned much more. This sort of evidence is important to recognize, but of course the details of the fires would be important (where ignitions started, which directions winds blew, etc.). More confidence in the effect of management on fire risk could be developed with more local comparisons of where fires burned and did not burn, perhaps coupled with measurements and modeling of fuels and fire behavior. The occurrence of one event (such as harvesting for forest products) likely influences the probability and impacts of other events (such as fires), but local details will always be important; any broad generalizations based on evidence would likely have very broad variation.

FIGURE 12.20   Severe fires in 1939 burned unmanaged forests across New South Wales, Australia, and about 60% occurred in the portions of the landscapes that would later be designated as managed state forests, and 20% in areas that would become unmanaged parks. This would suggest that managed forests were designated on lower, drier sites (more prone to fire) than were the park lands. Severe fires in 2009 burned parks more extensively than in 1939, and less extensively in managed forests than in 1939. This pattern would be consistent with reduced flammability of managed forests over time, and increased flammability of unmanaged forests. Confidence in this story would increase if detailed studies of the fires and landscapes (perhaps with modeling of fuels and fire behavior) supported the patterns. **Source:** data of Mark Adams and colleagues, in Binkley et al. 2018, used by permission.

Can Forests Remove Enough CO$_2$ from the Atmosphere to Save the Planet?

No, for two reasons. The first is that the rising CO$_2$ concentrations in the atmosphere will change many things on the planet, but the changes will not threaten the planet. At the scale of forest ecology, the effects of forests on the atmosphere are substantial, but the ability of trees to store C removed from the atmosphere is limited. The world's forests have been soaking up maybe 10% of the C released by burning of fossil fuels, and even a 10% increase in the effectiveness of forests would account for only a 1% reduction in

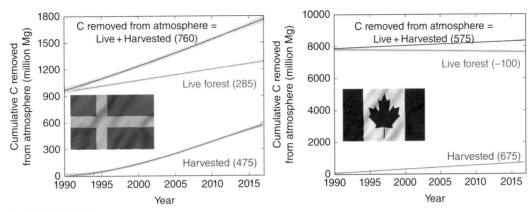

FIGURE 12.21   Rates of harvest increased in Sweden in the past 25 years, leading to a high cumulative amount of wood removed across the decades (left). The rate of forest growth increased even faster, leading to a large increase in the amount of C stored in living forests. Combining the C content of the forest with the C content of harvested forest products gives a measure of the net removal of C from the atmosphere (though of course some wood products would have decayed, with CO_2 returned to the atmosphere). Canada's more extensive boreal forests contain more C in living biomass than in Sweden, but growth rates did not increase, and wildfire and harvesting led to a slight decline in C storage. The sum of the change in live forest C and wood products shows a strong net removal from the atmosphere, but not as large as that of the intensively managed forests in Sweden. **Source:** Based on data from Högberg 2021.

the load of C added to the atmosphere each year from fossil fuels. Indeed, some scientists expect the ability of forests to remove C from the atmosphere may not continue to be as strong in the future as it was in the twentieth century, owing to shifts in the balance between deforestation, reforestation, and forest growth rates.

Advice can be offered on how forests might increase the removal of CO_2 from the atmosphere (Mayer et al. 2020; Ontl et al. 2020). Practices that minimize soil disturbance during logging might retain more C. Nitrogen-fixing tree species appear to be more effective at building up soil C than other species. It might seem that delaying harvests (or even stopping all harvests) might be the strongest way to ensure forests will continue to accumulate C from the atmosphere. This approach has two possible problems. Older forests generally grow more slowly than younger forests, and landscapes dominated by old forests probably sequester less C each year than those with a variety of forest ages. The second potential problem is that older forests don't last forever, and fires, insect outbreaks, and other events can reverse many decades of C accumulation. Perhaps the most important point is that complete accounting is needed for wise decision making. Some portion of the C in harvested trees is used in long-term products, essentially forming another sink for atmospheric CO_2. Buildings made of wood represent a C sink, not only from the C contained in the wood, but as a result of the CO_2 generation that was avoided by not using steel or concrete (both of which have massive CO_2 generation).

Full accounting of C in forests and forest products is a complex topic (MacKinley et al. 2011) beyond the domain of forest ecology. Perhaps the most useful, broad overview of the role of forest management is in the total amount of C contained in forests, and removed from forests, over recent decades (Figure 12.21). Most of the forests in Sweden have been managed with rotational forestry since the mid-twentieth century. Intensive investments in silviculture dramatically increased growth rates along with possible effects of warming climates (Figure 9.14). The amount of wood harvested from Swedish forests increased from 1990 to 2015, yet the total forest biomass continued to increase. The higher stocks of C in living forests combined with large amounts of C removed in wood products to account for a very large net removal of CO_2 from the atmosphere. Inclusion of the CO_2 that would have been released if steel or concrete had been used rather than Swedish wood would increase the size of the net removal from the atmosphere. The boreal forests of Canada cover much more area than the forests of Sweden, with more than six times the amount of C accumulated in trees. The amount of wood products harvested from Canadian boreal forests was somewhat larger than in Sweden. The boreal forests of Canada are managed less intensively than those in Sweden, and the total C in the living forests declined slightly over the decades. All of these details of forest ecology are relevant to the net effect of forests on atmospheric CO_2, but even intensive forest management would not be able to counter the massive amounts of CO_2 added to the atmosphere when fossil fuels are burned.

Temperatures across the land surface of Earth increased by 1 °C since 1960, or 0.5 °C globally (air over oceans warms more slowly). This magnitude of change will have widespread effects on forest ecology, but the changes will not be simple. A key question for this chapter would be how the ecological effects and legacies of forest management will change as the climate changes. Older forests that experienced more drought stress in a warmer or drier future might be at higher risk of beetle outbreaks, and so rotation lengths might be shortened as a counter-strategy. Increasing risks of fires might have the strongest effect on decisions about forest management, with consequences for forest ecology. The future of managed forests will not repeat the past, because of changes in management, changes in climate, and changes in the impacts of fires, pests and diseases (Chapter 14).

Ecological Afterthoughts: What's Next for These Forests?

The future of a forest depends in part on what happened in the past, as legacies of events such as fires and harvesting influence the success of various species and individuals of each species. Contingent events also shape the future of a forest, with droughts, insect outbreaks, invasions of new pests and diseases, fires, and storms. Consider two forests (Figure 12.22), one never harvested and almost unaffected by people, and another with a thousand-year history of human influence (including planting the current forest). Both forests have been shaped by legacies of past events. In pondering the future, would the likely future be more predictable for one than the other? Would legacies have a strong influence for one, and contingent events (which may or may not occur) be more likely for the other? Or would there be little if any value in thinking that unmanaged forests and managed forests would differ in the strength of influence of legacies versus events?

FIGURE 12.22   The unmanaged forest on the left has trees of a variety of ages, including some that were up to 300 years old (Englemann spruce, lodgepole pine, and subalpine fir in the Rocky Mountains, USA). The managed forest on the right is in the Czech Republic with a long history of human management, including planting the current forest about 70 years ago with Norway spruce and non-native Douglas-fir. Both forests have long legacies that led to their current composition, structure and function. Would it be likely that the future of one of the forests would be shaped more by contingent events than by the momentum of legacies from the past?

CHAPTER 13

Conservation, Sustainability and Restoration of Forests

We end, I think, at what might be called the standard paradox of the twentieth century: our tools are better than we are, and grow better faster than we do. They suffice to crack the atom, to command the tides. But they do not suffice for the oldest task in human history: to live on a piece of land without spoiling it.

Aldo Leopold 1938

Forest ecology is all about science, blending together physics, chemistry, and especially biology to understand how forests function, and how they change through time and across space. Humans strongly affect these ecological aspects of forests, and we often have preferences about how we would like a forest to be. People differ in their preferences, so bringing human values into forests creates an arena where values engage with the ecological realities of forests.

The future of forests is always open to change, and this includes landscapes that seem to have been spoiled. The conservationist Aldo Leopold purchased 240 ha of mostly treeless land in Wisconsin, USA in 1935. Historically the landscape would have been a matrix of low-density pine forests mixed with small meadows, but the land had been converted to agricultural uses that were not profitable in the long run. Leopold and his family set out to restore the landscape by planting tens of thousands of trees, and 80 years later the land is covered with forests of conifers, broadleaves, and small patches of prairie meadow (Figure 13.1). The legacy of that restoration work exists on the landscape, and in the form of a classic book (*Sand County Almanac*) that everyone should read, or reread.

Conservation, Sustainability and Restoration Build Values, Ethics, and Esthetics onto a Foundation of Forest Ecology

Conservation, sustainability and restoration overlay human values onto the ecology of forests. Misunderstandings about the nature of forests (over space and time) may lead to poor matches between expectations based on values and the way the future actually unfolds. The current composition and structure of a forest is a starting point, and sometimes insights about the historical condition of a forest is the focus for goals of conservation, sustainability and restoration. What's next for a forest depends on the legacies (or inertia) of historical factors shaping the forest, and the interventions that people use to shape the future forest. The term "inertia" has a precise definition in physics, but in ecology it could be used loosely to describe how factors that have shaped a forest recently (such as competition favoring growth of large trees at the expense of small trees) may continue a forest's trajectory into the future. Inertia could also describe factors that make it hard for a forest to change in response to treatments that aim to change a forest. The presence of a large numbers of plants or animals of undesirable species can present an inertia that is hard to overcome, increasing the challenge of fostering desirable species. Similarly, the loss of soil through erosion can have a high inertia that depends on restoring soils as a hurdle to restoring forests. The idea of inertia may be useful for thinking about forest ecology, though the value of the idea would decline if it led to a belief that it referred to a real, well-defined force as in physics.

FIGURE 13.1   A panoramic photo set was taken from a sycamore tree in 1939, looking across Aldo Leopold's land. A repeat photo in 2020 from the same tree gives an idea of how effective forest restoration can be. Leopold's "shack" is circled in both photos. **Source:** top photo by Carl Leopold, bottom photo by Arik Duhr.

Bringing human values into forest ecology creates unavoidable challenges for communication. The idea of a desirable or undesirable species makes sense from a perspective of human values, but these distinctions have no simple meaning in ecology. For example, a value statement could be that aspen trees are desirable on a landscape. This would not be an ecological imperative in any way, as forests without aspens clearly function well. Once the value is stated, some other ecological features become endowed with values. If aspen regeneration is held back by heavy browsing, then we can ask if predators have an ecological value. The ecological outcome for aspen might depend strongly on the predators, and the "value" of predators could be defined in clear ecological terms: doubling the predator population would increase aspen by some quantitative amount.

Values and ecology can also get tangled up when considering concepts such as "health." The idea of health applies relatively well to organisms, such as a brown bear or an oak tree. Health might be described based on either the presence of desirable things (good nutritional status, good rates of growth and reproduction), and by the absences of undesirable things (few diseases and injuries). Health has many dimensions for an organism, and it might be hard to provide a simple value for overall health. However, unhealthy conditions can be very clear, because the presence of diseases, malnutrition, and injuries link directly to poor growth, poor reproduction, and death.

What can health mean when applied to forests? Typing "forest health" into a search engine on the web returns about 1.5 million hits, so the idea is widespread. But what exactly does the term mean in forest ecology? A tree attacked by bark beetles may be in a very unhealthy state (and soon to be dead), but is a forest full of beetle-attacked trees an unhealthy forest? Trees of species that are not attacked by the beetles would likely increase in growth, so the death of some trees leads to improved growth (and better health) for other trees (Chapter 10). A more extreme example would be a large forest area of fast-growing trees with a high risk of severe wildfire that would convert the forest to a grassland for a long period into the future. Would the forest or grassland be healthier? The grassland would not have a large risk from fire, but now the word "risk" becomes complicated. The grassland would be more likely to reburn (a high risk), as dry grasses are more flammable than trees. However, the grasses would simply survive and regrow, and fire poses no risk to the persistence of the grass. The point here is that we have limited words available for describing complex stories, yet words shape how we think and how we understand other people. Careful attention is needed to minimize the fuzziness that words bring to our thinking and communication.

Human values can be a bit nebulous, and the words used to communicate about them may add more fuzziness (and challenge) to communication. When it comes to values in forest ecology, the general definition of terms also change over time. More than a century ago the concept of conservation was predominantly focused on use of resources, not on the long-term continuance of historical forest composition, structure, and function. Forest conservation largely viewed trees as raw materials in an engineering sense:

The first great fact about conservation is that it stands for development… Conservation demands the welfare of this generation first, and afterward the welfare of the generations to follow… The outgrowth of conservation, the inevitable result, is national efficiency…

Gifford Pinchot 1910

This perspective would fit the management of tree farms, where a clear goal of management guides the choices of options that increase efficiency and maximize profits. Modern ideas of conservation no longer focus so heavily on "command and control" of forests, but rather on broader and more flexible aims.

A secondary part of the conservation concept was sustainability, aiming to provide resources and raw materials generations to follow. The sustainability of resource production resulted from rates of use that did not exceed rates of renewal: timber harvests should not exceed the rate at which recovering forests grew new wood. Converting slower-growing old forests to young forests increased the amount of wood that could be sustainably harvested, which was good news for wood supply but bad news for old forests.

Conservation, Sustainability and Restoration Are About the Future

Modern ideas of forest conservation usually imply sustaining a forest's current, desired composition and structure from the present into the future. Sustainability might be used to mean the same thing, or it might imply keeping the flow of resources consistent (both meanings are common). Restoration deals with taking a forest that is currently judged to be undesirable, and moving it toward a more desirable condition. These related ideas bring values onto a foundation of forest ecology, and the future value of the forests can be sustained or increased with understanding of ecological sciences.

Why Do Species Go Extinct, and How Can This Be Prevented?

Why do species become extinct? Because they first become rare. Why do they become rare? Because of shrinkage in the particular environments which their particular adaptations enable them to inhabit. Can such shrinkage be controlled? Yes, once the specifications are known. How known? Through ecological research. How controlled? By modifying the environment with those same tools and skills already used in agriculture and forestry.

Leopold 1933

The idea that loss of habitat is the prime driver of species extinction is so ingrained in expectations that we might take it as a given, much like George Perkins Marsh thought it redundant to look at the evidence proving that tree removal dries up soils and streams (Chapter 6). But both expectations turn out to be strongly disproven by evidence. Habitat loss is not among the most important drivers of species extinctions.

If habitat loss is not the key driver of extinctions, how did the idea become so commonly accepted? As noted in Chapter 8, larger areas have higher numbers of species, and this species-area curve might be used in reverse: if habitat area decreased, the number of species supported in the area would decrease. Famous biologist E.O. Wilson (1992) used this idea to estimate that a 1% annual loss rate of forest area would lead to over 20 000 species extinctions annually, or over half a million in two decades. He guessed that larger birds and mammals would be more susceptible to extinction as a result of declining habitat area. This idea from Leopold, Wilson and many other sincerely concerned people can be tested with evidence. Wilson's extinction rate guess would mean about a dozen mammal species and two dozen bird species should have gone extinct each year since 1980. His guess was an overestimate by 100-fold, so the guess and the reasoning behind it were too wrong to provide insight.

Some species of unknown arthropods or plants might go extinct without ever being documented, but just about all the world's species of birds and mammals have been named and the loss of species would be documented (though a few might be missed). The International Union for Conservation of Nature and Natural Resources (IUCN 2020) maintains a red list of threatened, endangered, and extinct species. The actual rates of extinctions are far lower than commonly claimed, and loss of habitat is only the third or fourth most important reason. In the past 500 years, 159 species of birds became extinct (~0.8% of all bird species), with more than half occurring on islands or Australia. The estimate for mammal extinctions is 84 species (~1.3% of all mammal species), with two-thirds on islands or Australia (Loehle and Eschenbach 2011). More than half of the extinctions in the past 500 years happened before 1900. Perhaps the most important insight from tallying the known extinctions is the identification of the major causes. Almost all cases can be traced to a one major driver, though sometimes interactions between drivers was important. The two most important drivers have been introduction of exotic species (including competitors, predators, parasites and diseases) and direct hunting by humans (Szabo et al. 2012; Figure 13.2). Each of those drivers is three to four times more important in extinctions than habitat loss.

FIGURE 13.2   90% or more of known extinctions of mammals and birds resulted from hunting or exotic species introductions; loss of habitat may be important, but much less common. The last Tasmanian tiger (typical masses of 15 to 20 kg) in the wild was shot by Wilfred Batty in 1930 (left); extinction was driven by hunting for bounties, and maybe also by reduction in prey as a result of competition from exotic domestic sheep. The Little Swan Island hutia (upper right) was a 30-cm long rodent, exterminated by predation from exotic housecats by the mid-1900s (**Source:** photo by Simon J. Tonge). The Island of Hawaii lost many species of honeycreepers and other birds, as a result of predation by exotic rats. After 1900, more species extinctions resulted from exotic avian malaria, which was spread by exotic mosquitos already present on the islands (photo lower right of a stuffed extinct Hawaiian ʻōʻō at the Museum of Natural History in Paris, by Vassail). Warming climate may allow mosquitos to move up in elevation, threatening remaining honeycreeper species.

When evidence shows that predicted values were wrong by a factor of 100, what conclusions should be reached? There are two likely options:

- Something big was wrong with the idea, and the idea should be discarded (and a better one developed, consistent with the evidence); or
- The idea is correct, but the evidence was unreliable (after-the-evidence arguments could include a belief that hundreds of species are actually doomed to extinction even if they persist for now).

Both conclusions would be embraced by some scientists. How much does it matter when an idea is strongly disproven by evidence? In this case, the implications are existentially important. The spread of rats by humans around the world, especially onto islands, probably drove more species to extinction than all the human-driven changes in habitats in the past 10 000 years. A focus on habitat preservation to prevent species extinction may be useful and even important, but overlooking the factors behind 90% of the cases would be a poor use of evidence to avoid undesirable futures.

Conserving Old Forests Is Important, but Old Forests Do Not Last Forever

The conservation of forests might bring to mind images of large old trees, perhaps with decaying remains of trees that died in recent decades and centuries (Figure 13.3). The reality of living, dynamic forests always includes changes, either so slowly that it takes decades to notice or so rapidly the events may seem catastrophic. Major storms and fires can kill old trees across vast areas in a matter of hours, and outbreaks of insects (usually in association with diseases) can drastically change forest structure in a few years. It might seem that a forest conservation area could be large enough to handle such changes, with a uniform portion of the area occupied by each age class of forest. However, events that shape the age structure of forests often exist at scales that match or exceed the size of any designated conservation area, so major changes need to be expected even in conservation areas (see Chapters 8 and 9, and Figure 9.19).

Conservation and Sustainability Have Similarities

The science of forest ecology deals with things that can be measured, with evidence clearly (or not so clearly) indicating whether ideas are supported or refuted. Connecting forest ecology with people leads

FIGURE 13.3   These western hemlock and Douglas-fir trees are over 400 years old, and conservation plans could be developed to preserve the forest. However, forests always change, and expectations of preservation would need to be tempered by understanding likely changes that inevitably develop from background processes of growth, mortality, and recruitment of new trees, as well as the major changes that come with winds, fires, and other events. **Source:** photo by Paul Alaback.

to domains where values matter in addition to science. If forests cannot be preserved in unchanging reserves, then some sort of dynamic conservation goals become important. A common goal of conservation is sustainability, but like many human values sustainability defies clear definition. The National Forests in the United States are to be managed to provide a sustained flow of resources, including water and wood fiber. Sustainability might focus instead on sustaining quality habitat for a particular species of bird, mammal, or fish. Sustainability becomes particularly important for species that are listed as threatened or endangered with extinction. Suitable habitat for one species is of course unsuitable for many other species, so there is no clear path for a definition of sustainability that would satisfy people holding different values. Where sustainability cannot be defined objectively, perhaps a reverse perspective on *un*sustainability could be useful (see Chapter 14's discussion of undesirable futures as a useful guide).

Restoration Comes into Play When Conservation and Sustainability Have Not Been Achieved

The removal of large predators can lead to major population increases in herbivores and declines in plant species eaten by those herbivores (Chapter 5). The future of such forests might shift if large predators were restored to the landscapes. Yellowstone National Park in the Rocky Mountains, USA provides a good case study. In the 1800s and early 1900s, two major predators (humans and wolves) were removed from a 1 million ha region. Populations of elk and bison had been driven to near zero, but their populations recovered in the 1900s to very high levels. Most shoots of aspen and willow were being browsed in the late 1900s, with major changes in riparian areas and some upland areas. Wolves were brought back to the Park in the 1990s, and elk populations declined from highs of over 15 000 animals on the northern range to a fluctuating span from 4000 to 6000 (Figure 13.4). Aspen clones and riparian willows became more vigorous. This seems to be a clear case of

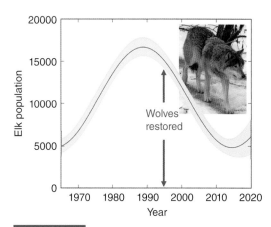

FIGURE 13.4   The population of elk was held low in the mid-1900s through hunting by managers in northern range of Yellowstone National Park. Predation by humans stopped in the 1960s and the elk population more than tripled, with large impacts on willow and aspen. The population had begun to decline before wolves were restored, but climate combined with increased hunting outside the Park (where elk migrate for winter habitat) may have had more effect on elk populations than wolves. **Source:** Data from US National Park Service.

simple cause and effect: wolves eat elk, more wolves means fewer elk, and fewer elk means more aspen and willows. The reality of forests at scales of thousands of hectares and several decades is actually more complicated. The wolves might affect the browsing of willows in riparian areas simply by chasing elk out of areas where they would otherwise congregate for long periods, so simply an increased *risk* of predation on elk might have consequences for aspen and willows. The number of elk killed by bears increased substantially in the period after wolves were introduced, so changes in elk populations might reflect impacts of bears, wolves, or bears plus wolves. Elk populations can decline during years with heavy snowpack or low summer precipitation. Hunting by humans does not occur within the Park, but elk migrate to low elevations outside the Park, where they are heavily hunted.

So, did the restoration of wolves in Yellowstone drive these "trophic dynamics" of elk, aspen, and willows? Maybe, but probably not so much. Restoration of wolves did not lead to simple, clearly demonstrable effects in the Park. Modeling of elk populations in Yellowstone indicated that the best evidence supported climate and human hunting (outside the Park in autumn) as the primary drivers of the decline in elk populations after wolves were restored. Any effect of wolves on the elk population was "primarily compensatory" (meaning the wolves killed elk that would not have survived, with little effect on elk populations; Vucetich et al. 2005). For a broader consideration of the evidence of predator control of foodwebs, see Allen et al. (2017). Simple stories of cause and effect can be compelling, but the reality of complex forests across time and space rarely fits simple expectations.

Digging deeper, was the Greater Yellowstone Ecosystem (an area of more than 1 million ha) restored by the return of wolves? This seems like a simple, scientific question, but of course forests don't provide simple answers to most simple questions. First, the historical conditions of the past were never constant, so there would be no particular "target" for assessing success. The impacts of hunting by Native Americans were likely important in historical times, as is modern hunting outside the Park boundaries. However, the continued presence of human hunting does not mean the impacts on elk would be the same.

In the absence of a single target for gauging restoration success across a dynamic forest landscape, perhaps a historic range of variation could be developed to gauge success. A great deal of work has been done to examine historical conditions in Yellowstone (see Figure 9.19 for example). Over periods of one or a few centuries, the range of forest age structure has been extremely variable. At longer time scales, shifts in climate substantially altered the extent of forests and grasslands, and of course human influences would have shifted too. These concerns undermine the value of a strictly scientific approach to gauging the success of a restoration effort. The nature of forests might be better suited to the reverse view mentioned at the end of the previous section. The forests of Yellowstone would clearly be "less distant" from historical conditions with the restoration of a major top predator. Whether the forests have been restored or not, the presence of wolves could be viewed as moving the landscape farther away from "*un*recovered." This distinction may seem a bit obtuse, and while it could be productive for shaping the understanding of a forest by an expert, the term "recovery" would undoubtedly be a good word for communicating with people broadly.

The History of a Forest Might Be Read in Reports, in Photographs, in Trees and Remnants of Trees

Human influences on forests can date back thousands of years (Chapter 12), though our best information about historical forest conditions usually goes back only a few centuries (Chapter 9). In some cases European explorers developed detailed sketches of landscapes that can be compared with modern conditions. John Wesley Powell led some explorations of the landscapes in Utah and Arizona, USA in the late 1800s, with sketches capturing the extent of trees before the widespread grazing of cattle stopped the frequent fires (Figure 13.5). A modern photograph of the same scene shows the vast expansion of tree cover. Comparing the two images give insights into how the landscape has changed, and what might need to be done if restoring historical conditions would be a goal of management.

Landscape photography has been common for more than a century, and many areas have photographic records of forest regions from many vantage points. A comparison of more than 100 historical photos of the San Juan Mountains in Colorado, USA gives a visual impression of where changes have been substantial or slight (Figure 13.6). A pair of images shows that spruce and fir trees increased in size over a century, and that locations on the landscape that had aspen a century earlier remained dominated by aspen (counter to any expectation that conifers would outcompete aspen at these sites). Across 100 photo pairs, it was clear that areas dominated by spruce/fir and by aspen showed substantial expansion, at the expense of former meadows. Less than 10% of the photos showed reductions in cover of these species. Why did the vegetation on these landscapes change? About one-third of the photos showing expansion of forests likely represented recovery after some major event (fires or beetle outbreaks). The other cases were expansion of trees into meadows. An investigation into possible causes would be needed to develop evidence, but the long-term dynamics on these landscapes might be expected to show fluctuations in vegetation types at time scales of centuries. Non-forest vegetation is a major part of many landscapes, and forest expansion can substantially reduce the presence (and value) of meadows, grasslands and shrublands. Restoration in these cases may focus on removing trees rather than fostering trees.

FIGURE 13.5   A lithograph was made of the view from Mount Trumbull in Arizona, USA, during an 1882 expedition led by John Wesley Powell (upper). The same scene in 2017 shows major forest expansion across most of the landscape, as a result of cattle grazing and the cessation of frequent surface fires. **Source:** photo by Todd Miller and Julianne Renner, www.thegreatbasininstitute.org/recreation-1882-mount-trumbull-lithograph-sheds-light-landscapeevolution.

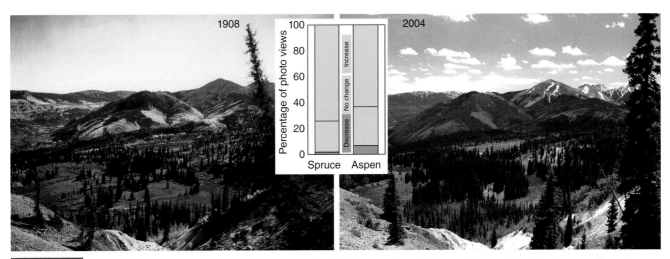

FIGURE 13.6   A century of change in the San Juan Mountains of Colorado, USA, showed substantial expansion of both spruce/fir and aspen trees. In some cases the trees may have been "reclaiming" area following major events (particularly fires and bark beetle outbreaks). In others, the trees were establishing in areas that had been meadows, and restoration in such cases might focus on restoring meadows lost to tree cover. **Source:** left photo by W. Cross, right photo by J. Zier, from Zier and Baker 2006, used by permission.

Changes visible in photographs can be more graphic and powerful than simple data, but data can also be represented in graphic formats that give strong visual impressions of forest change. The European settlement of central and western USA included government surveys that included some characterizations of forests. The information present in surveys from the mid-1800s can be translated into visualizations of forest composition and structure for comparison with current conditions (Figure 13.7). The historical forests of Wisconsin, USA differed substantially in response to the post-glacial development of soils. The forests developing on sandy soils (similar to those in Leopold's sand county area) were dominated by jack pine, with some red pine and northern

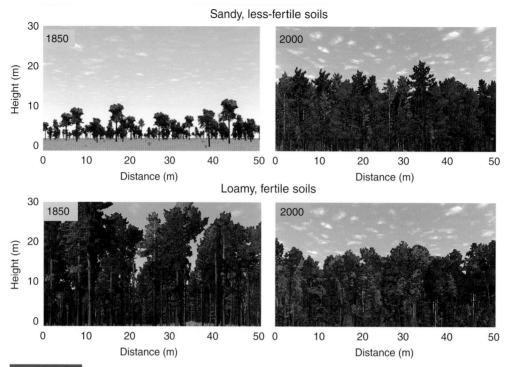

FIGURE 13.7 The historical forest composition and structure in northern Wisconsin, USA was recorded in Public Land Survey records, which can be compared with recent inventories. Computer visualization techniques use these numbers to create visual images of what the average forests would look like. Forests on less fertile sandy soils had short fire-free intervals that sustained sparse tree cover (upper left). With long fire-free periods, the density of the forests increased by 100-fold (upper right). Fertile, loamy soils had forests dominated by hemlock and yellow birch (lower left), and now these sites are dominated by sugar maple (lower right). Restoring these forests to historical composition and structure would be entirely different endeavors on each type of soil, underscoring the fundamental importance of soils in shaping forests. **Source:** Based on Stoltman et al. 2007, used by permission.

pin oaks (and an average density of only 10 trees ha^{-1}). Sites with loamy texture supported tall forests of eastern hemlock, white pine, and a wide variety of broadleaved species (with average density of 185 trees ha^{-1}). Both types of forests changed dramatically over 150 years. The sandy sites increased in tree density by 100-fold, without much change in species composition. The richer loamy sites increased tree densities by more than sixfold, with dominance shifting from the original eastern hemlock/yellow birch to sugar maple (with almost no hemlock).

Why did the forests in northern Wisconsin change so much? The low density of trees on the sandy soil in the 1800s resulted from short intervals between fires. Widespread conversion of forests to agricultural uses stopped the fires and led to increased densities of trees in the remaining forests. The change in species composition and density on more fertile soils resulted from intensive logging (first for conifers and then for broadleaved species), and post-logging conditions were better suited for different species. Restoration of the sandy-soil forest would involve more frequent fires, which of course might be socially unacceptable. Restoration of the forests on fertile loamy soils might require planting of vast numbers of hemlock trees, though with the arrival of the exotic hemlock woolly adelgid insect, hemlock dominated forests may no longer be possible.

Clues to the Past Structure of Forests Lurks in Tree Rings, Stumps, and Logs

Not all areas have good records or photographs of historical composition and structure of forests. Even where general information is available for a given forest type in a region, the variation in forests across space would mean the average story revealed from historical records may or may not apply to a specific forest within the region. The history of most temperate and boreal forests can be

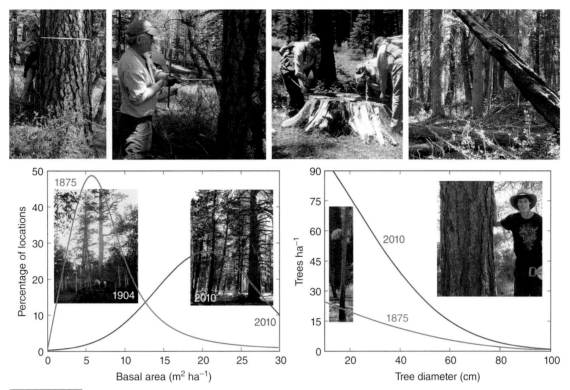

FIGURE 13.8   Historical composition and structure can be examined based on clues present in current forests. Reconstruction studies depend on estimating diameters of living trees for some past target date, based on tree rings (upper left). Not all historical trees currently survive in forests, but in many cases stumps and dead stems (upper right) provide information (sometimes with high precision, or only roughly). This reconstruction approach was used to characterize ponderosa pine forests in western Colorado, USA in 1875 (lower graphs). Across the landscape, the forests were mostly open (tree cover of 25%), with about 100 trees ha^{-1} and basal areas of 4–8 m^2 ha^{-1}. The modern forests had about four-times the historical density and basal area. Indeed, the most typical forest structures in 1875 were absent from the modern landscape. Sources: Based on Matonis et al. 2014; 1904 photo by S. Riley, from Baker 2017.

read in part based on clues present at each site. The history of tropical forests can be harder to figure out, owing to a lack of reliable annual tree rings and rapid decomposition of dead stems.

Reconstructing the history of a forest from on-site clues needs information from trees that are presently alive (how large were they at some historical time?), and from stumps and standing and fallen stems (Figure 13.8). The sizes of the now-dead trees back at the historical time can be determined from tree rings, if the wood remains solid. Otherwise, some estimation can be made based on the degree of decay. The sampling of historical clues can be done with a high precision for scientific research publications, or it can be done more simply as a rough guide for considering forest change and implications for restoration work.

What Does It Take to Restore a Forest?

The most frequent answer to that question might be "nothing," because tree species evolved in response to opportunities to establish (or reestablish) after major events removed the previous trees. The history of land-use changes also demonstrates the remarkable ability of forests to reestablish after forests were removed and soils were used for agricultural production. Beyond this general pattern, some important situations warrant consideration: when the reestablishing forest is limited by soil damage or loss; when undesirable species dominate; and when undesirable forest structures develop.

One of the most remarkable cases of a forest restoring itself comes from New Hampshire in the northeastern US. In the 1960s, some scientists were curious about what would happen to nutrient losses from a forest watershed if all the trees were cut and all the reestablishing vegetation was controlled by spraying of multiple herbicides for three years. Nutrient losses of course increased when uptake by vegetation plunged to near zero. More interesting is what happened once the herbicide treatments stopped: the forest recovered quickly. The accumulation of biomass over the next three decades was only a bit lower than the average trend expected for similar forests in the region (Figure 13.9).

Many Forests Have Reestablished Following Agricultural Land Use

Land use for agriculture has led to conversions from forests to crops or pastures, and back to forests. Sometimes the periods between forest clearing are only about a couple decades in length, such as shifting agriculture in tropical countries (which may cover about 250 million ha). Shifting cultivation entails cutting or burning of trees (Figure 13.10), followed by a few years of crop cultivation and then abandonment and forest reestablishment. The abandonment results from the increase of undesired weeds (which increase each year), and perhaps from declining nutrient supplies. The forest reestablishment period converts the vegetation from hard-to-control weedy species to trees, which are more easily controlled later by cutting or fire. The two key points here are that forest recovery is quite rapid, but the species composition and structure of forests in a shifting agriculture system are of course very different from those of older forests. Restoration work may not be needed to obtain a new forest, but a forest with a desired suite of species might require substantial work.

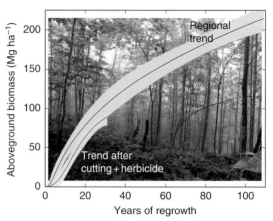

FIGURE 13.9   A broadleaved forest in New Hampshire, USA had all trees cut and all reestablishing plants sprayed with herbicides for three years. Over the next three decades, rapid forest recovery led to accumulation of biomass (red trend) that was only about 5% lower than the common trend for similar forests in the region (blue trend; **Source:** Based on data from Reiners et al. 2012).

Forest Reestablishment May Be Faster with Planting, and Contain More Desirable Species

The reestablishment of a forest can be accelerated by direct seeding or planting of seedlings, with opportunities for influencing the species that will comprise the forest (Brancalion et al. 2016; Crouzeilles et al. 2019). More than 90% of the Atlantic Forest of eastern Brazil was converted for agricultural uses. The species diversity of these forests rivaled that of the Amazon rainforests, and reestablishment of forests in the region usually focuses on development of high diversity of tree species. A long list of possible objectives, treatments, and economics influence restoration decisions (Brancalion et al. 2018). Tree species with smaller seeds, and seeds

FIGURE 13.10   A young reestablished forest was cleared at this site near Pemba in Northern Mozambique. Crops will be cultivated for a few years, and then the abandoned site will reestablish with a new forest. Shifting cultivation demonstrates how rapidly forests reestablish, but the short intervals between cropping phases leads to very different species composition and structure than in older forests. **Source:** photo by Ton Rulkens.

that remain viable for longer periods, are less expensive to produce. Many tropical tree species produce fruits with large seeds that lose viability in a period of weeks, and these are more challenging to use in a restoration program. However, large-seeded trees can provide crucial food for mammals and birds, so restoration goals benefit from investment in reestablishing these challenging species. Once young trees are established, new tree species arrive after dispersal of seeds by birds, bats, and other animals. This is an example of positive feedback, where an initial reestablishment of a forest leads to greater use by animals, which leads to input of new seeds of other species, which leads to further diversification of the forest. Use by animals tends to increase as young forests increase in biomass, and the initial rate of biomass accumulation can be enhanced with herbicides to control grasses (including very competitive exotic species) and fertilizers to increase nutrient supplies. Faster rates of biomass accumulation lead to higher animal use, as well as increases in rates of colonization by other tree species (Figure 13.11).

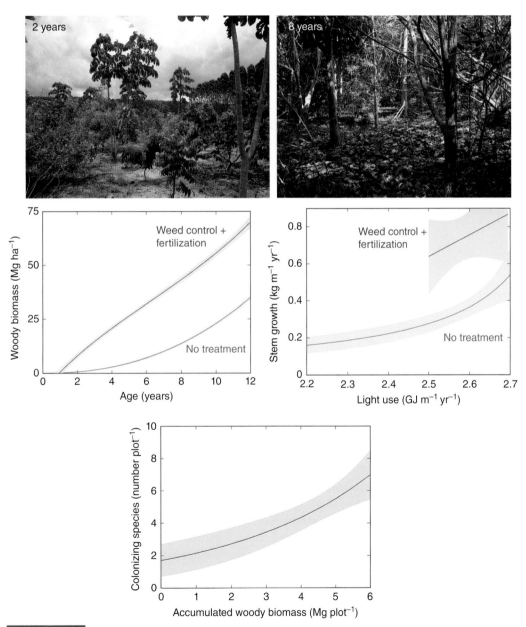

FIGURE 13.11   An experiment that mixed 20 native species (fast-growing, slow-growing, and legumes and non-legumes; **Source:** photos by Jose Luiz Stape) showed that biomass accumulated more than twice as fast when trees were aided by weed control and fertilization (middle left graph; in this example, based on data for mixtures that contained half fast-growing and half slow-growing species, at a planting density of 1600 seedlings ha^{-1}; **Source:** Data from Brancalion et al. 2019). The greater growth resulted primarily from higher efficiency of using light, rather than higher light capture (indicating that better growing conditions lowered partitioning to belowground production; **Source:** Based on data at eight years from Harjuniemi 2014). Across all treatments, colonization by other native tree species increased with increasing total plot biomass, likely as a result of increasing use by birds and mammals (**Source:** Data from Brancalion et al. 2019).

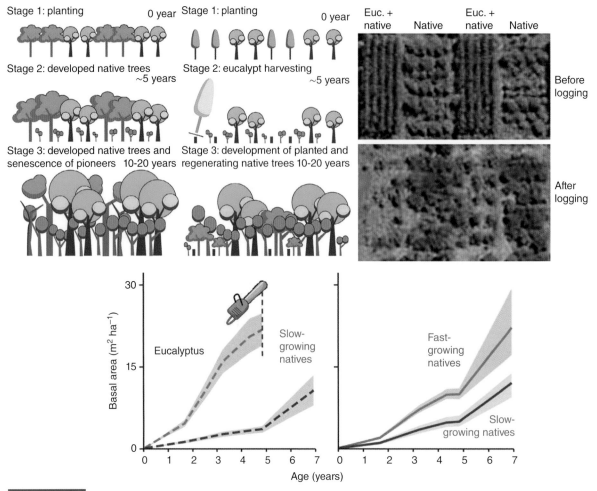

FIGURE 13.12   The cost of reestablishing native tree species might be reduced if the trees were initially planted in mixture with commercially valuable species, such as eucalyptus. The commercial trees would be harvested after about five years, releasing the native species trees to grow and develop into a forest. In this experiment, about half the trees in all plots were slow-growing native trees targeted to form the eventual forest. The other half of trees were eucalyptus in the commercial treatment, or fast-growing, short-lived native species in the native treatment. **Source:** from Brancalion et al. 2020, used by permission.

Reestablishing a forest on former agricultural cropland or pastureland is expensive, typically costing several thousand dollars per hectare. Native forest restoration is sometimes supported by funding tied to profits obtained from other forest activities. Forest policies may require a proportion of a forest company's lands be maintained in native forest, while the rest is used for intensive tree farming. The costs of reestablishment of native forest areas would be covered by income from the tree farms. Creative approaches might find ways to help cover the costs of native forest establishment by cropping a single rotation of commercial trees along with native species. An example in Brazil compared plots planted only with native species with plots that were planted with two rows of eucalyptus and two rows of native species (Figure 13.12). The eucalyptus trees were harvested at age 5, with profits covering about half of the cost of the native species planting. Some damage occurred to the native trees during logging, but few native trees died and the overall growth of the native trees in mixed plots was about half that of the pure plots of native species. Over the coming decades, this difference at age five would likely diminish, with the overall outcome of a reestablished native forest for half the cost (or the same cost, but with twice the area of restored forests).

Forest Reestablishment Leads to the Redevelopment of Forest Soils

Areas where forests were converted to agricultural fields have often returned to long-term forest cover, such as the examples in Figures 9.15–9.17. Forest reestablishment is often rapid, and again the species composition may not return to that of the forest that was cleared for agriculture. Soils under cultivation may have higher nutrient supplies for new trees, especially if soils were fertilized. In other cases, long-term agriculture may deplete soil nutrient supplies, particularly when erosion by water or wind was substantial. Abandonment of pastures may also result in forest reestablishment, and competition between tree seedlings and pasture vegetation might lead to different species composition than would develop after abandonment of cultivated fields. Soils are very dynamic, influencing the establishment success of species and growth rates, while forest development also changes soils (Figure 13.13). Soils that have been used for agriculture are generally sufficient for forest reestablishment, though rates of growth and long-term dynamics will be different than on soils that have remained forested.

It might seem that trees are such remarkable organisms that they can restore soil fertility following exploitative agriculture, erosion, or other problems. Trees are definitely capable of adding organic matter to soils, though most of the organic matter entering soils (from roots, mycorrhizal fungi, and aboveground litterfall) decomposes without leading to large accumulations. Most reestablished forests do accumulate a surface O horizon, and this may be where the majority of the new organic matter in a soil will be found. But unless the tree species are symbiotic nitrogen fixers, trees cannot increase soil N content, and the effects of trees on weathering of soil minerals tends to be slow. Well-established forests may develop higher supplies of available nutrients, as elements are taken from deeper in the soil and added to the near-surface soil, and perhaps through improved soil water status and vigorous communities of soil microbes and animals. An evidence-based conclusion about reestablishing forests would be that trees are good for soils, but they can't work miracles in a short period of time.

FIGURE 13.13 These soils are 25 m apart in the Duke Forest in North Carolina, USA. The forest was cleared for cultivated agriculture more than two centuries ago, and then the site was abandoned for the past century. The soil on the left was under the influence of grass cover (in a powerline right of way), and the one on the right developed under a pine forest. The trees reestablished a three-dimensional soil structure typical of forests, including an O horizon and legacies of root channels. Areas with higher concentrations of oxygen in the soil atmosphere are orange, and those with lower concentrations are gray. **Source:** from Binkley and Fisher 2020, used by permission, photo from Allan Bacon.

Reestablishing Forests in the Absence of Soils Is a Major Challenge, Requiring Insights and Money

The most severe challenges for forest reestablish are sites where soils have been lost. Agriculture may lead to erosion losses of fertile topsoil, and even subsoil where gullies or mass movement events develop. Trees can often establish and grow even in these situations. Limitations for trees are more severe following mining activities that strip the surface down to bedrock. Tree establishment can be even more difficult (or impossible) on some mine waste materials that are too acidic or too alkaline. Some of the challenges with forest reestablishment where soils are missing are illustrated in Figure 13.14. Acacia seedlings grown in mine wastes grew better when organic matter was added artificially, but the best growth depended on the addition of organic matter, phosphorus fertilizer and inoculum with arbuscular mycorrhizal fungus. The reestablishment of forests in all these cases depends

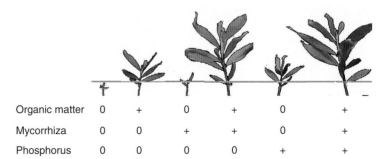

Organic matter	0	+	0	+	0	+
Mycorrhiza	0	0	+	+	0	+
Phosphorus	0	0	0	0	+	+

FIGURE 13.14 A greenhouse experiment demonstrated that acacia seedlings barely survived when planted in mine waste material, even though the seedlings were capable of fixing atmospheric nitrogen. Mixing organic matter with the mine waste greatly improved seedling growth, but the best growth resulted from addition of organic matter, phosphorus, and arbuscular mycorrhiza ("+" indicates a treatment was added; "0" indicates it was not; **Source:** based on information from Santana et al. 2020).

on treatments that bring the soil up to a condition suitable for trees, and then tree growth can contribute to the development of organic matter in soils and increased biological activity (for a general overview, see Adams 2017).

Management Can Shift Forests Away from Undesirable Conditions

The domain of forest restoration can go beyond situations where no forest is currently present to cases where a current forest has traits that are undesirable. Undesirable traits might include too many trees, trees of a range of sizes that could allow fire to move from the ground into tall crowns, too few large trees, and a lack of diversity of tree species or sizes. Each of these cases would call for different sorts of treatment to achieve restoration goals, and indeed each individual site calls for adapting general treatments to meet site-specific conditions.

Restoration in existing forests is a common priority for landscapes where the pattern of short intervals between fires was interrupted (frequent-fire landscapes). Unusually prolonged fire-free intervals allow increased tree density, development of trees that serve as ladder fuels, and loss of understory plants. Unmanaged ponderosa pine forests that have not experienced fire for many decades have a higher likelihood of severe crown fires than those that have been harvested or intentionally burned with surface fires (Figure 13.15). After a century without fire, unmanaged forests under severe weather conditions may experience twice as much extent of severe fire (in larger patch sizes), with only 25% as many trees surviving the fire (Finney et al. 2005; Shive et al. 2013). Restoration treatments in these types of forests may have a variety of goals, but two major goals would be developing a forest structure with lower risk of severe fire, and a structure that facilitates lower intensity surface fires (see Figure 11.14, and the Ecological Afterthoughts section at the end of this chapter).

A second situation where restoration treatments might be used to change an existing forest is fostering a faster development of structural characteristics of old-growth forests. For example, a typical plantation of Douglas-fir may have developed from plans for maximum wood yield, with high densities of a single age cohort of a single species. After 50 years, such a plantation would lack many of the key characteristics that would be typical for old-growth forests, such as several very large trees per hectare, large logs on the ground, and a multi-level canopy (Figure 13.16). A restoration treatment could be applied that would thin the forest (generating income), along with underplanting of seedlings that would diversify the height profile of the forest canopy (Newton and Cole 1987). Some logs could be left on the ground, or later storms might topple some trees to increase the pool of deadwood. After a couple decades, the treated forest would be less dissimilar from old-growth forests, and management approaches could be developed for continued harvesting of trees in a continuous-cover forestry system. A restored forest would not be identical to an old-growth forest, but old-growth forests are not all alike in any case. A key point for restoration would be careful identification of the undesirable condition of the current forest, and how those undesirable features could be reduced.

FIGURE 13.15 A large fire in Arizona, USA (Rodeo-Chediski) burned more severely through forests that were unmanaged (left, with no thinning or fire) than through forests that were harvested to reduce tree densities and burned to reduce risks of severe fire behavior (right).

FIGURE 13.16   A 50-year-old plantation of Douglas-fir (with some western hemlock) in Oregon, USA was treated to accelerate development of traits more common in old-growth forests. Thinning was followed by underplanting (along with volunteer seedlings). The treatment led to an increase in the size of large trees in the forest, and increased diversity in canopy structure. Windstorms toppled some trees, increasing the pool of deadwood on the ground. Within two decades, some habitat use was occurring by species expected to be found primarily in old-growth forests. **Source:** photos by Liz Cole.

Two Key Ideas Connect Forest Ecology with Conservation, Sustainability, and Restoration

There is no single way for a forest to be. If there was only a single, narrow condition for a viable forest, then goals of conservation, sustainability, and restoration would need to be equally narrow and precise. Fortunately all forests can be viable with a rather broad range of composition of tree species, other plant species, and animals, along with a great range in tree sizes and arrangement. Forests are so dynamic across space and through time that goals for wise stewardship do not (and should not) focus on a very narrow range of future forests. Narrow expectations are well-suited for tree farms, where intensive investments aim for precise outcomes, but diverse, less-managed forests demand broader potentials for future changes. Conservation, sustainability and restoration goals need to match the highly variable and dynamic realities of real forests. A group of old-growth forests across a landscape would show great variation, including variation in how things change over time in response to chronic interactions and major events. Therefore, goals for conserving or restoring such forests would not benefit from any simple, one-size-fits-all approach. Clear descriptions are needed for exactly how forests might be at risk, and then effective prescriptions become possible for reducing risks.

The second idea is that programs aiming to conserve, sustain, or restore forests need to recognize that dynamic forests will not stay unchanged. A program might aim to restore frequent fires to landscapes where historical structures supported frequent, low-intensity surface fires. Restoration treatments would only be the first step into the future. If frequent surface fires do not follow the initial treatments, the processes that led to undesirable conditions would simply recreate those undesired conditions. Plans that fail to account for the continuing dynamics of forests may simply fail.

A third idea for connecting forest ecology with conservation, sustainability and restoration is Pocket Science, an approach for learning and improving through experience. This idea is developed in Chapter 14.

Ecological Afterthoughts: Restoring Forests May Be About Restoring Non-Tree Vegetation

Lengthening the periods between fires can lead to undesirable changes in forests. This case study (Figure 13.17) of restoration aimed to reverse the changes that developed slowly over a century without fire, in a landscape where fires historically occurred in most decades. This restoration treatment occurred near the location of the historical photos in Figure 9.2. The restoration treatment could have been done in a variety of ways. The treatment might have just involved intentionally burning the site, at a time when weather conditions favored either a low- or high-intensity fire. The site might also have been thinned to remove the excess smaller trees, without the use of fire. The thinning could have retained trees with uniform spacings (for maximum growth), or in clumps (increasing understory vegetation). In this plot the full-restoration treatment entailed:

- removal of most of the medium and small trees;
- retention of trees in clumps;
- prescribed burning; and
- addition of seeds of native understory plants.

What are the key features that describe the differences in the restored treatment ("What's up with this forest?"): How would the outcomes have differed if:

a. the thinning had aimed to retain a uniform spacing between trees?

b. the thinning had retained twice as many trees?

c. the thinning had not been followed by burning? or

d. no seeding of understory plants was done?

FIGURE 13.17   A restoration treatment of this ponderosa pine forest in northern Arizona, USA reduced the number of trees from 1250 ha^{-1} to 60 trees ha^{-1}, close to the 1875 density of 25 trees ha^{-1}. Four years after treatment, perhaps the most ecological important change was the growth of understory plants. **Source:** photos by Doc Smith, for more information see Moore et al. 2008.

Another series of important questions consider the future of this forest ("What's next?"):

a. how would the forest develop if the interval between fires would be 10 years versus 50 or 100 years?

b. if an outbreak of bark beetles occurred, would the mortality risk for large, old trees be different as a result of the restoration treatments?

c. if a severe drought developed, how would the responses be different in the restored area?

d. if the climate does not repeat the historical climate before 1900, would the treatment that aimed to restore typical structure of landscapes in the 1800s be a rational thing to do?

e. Thirty years after the restoration treatments, few new seedlings of pines have established. What possible factors would be good to explore with some simple experiments?

CHAPTER 14

Forests of the Future

The first two questions in the core framework deal with the present and the past of forests. This chapter focuses on the third question: what does the future hold for forests? The future only becomes clear slowly, at the pace of real time. Some developments are more likely than others, in part because of the positive feedbacks in forests. Dominant trees grow faster than smaller trees, which can increase the share of site resources obtained by dominant trees, increasing the size disparities among trees. The outcome of positive feedbacks can be changed by events, such as lightning that is more likely to strike taller trees. These sorts of common trends were the topics of previous chapters. The focus of this chapter is on some of the very large, pervasive trends that are likely to shape forests in major ways, outside the historical norms of past centuries: species invasions, climate change, and human impacts. Predictions of the future of forests usually fall wide of what actually develops over time (D'Amato et al. 2017), so skepticism about predictions is always important.

Forests Have Already Changed, and Continue to Change

No forest stays the same for long. Over the typical lifespan of a tree, forests change in the numbers of trees, the size of trees, and often in species composition (Palik et al. 2021). It might seem that changes would be routine at the scale of a single forest, that averaging out over large landscapes would result in relatively constant forest conditions at regional scales. Evidence generally refutes the expectation of constancy at just about all scales (recall Figures 9.16, 9.17 and 9.19). Even when averaged over scales of 1000 km and more, forests do not remain the same for long periods.

Maples increased across most of the eastern United States since 1980, while pines increased somewhat, and oaks dramatically decreased (Figure 14.1). These changes resulted from a number of factors, including large increases in deer populations (changing the success of seedlings of different species), changes in soils (including legacies of high rates of N addition from air pollution), the occurrence of events (harvesting, fires, storms), and perhaps changing climate. These forest-shaping factors will be different in the future than in the past three decades. We can be confident that forests will change in response, but given differences in rates of change of driving factors, and major interactions between factors, the future details of forests can only be known as time moves slowly ahead. With cautious humility, what futures might be more likely?

Rates of processes in forests have also changed in recent decades, across very large areas. Forests across much of Europe and Scandinavia now grow substantially faster than a few decades ago (see Figure 9.14). Faster growth can also lead to higher rates of mortality, as a normal outcome of competition. The rates of mortality have also increased substantially in recent decades (Figure 14.2), but were these increases driven by increased growth, increased stress (such as drought), or other factors? The trends are clear, but the driving factors remain unclear (and would not likely be the same across forests and regions).

Rising temperatures have already changed the seasonality of the life cycles of plants in temperate regions. The naturalist Henry David Thoreau began recording in the 1850s when plants first began flowing in the spring, and when migrant bird species first arrived. After a century and a half, it's clear that the average timing of plant flowering in eastern Massachusetts, USA now arrives

Forest Ecology: An Evidence-Based Approach, First Edition. Dan Binkley.
© 2021 John Wiley & Sons Ltd. Published 2021 by John Wiley & Sons Ltd.

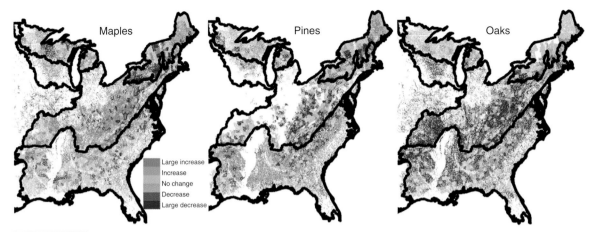

FIGURE 14.1 The forests of the eastern United States since 1980 experienced large increases in the number and sizes of maples, going up by about 35 stems ha^{-1}, and 1.2 m^2 ha^{-1} of basal area. The regional average increase was similar for pines, increasing by about 15 stems ha^{-1}, and 1.0 m^2 ha^{-1}, but the average change resulted from the combination of notable increases and decreases in different areas. Oaks declined across almost all the region, dropping by 30 stems ha^{-1}, and 0.6 m^2 ha^{-1}. **Source:** based on Knott et al. 2019, used by permission.

about two weeks sooner than in Thoreau's time (Figure 14.3), likely resulting from the 2.5 °C increase in average March temperatures. The timing of flowering of plants (and the breaking of buds to grow leaves, Figure 2.17) is sensitive to springtime temperatures. Interestingly, the timing of the arrival of migrating birds has not shifted over 150 years. What features of forest ecology might change in response to divergence of spring events for plants versus birds? Local insect populations might go up and down in response to flowering times, and insectivorous birds might encounter differences in food supply as climates warm. The patterns in Figure 14.3 reflect only a small part of the ecological story of plant flowering and bird migration, given the large number of species, likely interactions among the species, and cascading effects through foodwebs. The key point would be that plant phenology has already responded to warming, and such shifts hold the potential to resonate (or be muted?) across the forest ecosystems.

We must make no mistake: we are seeing one of the great historical convulsions in the world's fauna and flora.

Charles Elton (1958)

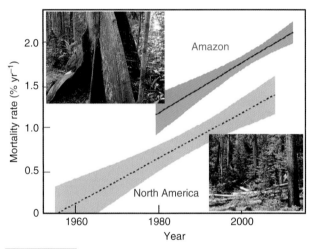

FIGURE 14.2 Across North America, annual mortality rates increased from near 0 in 1960 go over 1% in 2000. A similar trend was apparent in the Amazon after 1980. The evidence for increased mortality is very strong, but the importance of potential driving factors remains largely unknown (and likely varies among forests and regions; **Source:** Based on data from McDowell et al. 2018).

The establishment of non-native (invasive) species can change forests in two major ways: removing native species, and competing with native species. As noted in earlier chapters, the temperate forests of North America have been drastically altered by invasions of non-native insects and diseases, including the near extirpation of American chestnut, American elm, and on-going declines of ashes, eastern hemlock, and five-needled pines. In some other cases, non-native species have established and reshaped forests without extirpating native trees.

Species have tremendous potential for positive feedbacks, growing massive populations after a few individuals establish in novel territories. Coevolution of species can lead to a prominence of genes in host trees that limit the effectiveness of pathogens or harmful insects. Closely related species in novel areas may lack such genes, and host trees are very susceptible to the invading species.

Can Invasions Be Predicted?

Yes, but only in a limited way. Most species that arrive in novel areas either do not manage to establish, or they become established at low numbers without strong effects on forests. A minority of species have very large effects, but predicting which species will become major invasives is not generally possible. Once a species has established, however, some techniques have been effective

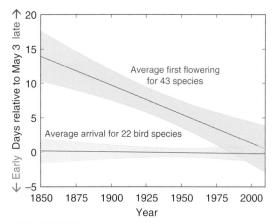

FIGURE 14.3   Henry David Thoreau recorded the timing of flower blooming and the arrival of migrating birds in the 1850s, and other people continued the records across the next century and a half. Plants now flower about two weeks sooner than in Thoreau's time (the declining red line indicates earlier flowering), but the arrival of migrant birds shows no clear trend (see Primack, 2014; **Sources:** flower dates based on data from Primack and Miller-Rushing 2012, bird dates based on data from Ellwood et al. 2010).

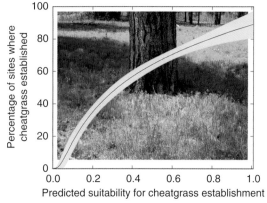

FIGURE 14.4   A MAXENT analysis of sites in Rocky Mountain National Park, USA led to predictions of probabilities that various types of sites would be suitable for new establishment of cheatgrass (the light-colored grass with dangling seedheads at the base of the ponderosa pine tree). Surveys six years later verified that the predictions were reasonably accurate. **Source:** based on data in West et al. 2016.

for predicting the spread of the invading species. The environmental characteristics of sites with successful invasion can be measured, and quantitative methods can identify the factors that are common among the sites. This calibration step can then be applied across uninvaded areas to identify where the species will have a higher probability of establishment. One approach for this sort of analysis is called "maximum entropy" (MAXENT), which essentially analyzes all the information available about potential factors, providing insights on which combination (and weighting) of factors has the best agreement with the information. The idea of entropy can be a complicated one to understand, but it's fundamentally important in physics, chemistry, biology, statistics, and information theory and technology.

This approach was used to predict where an invasive annual grass, cheatgrass, would establish in Rocky Mountain National Park in Colorado, USA. Field plots were spread across the Park, and the environmental characteristics of plots with and without current cheatgrass were tallied. Predictions were made for where cheatgrass would expand its range, and a resampling several years later clearly tracked the predicted patterns (Figure 14.4).

A similar predictive approach was used to examine the possible extent of invasive plant species establishment across the USA (Figure 14.5). The predictions were based on invasive species currently established somewhere in the country, omitting novel species that might arrive in the future. In some areas, less than 20% of the species that would be predicted to establish were currently present, so the number of invasive species in many areas might be expected to increase by fivefold. Most invasive species currently occupy less than 1% of the geographic range that might be suitable for them. All the evidence points to increasing invasions in the future of forests.

Some Forests Are More Invasible Than Others

Not all forests have the same presence of non-native species, so what factors might account for the pattern? A variety of possibilities could be tested against evidence. One of the earliest discussions of the possibilities came in Elton's 1958 book about invasions. He suggested that strong competition with established species would limit the success of invasives, and systems with large numbers of native species would somehow resist invasion better than simpler systems.

How would competition with native species limit the opportunity for invasives to establish? Technically, species cannot compete with each other, only organisms can. And how would a plant distinguish between a neighboring plant that was native and one that was non-native? Competition would likely challenge the establishment of any new plant, whether native or invasive, so does the simple number of species present in a forest have an influence on the invasibility of the forest?

Field evidence clearly indicates that forests with more native species actually have *more* invasive species, not fewer. In the Rocky Mountains, USA, forests with high cover of understory plants tended to have high numbers of native species, and relatively high numbers of invasive species (Figure 14.6). Having more native species present did not convey resistance to establishment of non-native species; sites that fostered high native diversity also fostered high invasive diversity. In this case, the sites with higher understory cover (and likely growth) occurred on more fertile soils.

What if an experiment held the soil constant, and the number of species was experimentally manipulated to give a range of species numbers? These sorts of experiments have been done for meadows, and the subsequent establishment of additional species has usually declined as the number of species present in the plots was higher: more resident species did lead to lower invasion by new species. However, details are always important in understanding the outcome of experiments. If the choice of species to sow

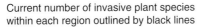

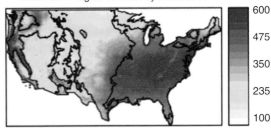
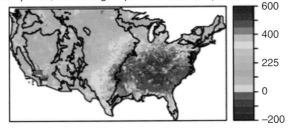

FIGURE 14.5 Regions of the USA have up to several hundred invasive (non-native) plant species (upper left), but most of the species occupy only a small portion of the area where suitable environments occur. Based on suitable environments (a MAXENT analysis), much of the country would be suitable for hundreds more invasive plant species. The current climate might favor some invasive species more than future climates; parts of the southwest might become too hot and dry for some species to thrive (relative to current climate, lower map; **Source:** based on Allen and Bradley, 2016, used by permission).

into experimental plots was done at random, then plots that received larger numbers of species would be likely to include one or more species with high growth rates. Rapid growth might then impede establishment of new plants later, and so the outcome of the experiment might relate to competition, not directly to diversity of species. Species differ in traits, including rates of growth and competitive ability. Questions about "diversity" that do not include the identities of species leave out important information.

Not all Invasive Species Are Alike: Identity Matters

Forests on isolated islands are often readily invaded, by plants, animals, and diseases. Three examples of invasive plants in Hawaii illustrate some of the features that can help species establish in new areas. Low elevation forests on the island of Hawaii have been invaded by falcataria (albizia) trees (Figure 14.7). The dominant native species, ohia, grows slowly, and falcataria trees have the capacity to fix nitrogen and grow rapidly. Litter from falcataria enriches the soil, which facilitates the growth of other invasive plants. One of the species that grows well under the influence of falcataria is clidemia, a fast-growing shrub with seeds that are widely dispersed by birds. The combination of albizia and clidemia can prevent the establishment of native plants. Strawberry guava is a third invasive plant, which prevents establishment of native koa trees. The guava fruits are eaten by non-native pigs which disseminate the seeds throughout the forests. The overall outcome of species invasions on islands is typically the conversion of native forests to forests of novel composition and often novel structure.

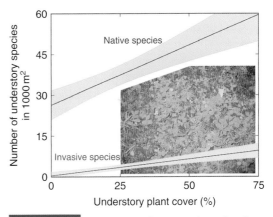

FIGURE 14.6 Forests in Rocky Mountain National Park, USA with high cover of understory plants also had high numbers of both native and invasive species. **Source:** based on data from Stohlgren et al. 1999.

Plantations of Non-Native Trees Can Lead to Invasions

The highest growth rates in tree farms often result from the use of non-native tree species. The non-native species are chosen in part because they grow faster than native tree species. The opportunity for planted tree species to invade native grasslands and forests can be very high indeed (Figure 14.8). Once an invading species has become widespread, the future forests will be quite different from historical forests. The strong biological potential for reproduction means that invasive species can become permanent components of forests; intensive management may be able to moderate impacts, but complete removal is not practical.

Biological Control May Help Limit Invasive Species

The opportunity for invasive species to affect forests may stem from the absence of other species that limit the success of the invasive in its native area. Insect populations are affected by an array of species and processes, including predation (by birds, mammals and arthropods), parasitism (by other insects and invertebrates), and diseases. Biological control can bring one or more of these species into invaded forests, impairing the success of the unwelcome invasive.

A classic example of using biological control to limit the success of an invasive comes from eastern North America, where winter moths were accidentally introduced from Eurasia in the mid-1900s. Larvae of winter moths eat leaves of broadleaved trees, including maples and oaks. A wide range of potential biological control species from Eurasia were considered, and a species of tachinid flies (*Cyzenis albicans*) was intentionally released in Nova Scotia. The flies lay eggs on leaves that are then eaten by moth larvae, and the pupae of the flies kill their hosts (a parasite that always kills its host may be called a parasitoid). The flies seemed to be remarkably successful at dropping the populations of winter moths and the defoliation of canopies (Figure 14.9). Or was this only a co-occurrence of two events in time? The fact that winter moths and defoliation declined after introduction of the parasitoid flies provides evidence, but the evidence might not be conclusive. The winter moth population could have declined as a result of other factors, such as bad weather, the accumulation of diseases in the winter moth population (unrelated to the parasitoid), or other factors. How might confidence in the effectiveness of the biological control be increased? More evidence could be gathered, including winter moth population dynamics in nearby locations where the biological control agent was not (yet) present. In Nova Scotia, the winter moth decline was observed only in the biological control area, increasing confidence that the local decline did not result from weather or other broad-scale

FIGURE 14.7 The falcataria trees in the upper picture were more than 40 m tall when killed by herbicide injection (**Source:** photo by J.B. Friday, University of Hawaii). The falcataria were killed because of their ability to enrich soil in N, furthering invasion of other species and threatening native species. Another exotic shrub that can prevent establishment of native plants is clidemia (middle photo, showing the extended hand of a standing person), which thrives on enriched soils under falcataria. Invasion by strawberry guava creates a dense woody understory (more than 5 m tall) that prevents establishment of native tree seedlings (lower picture). The seeds of guava are spread by exotic pigs.

factors. Another robust challenge to the effectiveness of biological control would be to repeat the experiment in another area at another time, and indeed the outcome from a trial 50 years later in Massachusetts, USA gave the same outcomes. High confidence is warranted in this case of biological control of invasive winter months.

Biological control will play a substantial role in limiting the impact of invasive species in many forests in the future, but not all cases can be successful. Biological control works well when a list of requirements is met: one or more control species are available, with large enough impact to control the invasive species; the control species are highly specific to the invasive species (and so do not threaten native species); and the control species can be successfully established for long-term persistence. Many examples are

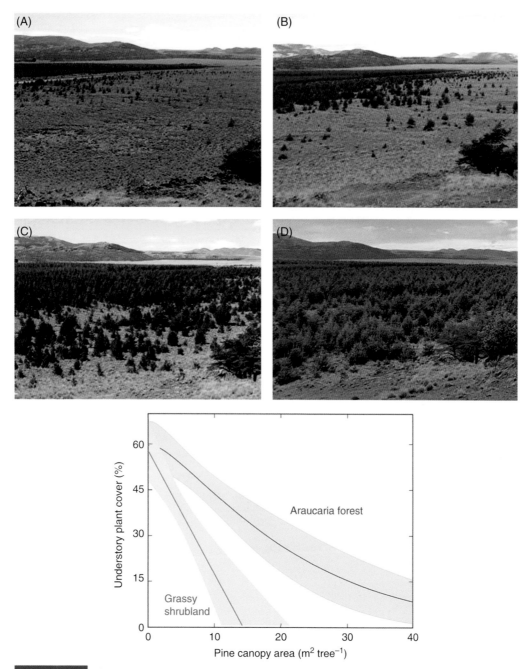

FIGURE 14.8 The pictures show the invasion of lodgepole pine (from North America) across a Patagonian grassland/shrubland in just 10 years (2007, 2011, 2015, and 2017). The ecological impact of pine invasion includes drastic losses of understory plant cover, especially in grassland/shrublands. **Source:** based on García et al. 2018, used by permission.

available of poorly conceived attempts at biological control, such as introduction of mongooses (a broad, generalist predator from Asia) onto the Hawaiian Islands to reduce rat consumption of sugar cane. The mongooses did not affect the rat population, and instead became a major predator of native (and exotic) birds.

Genetics Matter

Future forests will be shaped in part by a wide range of choices made about genetics. Forestry has typically promoted some species over others, which is a species-level form of genetic selection. For more than a century foresters have also paid attention to geographic regions in collecting seeds for use in plantations, as growth often depends on using a well-suited "provenance" of seeds (Figure 14.10).

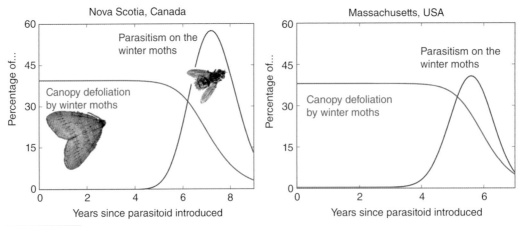

FIGURE 14.9   Before the introduction of parasitoid flies, invasive winter moths ate about 40% of the canopies of broadleaved trees in Nova Scotia, Canada. About four years after the flies were introduced, the fly population was large enough to affect the moths, reaching a peak of about 50% parasitism of moths after seven years. The moth population declined, and so did the percentage of parasitized moths. The same approach was used 50 years later in Massachusetts, US, with very similar results, increasing the confidence warranted in the story. **Source:** based on Elkinton et al. 2015, moth photo by D. Wagner, fly photo by Nicholas Condor.

FIGURE 14.10   The importance of provenance (the geographic location of seed collection) is evident in this row of birch trees at the Swedish University of Agricultural Sciences near Uppsala. The tall trees on the left come from southern locations, and those to the right from increasingly northerly locations (all trees are the same age). The southern provenances grow faster, and begin to turn color a few weeks later in the autumn.

Tree farms can be established with seedlings, with cuttings from trees (common for some poplar plantations), and with plantlets developed from tissues of trees (common for eucalyptus). Plantations of single genotypes (clones) typically outproduce seed-origin plantations because of higher uniformity in sizes of trees.

In the second half of the twentieth century, genetic selection intensified to foster genotypes with particular traits, such as straight stem form or resistance to diseases. Many species of eucalyptus can hybridize, and breeding programs focused on blending genotypes to achieve high growth rates, resistance to drought, and other objectives. In the last decades of the 1900s, about 75% of the genetic work on eucalyptus in Brazil focused on growth and tolerance of environmental factors, and 25% on resistance to pests and diseases. After the turn of the century, the focus flipped, reflecting rising challenges of pests and diseases (J.L. Stape, personal communication).

Hybridization between species is also being used to develop strains of species that resist invasive diseases. The chestnut blight from Asia decimated American chestnuts, and cross-breeding programs were developed in the 1900s to see if blending American genotypes with Asian chestnut genotypes could increase resistance to the blight. A first-generation cross might have the resistance of the Asian chestnut, but lack many of the key traits of the American chestnut (growth rate, size, etc.). The first-generation hybrids were "backcrossed" with other American chestnuts to end up with trees that were 15 parts American and 1 part Asian, with some of the blight-resistance of pure Asian chestnuts.

Could a single gene in the Asian chestnuts be identified and transferred to American chestnut trees, giving essentially "pure" American chestnuts that resist blight? Some aspects of organisms are indeed regulated by a single gene, but it's much more common for a suite of genes to have more diffuse regulation of traits, and this indeed is the case with resistance to chestnut blight. Hybrid breeding programs are the only way to extend blight resistance from Asian chestnuts to American chestnuts.

But what if a single gene could be found in some other plant that would confer resistance? The success of some fungal diseases is linked to the production of oxalate, which lowers the pH (increases the acidity) of tissues and leads to other conditions that favor the fungus over the plant. A gene was isolated from wheat that interferes with oxalate production by fungal diseases, and this gene has been successfully inserted into American chestnuts (Powell et al. 2019). The inserted gene can be passed on to progeny, which gives some interesting possibilities for chestnut restoration programs. Whereas a hybrid breeding program leads to a somewhat narrow gene pool (all resistant individuals are closely related), the insertion of a novel gene can spread through a genetically diverse population. Using a gene-insertion approach would allow pollen from a resistant tree to be shared with genes from a wide array of surviving wild genotypes of American chestnut, with half of the progeny having high resistance to blight.

The current state of technology allows for the complete enumeration of the genes of species, for deleting specific genes, and for adding specific genes. These are powerful new tools that go far beyond historical cross-breeding programs, or the common use of high intensity irradiation (which led to major developments in food crops, from wheat to pears). The future influence of genetic technology in forests will depend in part on the attitudes of people. Is a 15/16ths American chestnut really an American chestnut that should be widely planted in restoration programs? Would the 1/16th impurity of the genotype mean the trees would not be legitimate members of future forests? A chestnut with a blight-resistant gene from wheat would be over 99.9999% American chestnut, but would a single impurity in the genome be grounds for keeping chestnut out of the forests of the future? One of the strongest moral issues for many people is an idea of purity (or sanctity), and a perception of impurity can drive strong responses (Haidt 2012). Genetics will clearly have fundamental influences on futures forests, but the ecological aspects will also depend on social values that determine how (and even whether) powerful technology will be applied.

The Future Is Certain to Be Warmer, with More CO_2 in the Atmosphere

The atmospheric concentrations of CO_2 will continue to increase, with an unavoidable consequence of raising global temperature. Global average temperatures are affected by a wide variety of factors. Ice ages are influenced by variations in the Earth's tilt and orbit, with follow-on consequences for biotic activity that regulates atmospheric CO_2 concentrations, and changes in CO_2 further alter temperatures. The global average temperature will continue to rise, even if current emissions from fossil fuels decline, as a result of exchanges of CO_2 between the atmosphere and oceans (where much of the fossil-fuel CO_2 resides temporarily).

Warmer temperatures unavoidably mean an increase in global precipitation, but the spatial distribution of changes in temperature and precipitation depends on global circulation of air and oceans. Areas that become hotter may not be the same areas that receive more precipitation, with complex influences on forests. One key to forest response to changing climate would be any change in the vapor pressure deficit; warmer temperatures would not imply increased water stress if humidity increases enough that no change develops in vapor pressure deficit (Chapter 4). Warming might be greater at night than in the daytime, especially if humidity is higher. Seasonality in temperature, precipitation, and vapor pressure deficit could be important for routine tree growth, and for events such as droughts and fires.

These changes in temperature and water might be joined by the effects of CO_2 concentrations on photosynthesis. Higher CO_2 concentrations in the air might increase both photosynthesis and water use efficiency as more molecules of CO_2 enter leaves while water evaporates. These expectations could be wrong, if photosynthesis was too constrained by factors such as N supply, or if vapor pressure deficits increased water loss from leaves. The isotope ratios of ^{13}C to ^{12}C can be used to gauge water use efficiency (C uptake per unit water loss), and the record of isotope ratios in tree rings traces changes in water use efficiency across years and centuries. Analyses from forests around the world have shown the efficiency of water use increased by 50% since 1900 (Figure 14.11), in response to a 25% increase in the concentration of CO_2. If total transpiration by forests did not change, the wood production of forests should have increased by about half; if transpiration increased, this expected increase in growth would be somewhat larger. The two lines in Figure 14.11 represent some nuances that might interest scientists. When the trend is examined across tree

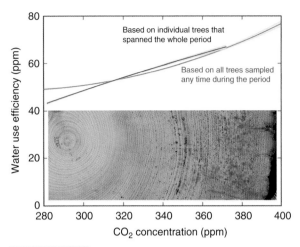

FIGURE 14.11   The patterns of ^{13}C isotopes in tree rings from over 400 forests around the world show that water use efficiency increased by about 50% with rising CO_2 concentrations since 1900 (green line). The representation of young and old trees would not have been consistent through the period, and the blue line is the subset of trees with data for the entire period. **Source:** based on data from Adams et al. 2020.

rings within individual trees, the rate of increase in water use efficiency slows slightly over time. This might be a result of declining response to further increases in CO_2, or it might be an age-related trend that results from older trees becoming taller trees (which can shift water use efficiency). When based on all available rings (include varying proportions of young trees and old trees), the increase in water use efficiency accelerated over time. The broad conclusion is robust in both cases, showing about a 50% increase in water use efficiency over the past century. The physics and biology of photosynthesis and growth are strong enough that future increases in CO_2 would seem likely to increase forest growth at a global scale, with much of the increase related to increased efficiency of using water.

If Droughts Increase, Which Forests and Trees Will Show Increased Mortality?

Tree species are well adapted to variations in water supply, and short periods of moderate drought may reduce photosynthesis without much effect on survival. Extended periods with substantial drought stress pose a larger challenge for both growth and survival. What generalizations about trees and drought might provide evidence for how future forests would change?

The forests in the Coweeta Experimental Forest (Chapter 1) experienced a severe drought between 1985 and 1988, with precipitation falling 24% below the long-term average. This magnitude of drought might occur about once in 200 years. When the severe drought occurred, did tree mortality increase evenly across the landscapes, or was it concentrated on drier sites or on wetter sites? Sites that are inherently dry (from a combination of low soil water storage capacity, and south-facing exposures) might be expected to show the highest mortality because of low water supplies even in good years. However, trees occurring on dry sites might be well adapted for dealing with drought, and trees on wetter sites might experience water limitations so rarely that they would be hit hardest by the severe drought. The evidence showed clearly that drought had a stronger effect on trees that normally experienced moister conditions than those on occurring on drier sites (Figure 14.12).

The same questions might be asked about which size of tree is most susceptible to drought: would mortality be higher for large, tall trees, or for short, suppressed trees? Tall trees have lower water potential, as water is raised to greater heights. Lower water potentials in tall trees might lead to greater challenges in dealing with droughts. On the other hand, small trees might be at a disadvantage in dealing with droughts because of low access to resources. The evidence clearly showed that

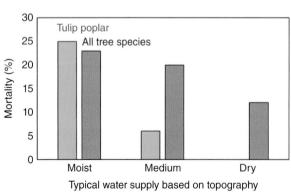

FIGURE 14.12   A strong, three-year drought in the Coweeta forest led to widespread mortality of trees. Mortality was higher at landscape positions that were moister (lower slopes), for tulip poplar trees and for all trees in general. **Source:** based on data from Elliott and Swank 1994.

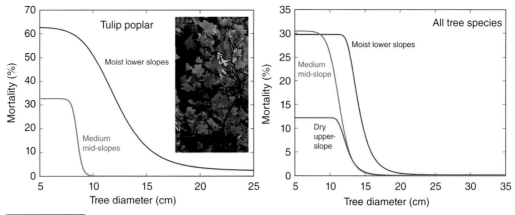

FIGURE 14.13   A much greater percentage of small tulip poplar trees died from the drought in the Coweeta Experimental forest than larger trees (left). The same trend occurred across all tree species (right; **Source:** Based on data from Elliott and Swank 1994).

mortality rates were higher for smaller trees, for tulip poplars and all trees in general, whether the trees grew on moister or drier sites (Figure 14.13).

Higher mortality in smaller trees was a very strong pattern at Coweeta, but would this be reliable evidence that droughts in the future would be more of a threat to small trees than larger trees? Given the great diversity of tree species across the planet, a pattern from one forest type might not apply broadly. The same questions could be asked about very different types of forests in locations with very different characteristics, and a consistent pattern across the broader sample would give a stronger basis for predicting general responses to future droughts. A severe drought in the mountains of California also led to high rates of mortality across landscapes, but in this case, mortality rates were about twice as high for taller trees than for short trees (Figure 14.14). This seems to indicate that a trend in drought mortality across tree sizes is unlikely to apply generally, and that local details are too important for simple trends. However, details about evidence can be important to consider, and in this case the details turn the story around. If the tree species comprising the taller trees showed higher drought mortality than the species with shorter trees, a pattern of species susceptibility to drought could be confused with a size-related risk. Indeed, the risk of mortality among broadleaved species, and among conifers other than pine, decreased for taller trees, just like at Coweeta. The overall pattern of high risk for the tallest trees resulted from the tallest trees being pines, and pines showed higher mortality than other groups. Within the group of pines, the taller trees did show higher mortality rates, but in this case the tree deaths resulted from bark beetle attacks (Chapter 10) rather than direct physiological problems with drought. Overall, it seems smaller trees are generally more at risk when droughts occur than larger trees.

Changing Climates Will Change the Distribution of Species

The current distribution of any species can be characterized as a function of climate; with maximum and minimum values for temperatures and precipitation associated with sites there the species occurs. Within such a climate envelope, the species would not be found on all sites, because of local details about topography that modify microenvironments, competition with other trees, impacts of browsers and diseases, and the time since major events. If the ecology of forests was very simple, then the future ranges of species could be predicted based on expected changes in regional climates. Each chapter in this book describes pieces of evidence about why such a simple view would be unlikely to work for real forests.

Three aspects of shifting ranges of species in response to changing climate might be worth keeping in mind. The first is that warming climates may be expected to extend viable ranges for species into areas where climate was previously too severe in winter. This would be the same

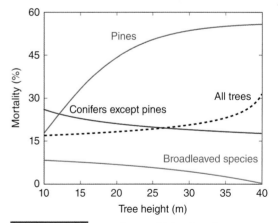

FIGURE 14.14   Mortality was higher for taller trees with drought in California, USA when averaged across all species (black dashed line), with trees taller than 30 m showing about double the mortality rate of trees less than 15 m. However, mortality risks actually decreased with size for broadleaved species and conifers other than pines. The overall trend was driven by increasing mortality of pines taller than 20 m, but these trees died as a result of bark beetle attack and not direct physiological problems of coping with drought. **Source:** based on data of Stephenson and Das 2019.

ecological phenomenon as invasion of a species into new territories, and quantitative predictions (such as MAXENT modeling) might prove to be accurate enough to predict extensions of species ranges.

Historical rates of species migration in response to changing climate have often been on the order of a 10–100 km in a century. Future rates of climate change might result in the climate characteristics of a given site "moving" at faster rate, perhaps faster than the potential dispersion rates of species. This might result in large areas of suitable conditions that remain colonized by migrating species. One option for approaching such open opportunities is "assisted migration" where people extend species ranges with active planting.

At the other end of the geographic range, the historical range of many species might be hotter and drier than a species would be expected to tolerate. This might lead to increased mortality and failure to regenerate, but in fact species often have a strong inertia for retaining current locations. Well-established species may have an advantage in holding onto sites relative to the establishment of new immigrants. Predictions of expansions of suitable ranges for a species might warrant more confidence than predictions about losses of current territories.

Fires Have Always Been Important in Forests, and Fires May Become More Important

Like many subjects in forest ecology, the occurrence and impacts of fires vary across space, and often vary across time as well. Random variation can produce illusions, generating patterns that appear to have deeper implications but in fact do not (Figure 11.17). Separating trends from random variation can be challenging, just as short-term trends may be in directions opposite from longer-term trends. For example, the area of forests burned in the US has more than doubled since 1980, but the trend in 1980 was for lower burned areas than in 1960 (Figure 14.15). Going farther back in time, every year in the 20 years between 1925 and 1945 had more forest area burned in the USA than the average value for the trend in 2020. Those mid-century fires tended to be widespread surface fires, not the high severity fires of recent decades, so comparisons of burned area may not provide simple insights about major changes over time.

Several factors combine to indicate that the influence of fire will in increase in future forests. A warmer climate with more frequent droughts would extend the period when forests can be ignited. Human populations have doubled in the past 50 years and will continue to increase, with increasing probabilities of fires ignited by people (accidentally or intentionally). Fire suppression efforts may lead to higher accumulations of fuels, which change the intensity and areal extent when fires do burn. These factors affect fires, along with the issues surrounding fire ecology and management. One example may illustrate the potential magnitude for future forests. The historical fires in the region of Yellowstone National Park, USA, can be related to the climate conditions in the years when large areas burned. Climate simulation models provide scenarios for future climate conditions, and these two sets of information can be combined to estimate how many years would go by between fires. The number of years between fires has averaged more than a century over recent centuries, but a warmer climate might drop that fire-free interval to 20 years or less, and do it very soon (Figure 14.16). Simulations may or may not warrant confidence, especially in predicting the unwritten future. However, they may provide the best available gauge for how large responses could be. The future forests in Yellowstone National Park seem very unlikely to be shaped by the sorts of fires that dominated the past few centuries.

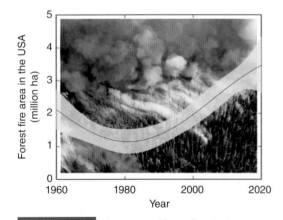

FIGURE 14.15   The extent of forest fires in the USA more than doubled after 1980. However, every year between 1925 and 1945 had more area burned than the 2020 peak in the recent trend, so the context of recent increases in fires needs to be considered carefully (**Source:** data from National Interagency Fire Center, https://www.nifc.gov; photo from US Forest Service, forest-http://atlas.fs.fed.us).

People Will Contribute to Shaping Future Forests

Tree farms in the future will continue to be managed with a classic approach of command and control: the factors influencing wood production will be considered and silvicultural and breeding techniques will be the key to high rates of growth. The future of managed forests and unmanaged forests will also be influenced by a wide variety of factors, including those considered in this chapter and throughout the book. Managed forests often have a

FIGURE 14.16   A shift to a warmer, drier climate with more frequent droughts would shorten the intervals between fires. Simulations for 8 million ha around Yellowstone National Park, US, showed that fire intervals would be likely to decline from more than 120 years in the twentieth century to only 20 years by the middle of the twenty-first century, well within the lifetimes of most readers of this book. **Source:** based on Westerling et al. 2011.

broader range of objectives than tree farms, and longer time frames have more potential for gradual change and for major events. Unmanaged forests may not have explicit goals for future conditions, but indeed people may have goals for unmanaged forests, such as prevention of severe fires, wildlife loss, and invasion by exotic species. The broad range of potential futures for managed and unmanaged forests spans a wide spectrum of potentially desirable and undesirable conditions.

The development of forests typically includes a large component of indeterminate factors: does a severe drought occur, does a fire burn through this particular watershed? Indeterminate factors combine with broad ranges of goals for future forests to frame a challenge that falls at the opposite end of the spectrum from command-and-control of tree farms. Tree farms can be managed with a clear focus on desired future conditions (high wood growth rates, and high profitability). Managed and unmanaged forests may be approached more productively with a focus on *undesirable* future conditions (Matonis et al. 2016). Among the many possible states of a future forest, some may be very undesirable, and thinking about forest management might focus productively on how to reduce those risks.

A good example of the value of thinking about undesirable conditions would be the restoration of the tree/meadow mosaic that would be characteristic of historical forests in frequent fire landscapes (such as Figures 13.15 and 13.17). These forests will experience climates and new invasive species that are outside the historical conditions that shaped vegetation across these landscapes in past centuries. A hope that returning to historical structure would ensure a desirable structure into the future may not be well-founded. However, the framing of goals might shift from asking, "What is most desirable?" to examining, "What risks would be worth reducing?" and "What undesirable futures would be good to avoid?" High densities of trees coupled with low cover of understory plants would predispose these forests to uncharacteristically severe fires. Restoring the forests to a structure more similar to historical times may be a good goal to avoid undesirable outcomes of severe fires. If future climates have more severe droughts, then "restoration" treatments to historical structures might be even more valuable for avoiding undesirable severe fires. Restoration ecology and management can often benefit strongly from a perspective of features that pose high risks of undesirable futures, even if precise desirable futures would be hard to define.

All These Factors Will Interact to Shape the Dynamics of Future Forests

Of course future forests will be shaped by changing climates, shifting species composition, and the occurrence of events. Forests are fostered by long periods between events, allowing trees to grow large under stable environmental conditions. Grasses cope very well with frequent events, including grazing and fire. A key difference between trees and grasses is that the growing tissues (meristems) of

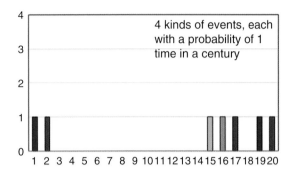

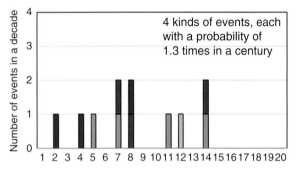

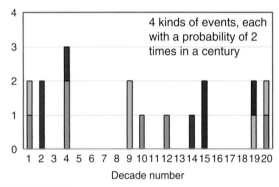

FIGURE 14.17 A random exercise illustrates how shortening the period between recurrence of events increases the likelihood that events will co-occur in the same decade. The top graph has four kinds of events (such as fire, drought, storms, and insect outbreaks) and each have a probability of occurring once in a century (a 1% probability each year). In a random run across two centuries, no decade experienced the occurrence of two events. The middle graph shortens the typical period between events, with a probability of occurring 1.3 times in a century, and three of the 20 decades had two kinds of events. A random run with a shorter return interval, with a probability of each event happening twice within a century, gave three decades with two types of events occurring, three decades where one kind of event happened twice, and one decade with three events. The point of this exercise is to show how the likelihood of types of events co-occurring goes up with the frequency of individual events.

trees are high in the crowns (and under bark), and when the tissue die (from being eaten or burned), regrowth may not be possible. The meristem cells for grasses are down at the root collar, so loss of grass leaves is like cutting off hair: the important growing cells are not harmed. A world with more events, such as increased frequency of severe fires, may not favor forests relative to shrublands and grasslands.

Two important points will be important for how events, such as storms, droughts, fires and insect outbreaks, will change forests. The first is that events often interact, such as beetle outbreaks after droughts, even though most of the examples in this book have focused on individual types of events (but see Figure 10.21). The second is that the intervals between events may shorten as climate changes, and this brings implications for the co-occurrence of events. As the intervals between events shorten, the random probability of two types of events happening close in time goes up. Forests are often dry enough to burn, but without an ignition no fire happens. Most forest fires that do ignite do not burn large areas, as fuels might be not quite dry enough, or windspeeds may be low. Severe fires tend to occur when these factors line up: dry fuels, ignitions, and strong winds. If the frequency of any one of these three factors increased, the incidence of severe fires would increase. If all factors increased, severe fires would become much more common.

The combinations of time span between events and the co-occurrence of multiple events is illustrated in a random exercise in Figure 14.17. Four different types of events are given a once-in-a-century probability of occurring. It's possible that an event may not actually come along in a given century, or it might randomly happen more than once. In a random run for two centuries, only one type of event happened in any single decade. When the probability of occurrence increased to 1.3 times in a century, the number of events of course increased, but so did the number of decades that experienced more than one event. A further increase in probability to twice in each century for each type of disturbance resulted in seven of twenty decades with more than one event, and one decade even had three events. Rerunning the random exercise would give different results, but co-occurrences of events become more frequent as the time between events shortens. If the broad driving factors of climate shorten the intervals between events, multiple events will stack up within the same decades. Given that some events are more likely to be associated than the random expectation in the exercise in Figure 14.17, future forests are likely to show greater dynamics (rates of change over time) than historical forests.

Rocket Science Can Get You to the Moon, but Pocket Science Leads to Better Outcomes in Forests

Back in the 1900s, it was common for forest management and science to be thought of as separate domains, with managers tasked with managing forests (sometimes learning new things from publications and at conferences), and scientists investigating details of processes (sometimes telling managers how to do a better job). Modern approaches recognize that these domains are more productive when connected, and collaborative science and management have become the norm for the best management (Franklin et al. 2018; Chapin 2020; Scott et al. 2020). One opportunity for keeping science and management connected is the idea of pocket science (Binkley et al. 2018). Every forest presents opportunities for learning, especially when a forest is intentionally changed by a management activity. The key point for scientific investigations is combining ideas with tests that have a chance of showing where the idea is wrong, and this approach can be applied routinely in projects that aim to manage forests.

For example, a forester may expect that controlling understory vegetation with herbicides will increase growth in a loblolly pine plantation, because this is a practice commonly used in the region. There's little reason to doubt that early growth of planted seedlings will do better with less competition, but it's possible that the early benefit disappears after a decade or two. If a 40-ha plantation will be treated with herbicide as a standard practice, a pocket-science approach would set aside one typical hectare without herbicide treatment. Only a small amount of information would need to be recorded, such as precise location of the plot along with a few representative photographs (for both the pocket-science plot and the general plantation). A few years later, it would be simple to retake some photographs, and maybe measure some trees. If the untreated hectare looked as good as the rest of the plantation, the expectation that herbicides lead to profitable returns on investment would be challenged. If the untreated hectare clearly showed poorer tree growth, confidence in herbicide treatment might go up. Unintended outcomes might also be spotted, such as high use of the untreated hectare by wildlife species.

Another example would be the benefit of creating brush fences after harvesting, aiming to reduce browsing on regenerating trees (Figure 14.18). If an entire harvested area were treated the same way, there would be little opportunity to learn whether the fencing was really necessary. If a portion of the cut forest remained outside the fenced area, then this pocket-science project would give a chance to see what might have happened if the whole area had not been fenced. The insights might not be conclusive, as the outcome for a small pocket-science plot might not give an accurate representation of what would have happened for a larger area. Perhaps a large area would disperse browsing pressure and not show the severe impacts that occurred on a small patch. A pocket science approach may be too rough and simple to provide a rigorous test, but it always provides food for thought (and perhaps learning and improved choices in the future).

Pocket science approaches support learning from the operational experiences of other land managers. If an entire forest is treated to restore historical forest structure and reduce risk of severe fire, how would the next set of land managers understand why the restoration treatment had been so important? A visit to the restored site would show a forest without very undesirable risks, and a retained, untreated patch would clearly show the risks without restoration treatments (Figure 14.18).

FIGURE 14.18   An aspen/conifer forest was harvested to promote development of a young aspen forest. Based on experience in other cases, the managers were concerned that elk browsing might prevent regeneration. An expensive fence was constructed from logging slash, and some of the harvested area was left outside the fence, giving a pocket-science opportunity for learning whether the expected outcomes actually occurred, and if any surprises developed (upper photo). A similar approach to a forest restoration project retained an unburned patch within a large landscape that was burned twice to restore low-density forest structure in a frequent-fire landscape. The educational value of this pocket-science approach is clear when visitors tour the restored forest (lower photo).

A manager who uses a pocket-science approach across multiple projects might need to allocate a couple days each year to revisiting past projects, repeating photographs and looking to see if the expected outcomes occurred, and if any surprises developed. Sets of pocket-science projects offer great opportunities for showing other people why a given treatment might be particularly important, or why it might not (perhaps showing where more detailed investigation would be particularly helpful). Perhaps the most valuable aspect of pocket science is the thinking it encourages. Routinely thinking of management actions as learning opportunities, and all forests as schoolhouses, fosters and feeds the curiosity that underlies responsible forest stewardship. The pocket science approach is a small, workable corner of the large discipline of adaptive management promoted by C.S. Holling and colleagues (Holling 1978).

The Core Framework Actually Needs a Fourth Question

As a science, forest ecology deals with measurable features of forests, using reasoning and evidence to explain the current and past state of the forest and to make some educated guesses about the future. The science of understanding forests is not about art, beauty, philosophy, or values. However, everyone who uses scientific approaches to understand forests also has ideas and values from domains outside science. If a survey were conducted about why people have interests in forests, the top responses would likely deal with ideas of beauty, the importance of nature, and deep and diverse values about trees, animals, and maybe even about soils.

The Core Framework can now be extended (Figure 14.19). The first three questions explicitly recognize that scientific understanding is fundamental and powerful, and a fourth question brings in all the other dimensions that matter to people: When entering a forest, what meaning will you find? This last question is outside and beyond the scope of this book, but insightful guidance is available in classic books (such as Leopold's *Sand County Almanac* (Leopold, 1949), and Kimmerer's *Braiding Sweetgrass* (Kimmerer 2013)), from interviews with diverse people, lectures, documentaries, and especially from directly experiencing forests.

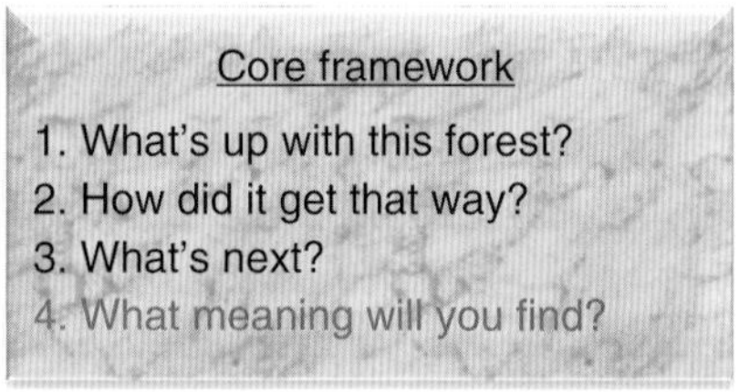

Core framework

1. What's up with this forest?
2. How did it get that way?
3. What's next?
4. What meaning will you find?

FIGURE 14.19 The core framework that applies to all forests, across space and time, may be more complete if a dimension is included that is outside science: what meaning will you find in this forest?

Ecological Afterthoughts: Growing Meaning in Forests

It's time to take the insights from this book and apply the four questions of the Core Framework to a forest. Pick a forest – any forest will do, because the questions are universal, even if the answers are always case-specific. The first three questions are objective, and can be answered with the evidence-based approaches emphasized in this book. The fourth question is personal and bound to differ from person to person, but it too is one that can be shared with other people, not only as science but also as art, beauty, meaning, and philosophy (Figure 14.20).

FIGURE 14.20 What insights, meanings, and beauty will you quilt together from your forest? **Source:** image of a quilt created by Penny Strockbine, inspired by the Pando aspen clone.

References

Aaltonen, **V.T.** (1919). Über die natürliche Verjungung der Heideväler im Finnischen Lappland. *Communicationes ex Instituto Quaestionum Forestalium Finlandiae* 1: 1–319. (In Finnish with German summary).

Abella, **S.R.** (2018). Forest decline after a 15-year "perfect storm" of invasion by hemlock woolly adelgid, drought, and hurricanes. *Biological Invasions* 20: 695–707.

Ackerman, **D.**, **Millet**, **D.B.**, and **Chen**, **X.** (2018). Global estimates of inorganic nitrogen deposition across four decades. *Global Biogeochemical Cycles* 33 https://doi.org/10.1029/2018GB005990.

Adair, **E.C.** and **Binkley**, **D.** (2002). Co-limitation of first year Fremont cottonwood seedlings by nitrogen and water. *Wetlands* 22: 425–429.

Adams, **M.B.** ed. (2017). The Forestry Reclamation Approach: Guide to Successful Reforestation of Mined Lands. *US Forest Service General Technical Report* NRS-169, Newtown Square, PA.

Adams, **M.A.**, **Buckley**, **T.N.**, and **Turnbull**, **T.L.** (2020). Diminishing CO_2-driven gains in water-use efficiency of global forests. *Nature Climate Change* https://doi.org/10.1038/s41558-020-0747-7.

Albrecht, **A.**, **Hanewinkel**, **M.**, **Bauhus**, **J.**, and **Kohnle**, **U.** (2012). How does silviculture affect storm damage in forests of south-western Germany? Results from empirical modeling based on long-term observations. *European Journal of Forest Research* 131: 229–247.

Aleksić, **J.M.** and **Geburek**, **T.** (2014). Quaternary population dynamics of an endemic conifer, *Picea omorika*, and their conservation implications. *Conservation Genetics* 15: 87–107.

Allen, **J.M.** and **Bradley**, **B.A.** (2016). Out of the weeds? Reduced plant invasion risk with climate change in the continental United States. *Biological Conservation* 203: 306–312.

Allen, **B.L.**, **Allen**, **L.R.**, **Andrén**, **H.** et al. (2017). Can we save large carnivores without losing large carnivore science? *Food Webs* 12: 64–75.

Ally, **D.**, **Ritland**, **K.**, and **Otto**, **S.P.** (2008). Can clone size serve as a proxy for clone age? An exploration using microsatellite divergence in *Populus tremuloides*. *Molecular Ecology* 17: 4897–4911.

Al-Rawi, **F.N.H.** and **George**, **A.R.** (2014). Back to the cedar forest: the beginning and end of Tablet V of the standard Babylonian epic of Gilgameš. *Journal of Cuneiform Studies* 66: 69–90.

Ammer, **C.** (2018). Tansley review: diversity and forest productivity in a changing climate. *New Phytologist* https://doi.org/10.1111/nph.15263.

Audley, **J.P.**, **Fettig**, **C.J.**, **Munson**, **A.S.** et al. (2020). Impacts of mountain pine beetle outbreaks on lodgepole pine forests in the Intermountain West, U.S., 2004–2019. *Forest Ecology and Management* 475: 118403.

Bailey, **R.** (1980). Description of the Ecoregions of the United States. Washington, D.C.: USDA Forest Service Miscellaneous Publication 1391.

Baker, **W.L.** (2017). The Landscapes They Are A-Changin'. *Technical Brief*. Colorado State University, Colorado Forest Restoration Institute, Fort Collins, CO, 72 pp.

Baker, **S.** (2020). Silviculture Survey: Forest Management Practices of Investors and Managers in the U.S. South. LLC, Athens, Gerogia, USA: Forisk Consulting.

Bär, **A.** and **May**, **S.** (2020). Bark insulation: ten central alpine tree species compared. *Forest Ecology and Management* 474: 118361.

Barczi, **J.F.**, **Rey**, **H.**, **Griffon**, **S.**, and **Jourdan**, **C.** (2018). DigR: a generic model and its open source simulation software to mimic three-dimensional root-system architecture diversity. *Annals of Botany* 121: 1089–1104.

Basile, **M.**, **Mikusiński**, **G.**, and **Storch**, **I.** (2019). Bird guilds show different responses to tree retention levels: a meta-analysis. *Global Ecology and Conservation* 18: e00615.

Basset, **Y.**, **Cizek**, **L.**, **Cuénoud**, **P.** et al. (2012). Arthropod diversity in a tropical forest. *Science* 338: 1481–1484.

Bateson, **G.** (1979). ind and Nature: A Necessary Unity. New York: Hampton Press.

Bauer, B. (2018). How tree rings tell time and climate history: www.climate.gov/news-features/blogs/beyond-data/how-tree-rings-tell-time-and-climate-history.

Bauerle, W.L., Hinckley, T.M., Cermack, J. et al. (1999). The canopy water relations of old-growth Douglas-fir trees. *Trees* 13: 211–217.

Beck, D.E. and **Della-Bianca, L.** (1972). Growth and yield of thinned yellow-poplar. *USDA Forest Service Research Paper* SE-101.

Beese, W.J. and **Sandford, J.S.** (2020). Understory vegetation response to alternative silvicultural treatments and systems: 26-year results. *Montane Alternative Silvicultural Systems (MASS) Project.* Vancouver Island University Research Services Contract File No.: 5615, Nanaimo.

Beese, W.J., Rollerson, T.P., and **Peters, C.M.** (2019a). Quantifying wind damage associated with variable retention harvesting in coastal British Columbia. *Forest Ecology and Management* 443: 117–131.

Beese, W.J., Deal, J., Dunsworth, B.G. et al. (2019b). Two decades of variable retention in British Columbia: a review of its implementation and effectiveness for biodiversity conservation. *Ecological Processes* 8: 33. https://doi.org/10.1186/s13717-019-0181-9.

Biederman, J.A., Somor, A.J., Harpold, A.A. et al. (2015). Recent tree die-off has little effect on streamflow in contrast to expected increases from historical studies. *Water Resources Research* 51: 9775–9789.

Bigelow, S.W., Runkle, J.R., and **Oswald, E.M.** (2020). Competition, climate, and size eects on radial growth in an old-growth hemlock forest. *Forests* 11: 52. https://doi.org/10.3390/f11010052.

Binkley, D. (2003). Seven decades of stand development in mixed and pure stands of conifers and nitrogen-fixing red alder. *Canadian Journal of Forest Research* 33: 2274–2279.

Binkley, D. and **Fisher, R.F.** (2020). Ecology and Management of Forest Soils. Hoboken: Wiley Blackwell.

Binkley, D., Dunkin, K.A., DeBell, D., and **Ryan, M.G.** (1992). Production and nutrient cycling in mixed plantations of Eucalyptus and Albizia in Hawaii. *Forest Science* 38: 393–408.

Binkley, D., Ryan, M.G., Stape, J.L., and **Fownes, J.** (2002). Age-related decline in forest ecosystem growth: an individual-tree, stand-structure hypothesis. *Ecosystems* 5: 58–67.

Binkley, D., Senock, R., Bird, S., and **Cole, T.G.** (2003). Twenty years of stand development in pure and mixed stands of *Eucalyptus saligna* and nitrogen-fixing *Facaltaria mollucana. Forest Ecology and Management* 182: 93–102.

Binkley, D., Moore, M., Romme, W., and **Brown, P.** (2006). Was Aldo Leopold right about the Kaibab deer herd? *Ecosystems* 9: 227–241.

Binkley, D., Stape, J.L., Bauerle, W.L., and **Ryan, M.G.** (2010). Explaining growth of individual trees: light interception and efficiency of light use by *Eucalyptus* at four sites in Brazil. *Forest Ecology and Management* 259: 1704–1713.

Binkley, D., Campoe, O.C., Gsalptl, M., and **Forrester, D.** (2013). Light absorption and use efficiency in forests: why patterns differ for trees and stands. *Forest Ecology and Management* 288: 5–13.

Binkley, D., Alsanousi, A., and **Romme, W.** (2014). Age structure of aspen forests on the Uncompahgre Plateau, Colorado. *Canadian Journal of Forest Research* 44: 836–841.

Binkley, D., Adams, M., Fredericksen, T. et al. (2018). Connecting ecological science and management in forests for scientists, managers and pocket scientists. *Forest Ecology and Management* 410: 157–163.

Binkley, D., Adams, M., Fredericksen, T. et al. (2018). Connecting ecological science and management in forests for scientists, managers and pocket scientists. *Forest Ecology and Management* 410: 157–163.

Binkley, D., Campoe, O.C., Alvares, C.A. et al. (2020). Variation in whole-rotation yield among Eucalyptus genotypes in response to water and heat stresses: the TECHS project. *Forest Ecology and Management* 462: 117953.

Bleby, T.M., McElrone, A.J., and **Jackson, R.B.** (2010). Water uptake and hydraulic redistribution across large woody root systems to 20 m depth. *Plant, Cell and Environment* 33: 2132–2148.

Bleiker, K.P. and **Smith, G.D.** (2019). Cold tolerance of mountain pine beetle (Coleoptera: Curculionidae) Pupae. *Environmental Entomology* 48: 1412–1417.

Blinn, C.E., Albaugh, T.J., Fox, T.R. et al. (2012). A method for estimating deciduous competition in pine stands using landsat. *Southern Journal of Applied Forestry* 36: 71–78.

Bohn, F.J. and **Huth, A.** (2017). The importance of forest structure to biodiversity–productivity relationships. *Royal Society Open Science* 4: 160521.

Bonner, F.T. and **Karrfalt, R.P.** (eds.) (2008). The Woody Plant Seed Manual. USDA Forest Service Agricultural Handbook 727. Washington, DC: USDA www.fs.usda.gov/treesearch/pubs/32626.

Booth, R.K., Brewer, S., Blaauw, M. et al. (2012). Decomposing the mid-Holocene Tsuga decline in eastern North America. *Ecology* 93: 1841–1852.

Botanic Gardens Conservation International. (2020). www.bgci.org/resources/bgci-databases/globaltreesearch, accessed January 2021.

Botkin, **D.** (1990). Discordant Harmonies: A New Ecology for the Twenty-First Century. New York: Oxford University Press.

Boyden, **S.**, **Binkley**, **D.**, and **Senock**, **R.** (2005). Competition and facilitation between Eucalyptus and nitrogen-fixing Falcataria in relation to soil fertility. *Ecology* 86: 992–1001.

Brancalion, **P.H.S.**, **Schweizer**, **D.**, **Gaudare**, **U.** et al. (2016). Balancing economic costs and ecological outcomes of passive and active restoration in agricultural landscapes: the case of Brazil. *Biotropical* 48: 856–867.

Brancalion, **P.H.S.**, **Bello**, **C.**, **Chazdon**, **R.L.** et al. (2018). Maximizing biodiversity conservation and carbon stocking in restored tropical forests. *Conservation Letters* 57: 55–66.

Brancalion, **P.H.S.**, **Campoe**, **O.**, **Medes**, **J.C.T.** et al. (2019). Intensive silviculture enhances biomass accumulation and tree diversity recovery in tropical forest restoration. *Ecological Applications* 29: e01847.

Brancalion, **P.H.S.**, **Amazonas**, **N.T.**, **Chazdon**, **R.L.** et al. (2020). Exotic eucalypts: from demonized trees to allies of tropical forest restoration? *Journal of Applied Ecology* https://doi.org/10.1111/1365-2664.13513.

Brando, **P.M.**, **Nepstad**, **D.C.**, **Balch**, **J.K.** et al. (2012). Fire-induced tree mortality in a neotropical forest: the roles of bark traits, tree size, wood density and fire behavior. *Global Change Biology* 18: 630–641.

Brown, **A.E.**, **Zhang**, **L.**, **McMahon**, **T.A.** et al. (2005). A review of paired catchment studies for determining changes in water yield resulting from alterations in vegetation. *Journal of Hydrology* 310: 28–61.

Buma, **B.**, **Bisbing**, **S.**, **Krapek**, **J.**, and **Wright**, **G.** (2017). A foundation of ecology rediscovered: 100 years of succession on the William S. Cooper plots in Glacier Bay, Alaska. *Ecology* 98: 1513–1523.

Burbaitė, **L.** and **Csányi**, **S.** (2009). Roe deer population and harvest changes in Europe. *Estonian Journal of Ecology* 58: 169–180.

Burns, **R.M.** and **Honkala**, **B.H.** (Technical coordinators). (1990). Silvics of North America: 1. Conifers; 2. Hardwoods. *Agriculture Handbook* 654. U.S. Department of Agriculture, Forest Service, Washington, DC. vol. 2, 877 p. www.srs.fs.usda.gov/pubs/misc/ag_654/table_of_contents.htm.

Busse, **M.D.**, **Shestak**, **C.J.**, **Hubbert**, **K.R.**, and **Knapp**, **E.E.** (2010). Soil physical properties regulate lethal heating during burning of woody residues. *Soil Science Society of America Journal* 74: 947–955.

Busse, **M.D.**, **Shestak**, **C.J.**, and **Hubbert**, **K.R.** (2013). Soil heating during burning of forest slash piles and wood piles. *International Journal of Wildland Fire* 22: 786–796.

Calcote, **R.** (2003). Mid-Holocene climate and the hemlock decline: the range limit of *Tsuga canadensis* in the western Great Lakes region, USA. *The Holocene* 13: 215–224.

Campoe, **O.C.**, **Stape**, **J.L.**, **Albaugh**, **T.J.** et al. (2013). Fertilization and irrigation effects on tree level aboveground net primary production, light interception and light use efficiency in a loblolly pine plantation. *Forest Ecology and Management* 288: 43–48.

Caprio, **A.C.** and **Lineback**, **P.** (2002). Pre-twentieth century fire history of Sequoia and Kings Canyon national parks: a review and evaluation of our knowledge. In: (ed. N.G. Sugihara, M. Morales, and T. Morales), 180–199. *Proceedings of the Symposium – Fire in California Ecosystems: Integrating Ecology, Prevention, and Management*. Association for Fire Ecology Miscellaneous Publication 1.

Carlson, **C.A.**, **Fox**, **T.R.**, **Colbert**, **S.R.** et al. (2006). Growth and survival of Pinus taeda in response to surface and subsurface tillage in the southeastern United States. *Forest Ecology and Management* 234: 209–217.

Cavaleri, **M.A.**, **Oberbauer**, **S.F.**, **Clark**, **D.B.** et al. (2010). Height is more important than light in determining leaf morphology in a tropical forest. *Ecology* 91: 1730–1739.

Chambers, **M.E.**, **Fornwalt**, **P.J.**, **Malone**, **S.L.**, and **Battaglia**, **M.A.** (2016). Patterns of conifer regeneration following high severity wildfire in ponderosa pine-dominated forests of the Colorado Front Range. *Forest Ecology and Management* 378: 57–67.

Chapin, **F.S.** III (2020). Grassroots Stewardship: Sustainability Within Our Reach. Oxford: Oxford University Press.

Chapin, **F.S.** III, **Conway**, **A.J.**, **Johnstone**, **J.F.** et al. (2016). Absence of net long-term successional facilitation by alder in a boreal Alaska floodplain. *Ecology* 97: 2986–2997.

Chen, **J.**, **Franklin**, **J.F.**, and **Spies**, **T.A.** (1995). Growing-season microclimate gradients form clearcut edges into old-growth Douglas-fir forests. *Ecological Applications* 5: 74–86.

Chen, **J.**, **Saunders**, **S.C.**, **Crow**, **T.R.** et al. (1999). Microclimate in forest ecosystem and landscape ecology. *Bioscience* 249: 288–297.

Chen, **J.**, **Song**, **B.**, **Rudnicki**, **M.** et al. (2004). Spatial relationship of biomass and species distribution in an old-growth Pseudotsuga-Tsuga forest. *Forest Science* 50: 364–375.

Chen, **J.**, **Hao**, **Z.**, **Guang**, **X.** et al. (2019). Liriodendron genome sheds light on angiosperm phylogeny and species-pair differentiation. *Nature Plants* 5: 18–25.

Chong, **G.W.**, **Simonson**, **S.E.**, **Stohlgren**, **T.J.**, and **Kalkhan**, **M.A.** (2001). Biodiversity: aspen stands have the lead, but will non-native species take over? In: *USDA Forest Service Proceedings* RMRS-P-18, Fort Collins, Colorado, pp. 261–271.

Christina, **M.**, **Laclau**, **J.-P.**, **Gonçalves**, **J.L.M.** et al. (2011). Almost symmetrical vertical growth rates above and below ground in one of the world's most productive forests. *Ecosphere* 2, art 27: 149–160.

Clements, **F.E.** (1916). Plant Succession; An Analysis of the Development of Vegetation. Carnegie Institution of Washington.

Clements, **F.E.** (1936). Nature and structure of the climax. *The Journal of Ecology* 24: 252–284.

Cleveland, **C.** (2017). Temperature and rainfall interact to control carbon cycling in tropical forests. *Knowledge Network for Biocomplexity* https://doi.org/10.5063/F19021QT.

Coble, **A.P.**, **Autio**, **A.**, **Cavaleri**, **M.A.** et al. (2014). Influence of height, shade index, and age on Eucalyptus leaf morphology and nitrogen. *Trees: Structure and Function* 28: 1–15.

Cook, **J.G.**, **Irwin**, **L.L.**, **Bryant**, **L.D.**, **Riggs**, **R.A.**, and **Thomas**, **J.W.** (2004). Thermal cover needs of large ungulates: a review of hypothesis tests. *Transactions of the 60th North American Wildlife and Natural Resources Conference*. Wildlife Management Institute, https://wildlifemanagement.institute/store/product/57.

Coomes, **D.A.** and **Grubb**, **P.J.** (2000). Impacts of root competition in forests and woodlands: a theoretical framework and review of experiments. *Ecological Monographs* 70: 171–207.

Coop, **J.D.**, **DeLory**, **T.J.**, **Downing**, **W.M.** et al. (2019). Contributions of fire refugia to resilient ponderosa pine and dry mixed-conifer forest landscapes. *Ecosphere* 10: e02809.

Cooper, **S.V.**, **Neiman**, **K.E.**, and **Roberts**, **D.W.** (1991). *Forest habitat types of northern Idaho: A second approximation*. USDA Forest Service, INT-GTR-236. Ogden, UT.

Croft, **H.** and **Chen**, **J.M.** (2017). Leaf pigment content. *Comprehensive Remote Sensing* https://doi.org/10.1016/B978-0-12-409548-9.10547-0.

Crouzeilles, **R.**, **Bodin**, **B.**, **Alexandre**, **N.S. Guariguata**, **M.R.**, **Beyer**, **H.**, and **Chazdon**, **R.L.** (2019). Giving Nature a Hand. *Conservation International Science Progress Report*, www.conservation.org/docs/default-source/publication-pdfs/progress_report_natural_regeneration.pdf.

D'Amato, **A.W.**, **Jokela**, **E.J.**, **O'Hara**, **K.L.**, and **Long**, **J.N.** (2017). Silviculture in the United States: an amazing period of change over the past 30 years. *Journal of Forestry* 116: 55–67.

D'Amato, **A.W.**, **Orwig**, **D.A.**, **Foster**, **D.R.** et al. (2017). Long-term structural biomass dynamics of virgin *Tsuga canadensis-Pinus strobus* forests after hurricane disturbance. *Ecology* 98: 721–733.

Dambrine, **E.**, **Dupouey**, **J.-L.**, **Laüt**, **L.** et al. (2007). Present forest biodiversitiy patterns in France related to former Roman agriculture. *Ecology* 88: 1430–1439.

Davis, **M.**, **Xue**, **J.**, and **Clinton**, **P.** (2015). Planted-Forest Nutrition. New Zealand Forest Research Institute, SCION Publication S0014 https://scionforestryfuture.files.wordpress.com/(2015)/05/plantationforestnutritionreport.pdf.

Dawkins, **R.** (1976). The Selfish Gene. Oxford: Oxford University Press.

Dawkins, **R.** (2010). The Greatest Show on Earth: The Evidence for Evolution. New York: Free Press.

De Jager, **N.R.**, **Rohweder**, **J.J.**, **Miranda**, **B.R.** et al. (2017). Modelling moose-forest interactions under different predation scenarios at Isle Royale National Park, USA. *Ecological Applications* 27: 1317–1337.

DeBano, **L.F.**, **Neary**, **D.G.**, and **Ffolliott**, **P.F.** (2008). Chapter 2: Soil physical properties. In: Wildland Fire in Ecosystems: Effects of Fire on Soils and Water, *Gen. Tech. Rep.* RMRS-GTR-42, vol. 4 (eds. **D.G. Neary**, **K.C. Ryan** and **L.F. DeBano**). Ogden.

Duffy, **C.J.** (2017). The terrestrial hydrologic cycle: an historical sense of balance. *WIREs Water* 4: 1–21.

Dupke, **C.**, **Bonenfant**, **C.**, **Reineking**, **B.** et al. (2017). Habitat selection by a large herbivore at multiple spatial and temporal scales is primarily governed by food resources. *Ecography* 40: 1014–1027.

Elkinton, **J.**, **Boettner**, **G.**, **Liebhold**, **A.**, and **Gwia**, **R.** (2015). Biology, spread, and biological control of winter moth in the eastern United States. *U.S. Forest Service report* FHTET-(2014)-07, www.fs.fed.us/foresthealth/technology/pdfs.

Elliott, **K.J.** and **Swank**, **W.T.** (1994). Impacts of drought on tree mortality and growth in a mixed hardwood forest. *Journal of Vegetation Science* 5: 229–236.

Elliott, **K.J.** and **Swank**, **W.T.** (2008). Long-term changes in forest composition and diversity following early logging (1919–1923) and the decline of American chestnut (*Castanea dentata*). *Plant Ecology* 197: 155–172.

Ellwood, **E.R.**, **Primack**, **R.B.**, and **Talmadge**, **M.L.** (2010). Effects of climate change on spring arrival times of birds in Thoreau's Concord from 1851 to (2007). *The Condor* 112: 754–762.

Elton, **C.S.** (1958). The Ecology of Invasions by Animals and Plants. Springer.

Elton, **C.** and **Nicholson**, **M.** (1942). The ten-year cycle in numbers of the lynx in Canada. *Journal of Animal Ecology* 11: 215–244.

Epron, **D.**, **Nouvellon**, **Y.**, **Mareschal**, **L.** et al. (2013). Partitioning of net primary production in *Eucalyptus* and *Acacia* stands and in mixed-species plantations: two case-studies in contrasting tropical environments. *Forest Ecology and Management* 301: 102–111.

Eyre, **F.H.** (2017). Forest Cover Types of the United States and Canada. Washington, DC: Society of American Foresters.

FAO. (2020). Global Forest Resources Assessment (2020). *Main report*. Rome. https://doi.org/10.4060/ca9825en.

Farjon, **A.** (2018). Conifers of the world. *Kew Bulletin* 73: 8.

Feng, **Y.**, **Negrón-Juárez**, **R.I.**, and **Chambers**, **J.Q.** (2020). Remote sensing and statistical analysis of the effects of hurricane María on the forests of Puerto Rico. *Remote Sensing of Environment* 247: 111940.

Fesenmyer, **K.A.** and **Christensen**, **N.L.** Jr. (2010). Reconstructing Holocene fire history in a southern Appalachian forest using soil charcoal. *Ecology* 91: 662–670.

Fierer, **N.**, **Breitbart**, **M.**, and **Nulton**, **J.** (2007). Metagenomic and small-subunit rRNA analyses reveal the genetic diversity of bacteria, archaea, fungi, and viruses in soil. *Applied and Environmental Microbiology* 73: 7059–7066.

Finney, **M.A.**, **McHugh**, **C.W.**, and **Grenfell**, **I.C.** (2005). Stand- and landscape-level effects of prescribed burning on two Arizona wildfires. *Canadian Journal of Forest Research* 35: 1714–1722.

Flaspohler, **D.J.**, **Giardina**, **C.P.**, **Asner**, **G.P.** et al. (2010). Long-term effects of fragmentation and fragment properties on bird species richness in Hawaiian forests. *Biological Conservation* 143: 280–288.

Floate, **K.D.**, **Godbout**, **J.**, **Lau**, **M.K.** et al. (2016). Plant–herbivore interactions in a trispecific hybrid swarm of *Populus*: assessing support for hypotheses of hybrid bridges, evolutionary novelty and genetic similarity. *New Phytologist* 209: 832–844.

Ford, **C.R.**, **Hubbard**, **R.M.**, and **Vose**, **J.M.** (2010). Quantifying structural and physiological controls on variationin canopy transpiration among planted pine and hardwood species in the southern Appalachians. *Ecohydrology* https://doi.org/10.1002/eco.136.

Ford, **C.R.**, **Elliott**, **K.J.**, **Clinton**, **B.D.** et al. (2012). Forest dynamics following eastern hemlock mortality in the southern Appalachians. *Oikos* 121: 523–536.

Foster, **D.R.** and **Motzkin**, **G.** (1998). Ecology and conservation in the cultural landscape of New England: lessons from nature's history. *Northeastern Naturalist* 5: 111–126.

Franklin, **J.F.** and **DeBell**, **D.S.** (1988). Thirty-six years of tree population change in an old-growth Pseudotsuga–Tsuga forest. *Canadian Journal of Forest Research* 18: 633–639.

Franklin, **C.M.A.** and **Harper**, **K.A.** (2016). Moose browsing, understorey structure and plant species composition across spruce budworm-induced forest edges. *Journal of Vegetation Science* 27: 524–534.

Franklin, **O.**, **Näsholm**, **T.**, **Högberg**, **P.**, and **Högberg**, **M.N.** (2014). Forests trapped in nitrogen limitation – an ecological market perspective on ectomycorrhizal symbiosis. *New Phytologist* 203: 657–666.

Franklin, **J.F.**, **Johnson**, **K.N.**, and **Johnson**, **D.L.** (2018). Ecological Forest Management. Long Grove, Illinois: Waveland Press.

Fridman, **J.** and **Nilsson**, **P.** (2017). Forest Statistics (2017), Official Statistics of Sweden. Umeå: Swedish University of Agricultural Sciences.

Fulé, **P.Z.**, **Crouse**, **J.E.**, **Cocke**, **A.E.** et al. (2004). Changes in canopy fuels and potential fire behavior 1880–2040: Grand Canyon, Arizona. *Ecological Modelling* 175: 231–248.

Fuller, **J.L.** (1998). Ecological impact of the mid-Holocene hemlock decline in southern Ontario, Canada. *Ecology* 79: 2337–2351.

Furukawa, **T.**, **Kayo**, **C.**, **Kadoya**, **T.** et al. (2015). Forest harvest index: accounting for global gross forest cover loss of wood production and an application of trade analysis. *Global Ecology and Conservation* 4: 150–159.

García, **R.A.**, **Franzese**, **J.**, **Policelli**, **N.** et al. (2018). Non-native pines are homogenizing the ecosystems of South America. In: From Biocultural Homogenization to Biocultural Conservation (eds. **R. Rozzi**, **R.H. May** and **F.S. Chapin**), 245–263. Switzerland AG: Springer Nature.

Gardiner, **B.**, **Schuck**, **A.**, **Schelhaas**, **M.-J.** et al. (eds.) (2013). Living with Storm Damage to Forests. What Science Can Tell Us #3. Joensuu, Finland: European Forest Institute.

Gauthier, **S.**, **Gagnon**, **J.**, and **Bergeron**, **Y.** (1993). Population age structure of *Pinus banksiana* at the southern edge of the Canadian boreal forest. *Journal of Vegetation Science* 4: 783–790.

Gergel, **S.E.** and **Turner**, **M.G.** (eds.) (2017). Learning Landscape Ecology: A Practical Guide to Concepts and Techniques, 2e. New York: Springer.

Giesler, **R.**, **Högberg**, **M.**, and **Högberg**, **P.** (1998). Soil chemistry and plants in fennoscandian boeal forest as exemplified by a local gradient. *Ecology* 79: 119–137.

Gill, J.L. (2014). Ecological impacts of the late quaternary megaherbivore extinctions. *New Phytologist* 201: 1163–1169.

Gill, N.S., **Jarvis**, D., **Veblen**, T.T. et al. (2017). Is initial post-disturbance regeneration indicative of longer-term trajectories? *Ecosphere* 8: e01924. https://doi.org/10.1002/ecs2.1924.

Gossner, M.M., **Pašalić**, E., **Lange**, M. et al. (2014). Differential responses of herbivores and herbivory to management in temperate European beech. *PLoS One* https://doi.org/10.1371/journal.pone.0104876.

Gragson, T.L. and **Bolstad**, P.V. (2006). Land use legacies and the future of Southern Appalachia. *Society and Natural Resources* 19: 175–190.

Graham, R.T., **Asherin**, L.A., **Jain**, T.B., **Baggett**, L.S., and **Battaglia**, M.A. (2019). Differing ponderosa pine forest structures, their growth and yield, and mountain pine beetle impacts: growing stock levels in the Black Hills. *US Forest Service General Technical Report* RMRS-GTR-393, Fort Collins.

Greer, B.T., **Still**, C., **Cullinan**, G.L. et al. (2018). Polyploidy influences plant-environment interactions in Quaking Aspen (*Populus tremuloides* Michx.). *Tree Physiology* 38: 630–640.

Gspaltl, M., **Bauerle**, W., **Binkley**, D., and **Sterba**, H. (2013). Leaf area and light use efficiency patterns of Norway spruce under different thinning regimes and age classes. *Forest Ecology and Management* 288: 49–59.

Hacket-Pain, A.J., **Lageard**, J.G.A., and **Thomas**, P.A. (2017). Drought and reproductive effort interact to control growth of a temperate broadleaved tree species (*Fagus sylvatica*). *Tree Physiology* 37: 744–754.

Haidt, J. (2012). The Righteous Mind: Why Good People Are Divided by Politics and Religion. New York: Pantheon.

Hall, B., **Motzkin**, G., **Foster**, D.R. et al. (2002). Three hundred years of forest and land-use change in Massachusetts, USA. *Journal of Biogeography* 29: 1319–1335.

Halliday, T.R. (1980). The extinction of the passenger pigeon, *Ectopistes migratorius*, and its relevance to contemporary conservation. *Biological Conservation* 17: 157–162.

Hammond, J.C., **Saavedra**, F.A., and **Kampf**, S.K. (2018). How does snow persistence relate to annual streamflow in mountain watersheds of the western U.S. with wet maritime and dry continental climates? *Water Resources Research* https://doi.org/10.1002/2017WR021899.

Hänninen, H. (2016). Boreal and Temperate Trees in a Changing Climate: Modelling the Ecophysiology of Seasonality. Dordrecht: Springer.

Hari, P., **Heliövaara**, K., and **Kulmala**, L. (eds.) (2013). Physical and Physiological Forest Ecology. Dordrecht: Springer.

Harjuniemi, A. (2014). *Effects of intensive silviculture in the restoration of northeastern Atlantic Forest in Brazil: survival, growth and carbon sequestered 8 years after planting*. MS thesis, North Carolina State University and University of Helsinki, Raleigh and Helsinki.

Harrington, C.A. and **Reukema**, D.L. (1983). Initial shock and long-term stand development following thinning in a Douglas-fir plantation. *Forest Science* 29: 33–46.

Hart, S.J., **Veblen**, T.T., **Eisenhart**, K.S. et al. (2017). Drought induces spruce beetle (*Dendroctonus rufipennis*) outbreaks across northwestern Colorado. *Ecology* 95: 930–939.

Hartmann, H. and **Trumbore**, S. (2016). Understanding the roles of nonstructural carbohydrates in forest trees – from what we can measure to what we want to know. *New Phytologist* 211: 386–403.

Harwood, C.E. and **Nambiar**, E.K.S. (2014). Sustainable plantation forestry in South-East Asia. *ACIAR Technical Reports* No. 84. Australian Centre for International Agricultural Research: Canberra.

Hasstedt, S.C.M. (2013). *Forest structure in unharvested old-growth: understanding the influence of soils on variability of long-term tree dynamics and fire history*. PhD dissertation, Colorado State University, Fort Collins.

Heard, M.J. and **Valente**, M.J. (2009). Fossil pollen records forecast response of forests to hemlock woolly adelgid invasion. *Ecography* 32: 881–887.

Heinonsalo, J., **Sun**, H., **Santalahti**, M. et al. (2015). Evidences on the ability of mycorrhizal genus *Piloderma* to use organic nitrogen and deliver it to Scots pine. *PLoS One* 10 (7): e0131561. https://doi.org/10.1371/journal.pone.0131561.

Henttonen, H.M., **Nöjd**, P., and **Mäkinen**, H. (2017). Environment-induced growth changes in the Finnish forests during 1971–2010 – an analysis based on National Forest Inventory. *Forest Ecology and Management* 386: 22–36.

Hicke, J.A., **Xua**, B., **Meddens**, A.J.H., and **Egan**, J.M. (2020). Characterizing recent bark beetle-caused tree mortality in the western United States from aerial surveys. *Forest Ecology and Management* 475: 118402.

Himes, A.J., **Turnblom**, E.C., **Harrison**, R.B. et al. (2014). Predicting risk of long-term nitrogen depletion under whole-tree harvesting in the coastal Pacific Northwest. *Forest Science* 60: 382–390.

Högberg, P. (2001). Interactions between hillslope hydrochemistry, nitrogen dynamics, and plants in Fennoscandian boreal forest. In: Global Biogeochemical Cycles in the Climate System (eds. **E.-D. Schulze**, **S.P. Harrison**, **M. Heimann**, et al.), 227–233. San Diego: Academic Press.

Högberg, **M.N.**, **Bååth**, **E.**, **Nordgren**, **A.** et al. (2003). Contrasting effects of nitrogen availability on plant carbon supply to mycorrhizal fungi and saprotrophs – a hypothesis based on field observations in boreal forest. *New Phytologist* 160: 225–238.

Högberg, **P.** (2021). *Sustainable boreal forest management – challenges and opportunities for climate change mitigation*. International Boreal Forest Research Association, www.ibfra.org.

Holling, **C.S.** (ed.) (1978). Adaptive Environmental Assessment and Management. London: Wiley.

Hollingsworth, **T.N.**, **Lloyd**, **A.H.**, **Nossov**, **D.R.** et al. (2010). Twenty-five years of vegetation change along a putative successional chronosequence on the Tanana River, Alaska. *Canadian Journal of Forest Research* 40: 1273–1287.

Holtslag, **A.A.M.** and **deBruin**, **H.A.R.** (1988). Applied modeling of the nighttime surface energy balance over land. *Journal of Applied Meteorology* 27: 689–704.

Honkaniemi, **J.**, **Lehtonen**, **M.**, **Väisänen**, **H.**, and **Peltola**, **H.** (2017). Effects of wood decay by Heterobasidion annosum on the vulnerability of Norway spruce stands to wind damage: a mechanistic modelling approach. *Canadian Journal of Forest Research* 47: 777–787.

Hubbard, **R.M.**, **Rhoades**, **C.C.**, **Elder**, **K.**, and **Negrón**, **J.** (2013). Changes in transpiration and foliage growth in lodgepole pine trees following mountain pine beetle attack and mechanical girdling. *Forest Ecology and Management* 289: 312–317.

Hungerford, **R.** (1980). Micro environmental response to harvesting and residue management. In: Environmental consequences of timber harvesting in Rocky Mountain coniferous forests. *USDA Forest Service General Technical Report* INT-90, Ogden, UT, pp. 37–73.

Huong, **V.D.**, **Nambiar**, **E.K.S.**, **Hai**, **N.X.** et al. (2020). Sustainable management of *Acacia auriculiformis* plantations for wood production over four successive rotations in South Vietnam. *Forests* 11: 550.

International Union for Conservation of Nature (IUCN) (2020). *ICUN Red List of Threatened Species*. ICUN, London. www.iucnredlist.org/search, accessed January 2021.

Itow, **S.** (1991). Species turnover and diversity patterns along an evergreen broad-leaved forest coenocline. *Journal of Vegetation Science* 2: 477–484.

Jackson, **S.T.** and **Weng**, **C.** (1999). Late Quaternary extinction of a tree species in eastern North America. *Proceedings of the National Academy of Sciences of the United States of America* 96: 13847–13852.

Jackson, **S.T.** and **Williams**, **J.W.** (2004). Modern analogs in quaternary paleoecology: here today, gone yesterday, gone tomorrow? *Annual Review of Earth and Planetary Sciences* 32: 495–537.

Jackson, **S.T.**, **Betancourt**, **J.L.**, **Lyford**, **M.E.** et al. (2005). A 40,000-year woodrat-midden record of vegetational and biogeographical dynamics in North-Eastern Utah, USA. *Journal of Biogeography* 32: 1085–1106.

Jackson, **C.R.**, **Webster**, **J.R.**, **Knoepp**, **J.D.** et al. (2018). Unexpected ecological advances made possible by long-term data: a Coweeta example. *WIREs Water*: e1273. https://doi.org/10.1002/wat2.1273.

Jensen, **J.R.** (2000). Remote Sensing of the Environment: An Earth Resource Perspective. New Jersey: Prentice-Hall.

Johnson, **K.P.**, **Clayton**, **D.H.**, **Dumbacher**, **J.P.**, and **Fleischer**, **R.C.** (2010). The flight of the passenger pigeon: phylogenetics and biogeographic history of an extinct species. *Molecular Phylogenetics and Evolution* 57: 455–458.

Kane, **V.R.**, **Gersonde**, **R.F.**, **Lutz**, **J.A.** et al. (2011). Patch dynamics and the development of structural and spatial heterogeneity in Pacific Northwest forests. *Canadian Journal of Forest Research* 41: 2276–2291.

Kardell, **O.** (2016). Swedish forestry, forest pasture grazing by livestock, and game browsing pressure since 1900. *Environmental History* 22: 561–587.

Kardol, **P.**, **Todd**, **D.E.**, **Hanson**, **P.J.**, and **Mulholland**, **P.J.** (2010). Long-term successional forest dynamics: species and community responses to climatic variability. *Journal of Vegetation Science* 21: 627–642.

Kashian, **D.M.**, **Romme**, **W.H.**, and **Regan**, **C.M.** (2007). Reconciling divergent interpretations of quaking aspen decline on the Northern Colorado front range. *Ecological Applications* 5: 1296–1311.

Kashian, **D.M.**, **Romme**, **W.H.**, **Tinker**, **D.B.** et al. (2013). Postfire changes in forest carbon storage over a 300-year chronosequence of *Pinus contorta*-dominated forests. *Ecological Monographs* 83: 49–66.

Kayes, **L.J.** and **Tinker**, **D.B.** (2012). Forest structure and regeneration following a mountain pine beetle epidemic in southeastern Wyoming. *Forest Ecology and Management* 263: 57–66.

Kelly, **R.L.**, **Surovell**, **T.A.**, **Shuman**, **B.N.**, and **Smith**, **G.M.** (2013). A continuous climatic impact on Holocene human population in the Rocky Mountains. *Proceedings of the National Academy of Sciences of the United States of America* 110: 443–447.

Kenzo, **T.**, **Ichie**, **T.**, **Watanabe**, **Y.** et al. (2006). Changes in photosynthesis and leaf characteristics with tree height in five dipterocarp species in a tropical rain forest. *Tree Physiology* 26: 865–873.

Kimmerer, **R.W.** (2013). Braiding Sweetgrass: Indigenous Wisdom, Scientific Knowledge and the Teachings of Plants. Minneapolis: Milkweed Editions.

Kittredge, **J.** (1948). Forest Influences. New York: McGraw-Hill.

Knight, **K.S.**, **Oleksyn**, **J.**, **Jagodzinski**, **A.M.** et al. (2008). Overstorey tree species regulate colonization by native and exotic plants: a source of positive relationships between understorey diversity and invasibility. *Diversity and Distributions* 14: 666–675.

Knoerr, **K.R.** and **Gay**, **L.W.** (1965). The leaf energy balance. *Ecology* 46: 17–24.

Knott, **J.A.**, **Desprez**, **J.M.**, **Oswalt**, **C.M.**, and **Fei**, **S.** (2019). Shifts in forest composition in the eastern United States. *Forest Ecology and Management* 433: 176–183.

Koch, **P.L.** and **Barnosky**, **A.D.** (2006). Late quaternary extinctions: state of the debate. *Annual Review of Ecology and Systematics* 37: 215–250.

Koch, **G.W.**, **Sillett**, **S.C.**, **Jennings**, **G.M.**, and **Davis**, **S.D.** (2004). The limits to tree height. *Nature* 428: 851–854.

Krebs, **C.J.**, **Boonstra**, **R.**, **Boutin**, **S.** et al. (1995). Impact of food and predation on the snowshoe hare cycle. *Science* 269: 1112–1115.

Krebs, **C.J.**, **Boonstra**, **R.**, and **Boutin**, **S.** (2017). Using experimentation to understand the 10-year snowshoe hare cycle in the boreal forest of North America. *Journal of Animal Ecology* https://doi.org/10.1111/1365-2656.12720.

Kucbel, **S.**, **Jaloviar**, **P.**, and **Špišák**, **J.** (2011). Quantity, vertical distribution and morphology of fine roots in Norway spruce stands with different stem density. *Plant Roots* 5: 46–55.

Kuijper, **D.P.J.**, **Jedrzejewska**, **B.**, **Brzeziecki**, **B.** et al. (2010). Fluctuating ungulate density shapes tree recruitment in natural stands of the Białowieża Primeval Forest, Poland. *Journal of Vegetation Science* 21: 1082–1098.

Kulakowski, **D.**, **Veblen**, **T.T.**, and **Bebi**, **P.** (2016). Fire severity controlled susceptibility to a 1940s spruce beetle outbreak in Colorado, USA. *PLoS One* https://doi.org/10.1371/journal.pone.0158138.

Kumar, **P.**, **Chen**, **H.Y.H.**, **Searle**, **E.B.**, and **Shahi**, **C.** (2018). Dynamics of understorey biomass, production and turnover associated with long-term overstorey succession in boreal forest of Canada. *Forest Ecology and Management* 427: 152–161.

Kuuluvainen, **T.** and **Ylläsjärvi**, **I.** (2011). On the natural regeneration of dry heath forests in Finnish Lapland: a review of V.T. Aaltonen (1919). *Scandinavian Journal of Forest Research* 26: 34–44.

Langvall, **O.** and **Löfvenius**, **M.O.** (2019). Long-term standardized forest phenology in Sweden: a climate change indicator. *International Journal of Biometeorology* https://doi.org/10.1007/s00484-019-01817-8.

Laughlin, **D.C.**, **Moore**, **M.M.**, and **Fulé**, **P.Z.** (2011). A century of increasing pine density and associated shifts in understory plant strategies. *Ecology* 92: 556–561.

Le Roux, **J.J.**, **Strasberg**, **D.**, **Rouget**, **M.** et al. (2014). Relatedness defies biogeography: the tale of two island endemics (*Acacia heterophylla* and *A. koa*). *New Phytologist* 204: 230–242.

Leopold, **A.** (1933). The conservation ethic. *Journal of Forestry* 31: 634–643.

Leopold, **A.** (1938). Engineering and conservation. In: The River of the Mother of God: And Other Essays by Aldo Leopold (eds. **J.B. Callicott** and **S.L. Flader**). Madison: University of Wisconsin Press.

Leopold, **A.** (1949). A Sand County Almanac and Sketches from Here and There. New York: Oxford University Press.

Lertzman, **K.**, **Fall**, **J.**, and **Dorner**, **B.** (1998). Three kinds of heterogeneity in fire regimes: at the crossroads of fire history and landscape ecology. *Northwest Science* 72: 4–23.

Leuzinger, **S.** and **Körner**, **C.** (2007). Tree species diversity affects canopy leaf temperatures in a mature temperate forest. *Agricultural and Forest Meteorology* 146: 29–37.

Linden, **D.W.**, **Roloff**, **G.J.**, and **Kroll**, **A.J.** (2012). Conserving avian richness through structure retention in managed forests of the Pacific Northwest, USA. *Forest Ecology and Management* 284: 174–184.

Locatelli, **T.**, **Gardiner**, **B.**, **Tarantola**, **S.** et al. (2016). Modelling wind risk to Eucalyptus globulus (Labill.) stands. *Forest Ecology and Management* 365: 159–173.

Loehle, **C.** and **Eschenbach**, **W.** (2011). Historical bird and terrestrial mammal extinction rates and causes. *Diversity and Distributions* 18: 84–91.

Löfvenius, **M.O.** (1993). *Temperature and radiation regimes in pine shelterwood and clear-cut areas*. PhD thesis, Swedish University of Agricultural Sciences, Umeå.

Lundqvist, **L.** (2017). Tamm review: selection system reduces long-term volume growth in Fennoscandic uneven-aged Norway spruce forests. *Forest Ecology and Management* 391: 362–375.

MacKinley, **D.C.**, **Ryan**, **M.G.**, **Birdsey**, **R.A.** et al. (2011). A synthesis of current knowledge on forests and carbon storage in the United States. *Ecological Applications* 21: 1902–1924.

Marsh, **G.P.** (1864). Man and Nature: Or, Physical Geography as Modified by Human Action. New York: Scribner.

Martin, **P.H.** and **Canham**, **C.D.** (2020). Peaks in frequency, but not relative abundance, occur in the center of tree species distributions on climate gradients. *Ecosphere* 11: e03149.

Martin, **T.G.**, **Arcese**, **P.**, and **Scheerder**, **N.** (2011). Browsing down our natural heritage: deer impacts on vegetation structure and songbird populations across an island archipelago. *Biological Conservation* 144: 459–469.

Matonis, **M.S.**, **Binkley**, **D.**, **Tuten**, **M.**, and **Cheng**, **T.** (2014). The forests they are a-changin' – Ponderosa pine and mixed conifer forests on the Uncompahgre Plateau in 1875 and 2010–13. *Technical Brief.* Colorado State University, Colorado Forest Restoration Institute, Fort Collins, CO, 27 pp.

Matonis, **M.S.**, **Binkley**, **D.**, **Franklin**, **J.**, and **Johnson**, **K.N.** (2016). Benefits of an "undesirable" approach to natural resource management. *Journal of Forestry* 114: 658–665.

de **Mattos**, **E.M.**, **Binkley**, **D.**, **Campoe**, **O.C.** et al. (2020). Variation in canopy structure, leaf area, light interception and light use efficiency among Eucalyptus clones. *Forest Ecology and Management* 463: 118038.

Mayer, **M.**, **Prescott**, **C.E.**, **Abaker**, **W.E.A.** et al. (2020). Tamm review: influence of forest management activities on soil organic carbon stocks: a knowledge synthesis. *Forest Ecology and Management* 466: 118127.

McArdle, **R.E.** (1930). The Yield of Douglas Fir in the Pacific Northwest. *USDA Technical Bulletin* 201, Washington, DC.

McDowell, **N.**, **Allen**, **C.D.**, **Anderson-Teixeira**, **K.** et al. (2018). Drivers and mechanisms of tree mortality in moist tropical forests. *New Phytologist* 219: 851–869.

McKinney, **S.T.** (2019). Fire regimes of ponderosa pine (Pinus ponderosa) ecosystems in Colorado: a systematic review and meta-analysis. In: *Fire Effects Information System.* U.S. Forest Service, Rocky Mountain Research Station, www.fs.fed.us/database/feis/fire_regimes/CO_ponderosa_pine/all.pdf.

McLaren, **K.**, **Luke**, **D.**, **Tanner**, **E.** et al. (2019). Reconstructing the effects of hurricanes over 155 years on the structure and diversity of trees in two tropical montane rainforests in Jamaica. *Agricultural and Forest Meteorology*: 276–277. https://doi.org/10.1016/j.agrformet.2019.107621.

McMurtrie, **R.E.** and **Näsholm**, **T.** (2017). Quantifying the contribution of mass flow to nitrogen acquisition by an individual plant root. *New Phytologist* https://doi.org/10.1111/nph.14927.

Messier, **C.**, **Posada**, **J.**, **Aubin**, **I.**, and **Beaudet**, **M.** (2009). Functional relationships between old-growth forest canopies, understorey light and vegetation dynamics. In: Old-Growth Forests (eds. **C.G.W. Gleixner** and **M. Heimann**), 115–139. Berlin: Springer.

Mietkiewicz, **N.** and **Kulakowski**, **D.** (2016). Relative importance of climate and mountain pine beetle outbreaks on the occurrence of large wildfires in the western USA. *Ecological Applications* 26: 2523–2535.

Moore, **M.M.**, **Huffman**, **D.W.**, **Bakker**, **J.D.**, et al. (2004). Quantifying Forest Reference Conditions for Ecological Restoration: The Woolsey Plots. *Final Report to the Ecological Restoration Institute for the Southwest Fire Initiative.* http://openknowledge.nau.edu/2527/2/Moore_M_etal_(2004)_QuantifyingForestReferenceConditionsFor%281%29.pdf.

Moore, **M.M.**, **Covington**, **W.W.**, **Fulé**, **P.Z.** et al. (2008). Ecological restoration experiments (1992–2007) at the G. A. Pearson natural area, Fort Valley experimental forest. In: Fort Valley Experimental Forest-a Century of Research 1908–2008 (eds. **S.D. Olberding** and **M.M. Moore**), 209–218. Fort Collins, CO: US Forest Service RMRS-P-55.

Moreira, **F.**, **Duarte**, **I.**, **Catry**, **F.**, and **Acácio**, **V.** (2007). Cork extraction as a key factor determining post-fire cork oak survival in a mountain region of southern Portugal. *Forest Ecology and Management* 253: 30–37.

Morris, **J.L.**, **Puttick**, **M.N.**, **Clark**, **J.W.** et al. (2018). The timescale of early land plant evolution. *Proceedings of the National Academy of Sciences of the United States of America* https://doi.org/10.1073/pnas.1719588115.

Moser, **W.K.**, **Hansen**, **M.H.**, **Nelson**, **M.D.**, et al. (2007). After the blowdown: a resource assessment of the Boundary Waters Canoe Area Wilderness, 1999–2003. *US Forest Service General Technical Report* NRS-7, Newtown, Pennsylvania.

Muir, **J.** (1894). The Mountains of California. New York: The Century Co.

NADP, National Atmospheric Deposition Program (NRSP-3) (2020). NADP Program Office. Madison, WI: Wisconsin State Laboratory.

Näsholm, **T.**, **Högberg**, **P.**, **Franklin**, **O.** et al. (2013). Are ectomycorrhizal fungi alleviating or aggravating nitrogen limitation of tree growth in boreal forests? *New Phytologist* 198: 214–221.

Negrón, **J.F.** (2020). Within-stand distribution of tree mortality caused by mountain pine beetle, *Dendroctonus ponderosae* Hopkins. *Insects* 11: 112.

Negrón, **J.F.** and **Huckaby**, **L.** (2020). Reconstructing historical outbreaks of mountain pine beetle in lodgepole pine forests in the Colorado front range. *Forest Ecology and Management* 473: 118270.

Neustadt, **R.E.** and **May**, **E.R.** (1986). Thinking in Time: The Uses of History for Decision Makers. New York: The Free Press.

Newton, **M.** and **Cole**, **E.C.** (1987). A sustained-yield scheme for old-growth Douglas-fir. *Western Journal of Applied Forestry* 2: 22–25.

Nicholson, **A.J.** (1933). The balance of animal populations. *Journal of Animal Ecology* 2: 132–178.

Nieminen, **M.**, **Piirainen**, **S.**, **Sikström**, **U.** et al. (2018). Ditch network maintenance in peat-dominated boreal forests: review and analysis of water quality management options. *Ambio* https://doi.org/10.1007/s13280-018-1047-6.

Nilsson, **P.**, **Neil**, **C.**, and **Sören**, **W.** (2014). Forest Data (2014): Information on Swedish Forests. Department of Forest Resource Management. Umeå: Swedish University of Agricultural Sciences.

Norris, **J.R.**, **Betancourt**, **J.L.**, and **Jackson**, **S.T.** (2016). Late Holocene expansion of ponderosa pine (*Pinus ponderosa*) in the central Rocky Mountains, USA. *Journal of Biogeography* 43: 778–790.

Oli, **M.K.**, **Krebs**, **C.J.**, **Kenney**, **A.J.** et al. (2020). Demography of snowshoe hare population cycles. *Ecology* https://doi.org/10.1002/ecy.2969.

Oliver, **C.D.** and **Larson**, **B.C.** (1990). Forest Stand Dynamics. New York: Wiley.

Ontl, **T.A.**, **Janowiak**, **M.K.**, **Swanston**, **C.W.** et al. (2020). Forest management for carbon sequestration and climate adaptation. *Journal of Forestry* 118: 86–101.

Orgiazzi, **A.**, **Bardgett**, **R.D.**, **Barrios**, **E.** et al. (eds.) (2016). Global Soil Biodiversity Atlas. Luxembourg: European Commission, Publications Office of the European Union http://atlas.globalsoilbiodiversity.org.

Oswald, **W.W.**, **Doughty**, **E.D.**, **Foster**, **D.R.** et al. (2017). Evaluating the role of insects in the middle-Holocene Tsuga decline. *Journal of the Torrey Botanical Society* 144: 35–39.

Palik, **B.J.**, **D'Amato**, **A.W.**, **Franklin**, **J.F.**, and **Johnson**, **K.N.** (2021). Ecological Silviculture: Foundations and Applications. Long Grove: Waveland Press.

Pan, **Y.**, **Birdsey**, **R.A.**, **Phillips**, **O.L.**, and **Jackson**, **R.B.** (2013). The structure, distribution, and biomass of the world's forests. *Annual Review of Ecology and Systematics* 44: 593–622.

Pausas, **G.** and **Keeley**, **K.E.** (2017). Epicormic resprouting in fire-prone ecosystems. *Trends in Plant Science* 22: 1008–1015.

Payn, **T.**, **Carnus**, **J.-M.**, **Freer-Smith**, **P.** et al. (2015). Changes in planted forests and future global implications. *Forest Ecology and Management* 352: 57–67.

Payne, **C.J.** (2018). *Long-term temporal dynamics of the Duke Forest*. PhD dissertation, University of North Carolina, Chapel Hill.

Pedelty, **J.A.**, **Morisette**, **J.T.**, **Schnase**, **J.L.** et al. (2003). High Performance Geostatistical Modeling of Biospheric Resources in the Cerro Grande Wildfire Site. New Mexico and Rocky Mountain National Park, Colorado: Los Alamos www.researchgate.net/publication/237268759_High_Performance_Geostatistical_Modeling_of_Biospheric_Resources_in_the_Cerro_Grande_Wildfire_Site_Los_Alamos_New_Mexico_and_Rocky_Mountain_National_Park_Colorado.

Petriţan, **I.C.**, von **Lüpke**, **B.**, and **Petriţan**, **A.M.** (2011). Effects of root trenching of overstorey Norway spruce (*Picea abies*) on growth and biomass of underplanted beech (*Fagus sylvatica*) and Douglas fir (*Pseudotsuga menziesii*) saplings. *European Journal of Forest Research* 130: 813–828.

Pinchot, **G.** (1910). The Fight for Conservation. New York: Doubleday, Page and Co.

Pommerening, **A.** and **Grabarnik**, **P.** (2019). Individual-Based Methods in Forest Ecology and Management. New York: Springer.

Pommerening, **A.** and **Sánchez Meador**, **A.J.** (2018). Tamm review: tree interactions between myth and reality. *Forest Ecology and Management* 424: 164–176.

Powell, **W.A.**, **Newhouse**, **A.E.**, and **Coffey**, **V.** (2019). Developing blight-tolerant American chestnut trees. *Cold Spring Harbor Perspective in Biology* https://doi.org/10.1101/cshperspect.a034587.

Prescott, **C.E.**, **Grayston**, **S.J.**, **Helmisaari**, **H.-S.** et al. (2020). Surplus carbon drives allocation and plant–soil interactions. *Trends in Ecology & Evolution* https://doi.org/10.1016/j.tree.(2020).08.007.

Pretzsch, **H.** (2020). Density and growth of forest stands revisited. Effect of the temporal scale of observation, site quality, and thinning. *Forest Ecology and Management* 460: 117879.

Pretzsch, **H.**, **Block**, **J.**, **Dieler**, **J.** et al. (2010). Comparison between the productivity of pure and mixed stands of Norway spruce and European beech along an ecological gradient. *Annals of Forest Science* 67: 712.

Pretzsch, **H.**, **Biber**, **P.**, **Schütze**, **G.** et al. (2014). Forest stand growth dynamics in Central Europe have accelerated since 1870. *Nature Communications* 5: 4967. https://doi.org/10.1038/ncomms5967.

Pretzsch, **H.**, **Forrester**, **D.I.**, and **Bauhus**, **J.** (eds.) (2017). Mixed-Species Forests, Ecology and Management. New York: Springer.

Primack, **R.B.** (2014). Walden Warming: Climate Change Comes to Thoreau's Woods. Chicago: University of Chicago Press.

Primack, **R.B.** and **Miller-Rushing**, **A.J.** (2012). Uncovering, collecting, and analyzing records to investigate the ecologicai impacts of climate change: a template from Thoreau's Concord. *Bioscience* 62: 170–181.

Räim, **O.**, **Kaurilind**, **E.**, **Hallik**, **L.**, and **Merilo**, **E.** (2012). Why does needle photosynthesis decline with tree height in Norway spruce. *Plant Biology* 14: 306–314.

Ramankutty, **N.**, **Graumlich**, **L.**, **Achard**, **F.** et al. (2006). Global land-cover change: recent progress, remaining challenges. In: Land-Use and Land-Cover Change (eds. **E.F. Lambin** and **H. Geist**), 9–39. Berlin: Springer.

Rao, **S.J.** (2017). Effect of reducing red deer *Cervus elaphus* density on browsing impact and growth of scots pine *Pinus sylvestris* seedlings in semi natural woodland in the Cairngorms, UK. *Conservation Evidence* 14: 22–26.

Reed, **D.E.**, **Ewers**, **B.E.**, **Pendall**, **E.** et al. (2018). Bark beetle-induced tree mortality alters stand energy budgets due to water budget changes. *Theoretical and Applied Climatology* 131: 153–165.

Rehfeldt, **G.E.**, **Ferguson**, **D.E.**, and **Crookston**, **N.L.** (2009). Aspen, climate, and sudden decline in western USA. *Forest Ecology and Management* 258: 2353–2364.

Reich, **P.B.**, **Oleksyn**, **J.**, **Modrzynski**, **J.** et al. (2005). Linking litter calcium, earthworms and soil properties: a common garden test with 14 tree species. *Ecology Letters* 8: 811–818.

Reilly, **M.J.** and **Spies**, **T.A.** (2016). Disturbance, tree mortality, and implications for contemporary regional forest change in the Pacific Northwest. *Forest Ecology and Management* 374: 102–110.

Reimer, **P.J.**, **Bard**, **E.**, **Bayliss**, **A.** et al. (2013). INTCAL13 and MARINE13 radiocarbon age calibration curves 0-50,000 years cal BP. *Radiocarbon* 55: 1869–1887.

Reiners, **W.A.**, **Driese**, **K.L.**, **Fahey**, **T.J.**, and **Gerow**, **K.G.** (2012). Effects of three years of regrowth inhibition on the resilience of a clear-cut northern hardwood forest. *Ecosystems* 15: 1351–1362.

Rhoades, **C.C.**, **Hubbard**, **R.M.**, **Elder**, **K.** et al. (2020). Tree regeneration and soil responses to management alternatives in beetle-infested lodgepole pine forests. *Forest Ecology and Management* 468: 118182.

Ripple, **W.J.** and **Beschta**, **R.L.** (2006). Linking a cougar decline, trophic cascade, and catastrophic regime shift in Zion National Park. *Conservation Biology* 133: 397–408.

Robichaud, **P.R.**, **Beyers J.L.**, and **Neary**, **D.** (2000). Evaluating the effectiveness of post-fire rehabilitation treatments. *USDA Forest Service General Technical Report* RMRS-GTR-63, Fort Collins.

Romme, **W.H.** (1982). Fire and landscape diversty in subalpine forests of Yellowstone National Park. *Ecological Monographs* 52: 199–221.

Romme, **W.H.** and **Knight**, **D.H.** (1982). Landscape diversity: the concept applied to Yellowstone Park. *Bioscience* 32: 664–670.

Ruefenacht, **B.**, **Finco**, **M.V.**, **Nelson**, **M.D.** et al. (2008). Conterminous U.S. and Alaska forest type mapping using forest inventory and analysis data. *Photogrammetric Engineering & Remote Sensing* 74: 1379–1388.

Running, **S.W.**, **Nemani**, **R.R.**, **Heinsch**, **F.A.** et al. (2004). A continuous satellite-derived measure of global terrestrial primary production. *Bioscience* 54: 547–560.

Ryan, **M.G.**, **Binkley**, **D.**, **Fownes**, **J.** et al. (2004). An experimental test of the causes of age-related decline in forest growth. *Ecological Monographs* 74: 393–414.

Ryan, **M.G.**, **Stape**, **J.L.**, **Binkley**, **D.**, and **Alvares**, **C.A.** (2020). Cross-site patterns in the response of Eucalyptus plantations to irrigation, climate and intra-annual weather variation. *Forest Ecology and Management* 475: 118444.

Sandford, **J.**, **Lejour**, **D.**, **Beese**, **W.J.**, and **Filipescu**, **C.** (2020). Montane Alternative Silvicultural Systems (MASS): Twenty-Five Year Growth and Survival of Planted and Naturally-Regenerated Conifers. Nanaimo: Report prepared for Mosaic Forest Management.

Santana, **M.C.**, **Pereira**, **A.P.A.**, **Lopes**, **B.A.B.** et al. (2020). Mycorrhiza in mixed plantations. In: Mixed Plantations of Eucalyptus and Leguminous Trees (eds. **E.J.B.N. Cardoso** et al.), 137–154. Cham, Switzerland: Springer.

Schietti, **J.**, **Martins**, **D.**, **Emilio**, **T.** et al. (2016). Forest structure along a 600 km transect of natural disturbances and seasonality gradients in Central-Southern Amazonia. *Journal of Ecology* 104: 1335–1346.

Schmidt, **M.**, **Hanewinkel**, **M.**, **Kändler**, **G.** et al. (2010). An inventory-based approach for modeling singletree storm damage – experiences with the winter storm of (1999) in Southwestern Germany. *Canadian Journal of Forest Research* 40: 1636–1652.

Schoenecker, **K.**, **Singer**, **F.J.**, **Zeigenfuss**, **L.C.** et al. (2004). Effects of elk herbivory on vegetation and nitrogen processes. *Journal of Wildlife Management* 68: 835–847.

Schoennagel, **T.**, **Sherriff**, **R.L.**, and **Veblen**, **T.T.** (2011). Fire history and tree recruitment in the Colorado front range upper montane zone: implications for forest restoration. *Ecological Applications* 21: 2210–2222.

Schowalter, **T.D.** and **Crossley**, **D.A.** (1988). Canopy arthropods and their response to forest disturbance. In: Forest Hydrology and Ecology at Coweeta (eds. **W.T. Swank** and **D.A. Crossley**), 207–218. New York: Springer.

Schulte, **L.A.** and **Mladenoff**, **D.J.** (2005). Severe wind and fire regimes in northern forests: historical variability at the regional scale. *Ecology* 86: 431–445.

Scott, **J.M.**, **Wiens**, **J.A.**, **Van Horne**, **B.**, and **Goble**, **D.D.** (2020). Shepherding Nature: The Challenge of Conservation Reliance. Cambridge: Cambridge University Press.

Seastedt, **T.R.** and **Crossley**, **D.A.** Jr. (1988). Soil arthropods and their role in decomposition and mpneralization processes. In: Forest Hydrology and Ecology at Coweeta (eds. **W.T. Swank** and **D.A. Crossley**), 233–243. New York: Springer.

Seidel, **D.**, **Leuschner**, **C.**, **Müller**, **A.**, and **Krause**, **B.** (2011). Crown plasticity in mixed forests – quantifying asymmetry as a measure of competition using terrestrial laser scanning. *Forest Ecology and Management* 261: 2123–2132.

Shive, **K.L.**, **Kuenzi**, **A.M.**, **Sieg**, **C.H.**, and **Fulé**, **P.Z.** (2013). Pre-fire fuel reduction treatments influence plant communities and exotic species 9 years after a large wildfire. *Applied Vegetation Science* 16: 457–469.

Sillett, **S.C.**, **Van Pelt**, **R.**, **Kramer**, **R.D.** et al. (2015a). Biomass and growth potential of *Eucalyptus regnans* up to 100 m tall. *Forest Ecology and Management* 348: 78–91.

Sillett, **S.C.**, **Van Pelt**, **R.**, **Kramer**, **R.D.** et al. (2015b). How do tree structure and old age affect growth potential of California redwoods? *Ecological Monographs* 85: 181–212.

Sillett, **S.C.**, **Van Pelt**, **R.**, **Freund**, **J.A.** et al. (2018). Development and dominance of Douglas-fir in North American rainforests. *Forest Ecology and Management* 429: 93–114.

Sillett, **S.C.**, **Van Pelt**, **R.**, **Kramer**, **R.D.** et al. (2020). Aboveground biomass dynamics and growth efficiency of *Sequoia sempervirens* forests. *Forest Ecology and Management* 458: 177740.

Sillett, **S.C.**, **Kramer**, **R.D.**, **Van Pelt**, **R.** et al. (2021). Comparative Development of the Four Tallest Conifer Species. *Forest Ecology and Management* 480: 118688.

Silva, **S.S.**, **Fearnside**, **P.M.**, **Graça**, **P.L.M.A.** et al. (2018). Dynamics of forest fires in the southwestern Amazon. *Forest Ecology and Management* 424: 312–322.

Similä, **M.**, **Aapala**, **K.**, and **Penttinen**, **J.** (eds.) (2014). Ecological Restoration in Drained Peatlands – Best Practices from Finland. Natural Heritage Services, Vantaa: Metsähallitus.

Skłodowski, **J.** (2020). Two directions of regeneration of post-windthrow pine stands depend on the composition of the undergrowth and the soil environment. *Forest Ecology and Management* 461: 117950.

Slinski, **K.M.**, **Hogue**, **T.S.**, **Porter**, **A.T.**, and **McCray**, **J.E.** (2016). Recent bark beetle outbreaks have little impact on streamflow in the Western United States. *Environmental Research Letters* 11: 074010.

Slot, **M.** and **Winter**, **K.** (2017). *In situ* temperature response of photosynthesis of 42 tree and liana species in the canopy of two Panamanian lowland tropical forests with contrasting rainfall regimes. *New Phytologist* 214: 1103–1117.

Smith, **S.** and **Read**, **D.** (2008). Mycorrhizal Symbioses, 3e. Cambridge: Academic Press.

Sobachkin, **R.S.**, **Sobachkin**, **D.S.**, and **Buzykin**, **A.I.** (2005). The influence of stand density on growth of three conifer species. In: Tree Species Effects on Soils: Implications for Global Change (eds. **D. Binkley** and **O. Menyailo**), 247–255. Amsterdam: Kluwer.

Song, **B.**, **Chen**, **J.**, and **Silbernagel**, **J.** (2004). Three-dimensional canopy structure of an old-growth Douglas-Fir forest. *Forest Science* 50: 376–386.

Song, **X.-P.**, **Hansen**, **M.C.**, **Stehman**, **S.V.** et al. (2018). Global land change from 1982 to 2016. *Nature* 560: 639–643.

Stephens, **S.L.**, **Collins**, **B.M.**, **Fettig**, **C.J.** et al. (2018). Drought, tree mortality, and wildfire in forests adapted to frequent fire. *Bioscience* 68: 77–88.

Stephenson, **N.L.** and **Das**, **A.J.** (2019). Height-related changes in forest composition explain increasing tree mortality with height during an extreme drought. *Nature Communications* https://doi.org/10.1038/s41467-019-12380-6.

Stephenson, **N.L.**, **Das**, **A.J.**, **Condit**, **R.** et al. (2014). Rate of tree carbon accumulation increases continuously with tree size. *Nature* 507: 90–93.

Stevens, **J.T.**, **Kling**, **M.M.**, **Schwilk**, **D.W.** et al. (2020). Biogeography of fire regimes in western U.S. conifer forests: a trait-based approach. *Global Ecology and Biogeography* 29: 944–955.

Stohlgren, **T.** (2006). Measuring Plant Diversity: Lessons from the Field. Oxford: Oxford University Press.

Stohlgren, **T.**, **Binkley**, **D.**, **Chong**, **G.** et al. (1999). Exotic plant species invade hotspots of native plant diversity. *Ecological Monographs* 69: 25–46.

Stohlgren, **T.J.**, **Schell**, **L.D.**, and **Vanden Heuvel**, **B.** (1999). How grazing and soil quality affect native and exotic plant diversity in Rocky Mountain grasslands. *Ecological Applications* 9: 45–64.

Stoltman, **A.M.**, **Radeloff**, **V.C.**, and **Mladenoff**, **D.J.** (2007). Computer visualization of pre-settlement and current forests in Wisconsin. *Forest Ecology and Management* 246: 135–143.

Stolzenburg, **W.** (2008). Wild Things Were: Life, Death, and Ecological Wreckage in a Land of Vanishing Predators. New York: Bloomsbury USA.

Strahan, **R.T.**, **Laughlin**, **D.C.**, **Bakker**, **J.D.**, and **Moore**, **M.M.** (2015). Long-term protection from heavy livestock grazing affects ponderosa pine understory composition and functional traits. *Rangeland Ecology & Management* 68: 257–265.

Sullivan, **M.J.P.** (2017). An estimate of the number of tropical tree species diversity and carbon storage across the tropical forest biome. *Scientific Reports* https://doi.org/10.1038/srep39102.

Swift, **L.W.** Jr., **Cunningham**, **G.B.**, and **Douglass**, **J.E.** (1988). Climatology and hydrology. In: Ecological Studies: Forest Hydrology and Ecology at Coweeta, vol. 66 (eds. **W.T. Swank** and **D.A. Crossley**), 35–55. New York, NY: Springer-Verlag.

Szabo, **J.K.**, **Khwaja**, **N.**, **Garnett**, **N.T.**, and **Butchart**, **S.H.M.** (2012). Global patterns and drivers of avian extinctions at the species and subspecies level. *PLoS One* 7: e47080.

Tateno, **R.**, **Hishi**, **T.**, and **Takeda**, **H.** (2004). Above- and belowground biomass and net primary productivity in a cool-temperate deciduous forest in relation to topographical changes and soil nitrogen. *Forest Ecology and Management* 193: 297–306.

Taylor, **P.G.**, **Cleveland**, **C.C.**, **Wieder**, **W.R.** et al. (2017). Temperature and rainfall interact to control carbon cycling in tropical forests. *Ecology Letters* 20: 779–788.

Thoreau, **H.D.** (1860). The Journals of Henry D. Thoreau, vol. 14. Republished 1906. New York: Houghton-Mifflin.

Tinkham, **W.T.**, **Dickinson**, **Y.**, **Hoffman**, **C.M.**, **Battaglia**, **M.A.**, **Ex**, **S.**, and **Underhill**, **J.** (2017). Visualization of heterogeneous forest structures following treatment in the southern Rocky Mountains. *US Forest Service General Technical Report* RMRS-GTR-365. Fort Collins, CO.

Torresan, **C.**, del **Río**, **M.**, **Hilmers**, **T.** et al. (2020). Importance of tree species size dominance and heterogeneity on the productivity of spruce-fir-beech mountain forest stands in Europe. *Forest Ecology and Management* 457: 117716.

Trung, **C.L.**, **Binkley**, **D.**, and **Stape**, **J.L.** (2013). Neighborhood uniformity increases growth of individual Eucalyptus trees. *Forest Ecology and Management* 289: 90–97.

Tudge, **C.** (2005). The Tree: A Natural History of What Trees Are, How They Live, and Why They Matter. New York: Crown Publishers.

Tulau, **M.J.**, **McInnes-Clarke**, **S.K.**, **Yang**, **X.** et al. (2019). The Warrumbungle post- fire recovery project – raising the profile of soils. *Soil Use and Management* 35: 63–74.

Turner, **M.G.** and **Gardner**, **R.H.** (2015). Landscape Ecology in Theory and Practice, 2e. New York: Springer.

Turner, **M.G.**, **Dale**, **V.H.**, and **Everham**, **E.H.** (1997). Fires, hurricanes, and volcanoes: comparing large disturbances. *Bioscience* 47: 758–768.

Turner, **M.G.**, **Whitby**, **T.G.**, **Tinker**, **D.B.**, and **Romme**, **W.H.** (2016). Twenty-four years after the Yellowstone fires: are postfire lodgepole pine stands converging in structure and function? *Ecology* 97: 1260–1273.

Urban, **D.L.**, **Miller**, **C.**, **Halpin**, **P.N.**, and **Stephenson**, **N.L.** (2000). Forest gradient response in Sierran landscapes: the physical template. *Landscape Ecology* 15: 603–620.

Valinger, **E.**, **Kempe**, **G.**, and **Fridman**, **J.** (2019). Impacts on forest management and forest state in southern Sweden 10 years after the storm Gudrun. *Forestry* 92: 481–489.

Van Derwarker, **A.M.** and **Detwiler**, **K.R.** (2000). Plant and animal subsistence at the Coweeta Creek Site (31MA34), Macon County, North Carolina. *North Carolina Archaeology* 49: 59–76.

Van Valen, **L.** (1975). Life, death, and energy of a tree. *Biotropica* 7: 260–269.

Vanden Broeck, **A.**, **Cox**, **K.**, **Brys**, **R.** et al. (2018). Variability in DNA methylation and generational plasticity in the Lombardy poplar, a single genotype worldwide distributed since the eighteenth century. *Frontiers in Plant Science* 13: 1635.

Veintimilla, **D.**, **Ngo Bieng**, **M.A.**, **Delgado**, **D.** et al. (2019). Drivers of tropical rainforest composition and alpha diversity patterns over a 2,520 m altitudinal gradient. *Ecology and Evolution* 2019: 5720–5730.

Votoroantim Instituto Memória. (2006). Guia de Acervo 88 anos de história.

Vucetich, **J.A.** and **Peterson**, **R.O.** (2020). The population biology of Isle Royale wolves and moose: an overview. URL: www.isleroyalewolf.org.

Vucetich, **J.A.**, **Smith**, **D.W.**, and **Stahler**, **D.R.** (2005). Influence of harvest, climate and wolf predation on Yellowstone elk, 1961–2004. *Oikos* 111: 259–270.

Wang, **Z.**, **Brown**, **J.H.**, **Tang**, **Z.**, and **Fang**, **J.** (2009). Temperature dependence, spatial scale, and tree species diversity in eastern Asia and North America. *Proceedings of the National Academy of Sciences of the United States of America* 106: 13388–13392.

Wang, **X.**, **Stenström**, **E.**, **Boberg**, **J.** et al. (2017). Outbreaks of *Gremmeniella abietina* cause considerable decline in stem growth of surviving scots pine trees. *Dendrochronologia* 44: 39–47.

Warman, **R.D.** (2014). Global wood production from natural forests has peaked. *Biodiversity and Conservation* 23: 1063–1078.

Webb, **G.E.** (1979). A.E. Douglass and the canals of Mars. *The Astronomy Quarterly* 3: 27–37.

Wehner, **C.E.** and **Stednick**, **J.D.** (2017). Effects of mountain pine beetle-killed forests on source water contributions to stream-flow in headwater streams of the Colorado Rocky Mountains. *Frontiers of Earth Science* 11: 496–504.

Wei, **L.**, **Zhou**, **H.**, **Link**, **T.E.** et al. (2018). Are temperature inversions the key driver of forest productivity in a small montane watershed? *Agricultural and Forest Meteorology* 259: 211–221.

West, **A.M.**, **Kumar**, **S.**, **Brown**, **C.S.** et al. (2016). Field validation of an invasive species Maxent model. *Ecological Informatics* 36: 126–134.

Westerling, **A.L.**, **Turner**, **M.G.**, **Smithwick**, **E.A.H.** et al. (2011). Continued warming could transform greater Yellowstone fire regimes by mid-21st century. *Proceedings of the National Academy of Sciences of the United States of America* https://doi.org/10.1073/pnas.1110199108.

Wilson, **E.O.** (1992). The Diversity of Life. Cambridge, MA: Harvard University Press.

Winter, **T.C.**, **Harvey**, **J.W.**, **Franke**, **O.L.**, and **Alley**, **W.M.** (1998). Ground Water and Surface Water: A Single Resource. Washington, DC: *U.S. Geological Survey Circular* 1139.

Woods, **S.W.**, **Birkas**, **A.**, and **Ahl**, **R.** (2007). Spatial variability of soil hydrophobicity after wildfires in Montana and Colorado. *Geomorphology* 86: 465–479.

Wright, **C.J.** and **Coleman**, **D.C.** (2000). Cross-site comparison of soil microbial biomass, soil nutrient status, and nematode trophic groups. *Pedobiologia* 44: 2–23.

Yarranton, **M.** and **Yarranton**, **G.A.** (1975). Demography of a jack pine stand. *Canadian Journal of Botany* 53: 310–314.

Yong, **E.** (2016). I Contain Multitudes: The Microbes Within us and a Grander View of Life. New York: Ecco Press.

Yu, **Y.**, **Chen**, **J.M.**, **Yang**, **X.** et al. (2017). Influence of site index on the relationship between forest net primary productivity and stand age. *PLoS One* 12 (5): e0177084.

Zhang, **L.**, **Dawes**, **W.R.**, and **Walker**, **G.R.** (2001). Response of mean annual evapotranspiration to vegetation changes at catchment scale. *Water Resources Research* 37: 701–708.

Zier, **J.L.** and **Baker**, **W.L.** (2006). A century of vegetation change in the San Juan Mountains, Colorado: an analysis using repeat photography. *Forest Ecology and Management* 228: 251–262.

Zimmerman, **J.K.**, **Everham**, **E.M.** III, **Waide**, **R.B.** et al. (1994). Responses of tree species to hurricane winds in subtropical wet forest in Puerto Rico: implications for tropical tree life histories. *Journal of Ecology* 82: 911–922.

Index

Forest Ecology: An Evidence-Based Approach, First Edition. Dan Binkley.
© 2021 John Wiley & Sons Ltd. Published 2021 by John Wiley & Sons Ltd.